MW00844818

Spark
Discharge

E.M. Bazelyan
Yu. P. Raizer

Spark
Discharge

CRC Press
Boca Raton   New York

Library of Congress Cataloging-in-Publication Data

Bazelian, Eduard Meerovich
    Spark discharge / E.M. Bazelyan and Yu. P. Raizer ; translated by
L.N. Smirnova.
       p. cm.
    Includes bibliographical references and index.
    ISBN 0-8493-2868-3
    1. Breakdown (Electricity) 2. Electric spark. 3. Electric
discharges through gases. I. Raizer, IU. P. (IUril Petrovich.)
II. Title.
    [DNLM: 1. Hepatitis B virus. QW 710 G289h]
    QC703.5.B39   1997
    537.5'2—dc21
                                                         97-14858
                                                                CIP

This book contains information obtained from authentic and highly regarded sources. Reprinted material is quoted with permission, and sources are indicated. A wide variety of references are listed. Reasonable efforts have been made to publish reliable data and information, but the author and the publisher cannot assume responsibility for the validity of all materials or for the consequences of their use.

Neither this book nor any part may be reproduced or transmitted in any form or by any means, electronic or mechanical, including photocopying, microfilming, and recording, or by any information storage or retrieval system, without prior permission in writing from the publisher.

All rights reserved. Authorization to photocopy items for internal or personal use, or the personal or internal use of specific clients, may be granted by CRC Press LLC, provided that $.50 per page photocopied is paid directly to Copyright Clearance Center, 27 Congress Street, Salem, MA 01970 USA. The fee code for users of the Transactional Reporting Service is ISBN 0-8493-2868-3/98/$0.00+$.50. The fee is subject to change without notice. For organizations that have been granted a photocopy license by the CCC, a separate system of payment has been arranged.

The consent of CRC Press LLC does not extend to copying for general distribution, for promotion, for creating new works, or for resale. Specific permission must be obtained in writing from CRC Press LLC for such copying.

Direct all inquiries to CRC Press LLC, 2000 Corporate Blvd., N.W., Boca Raton, Florida 33431.

© 1998 by CRC Press LLC

No claim to original U.S. Government works
International Standard Book Number 0-8493-2868-3
Library of Congress Card Number 97-14858
Printed in the United States of America 1 2 3 4 5 6 7 8 9 0
Printed on acid-free paper

Preface

This book is concerned with spark breakdown of long gas gaps. When we started writing this book, we assumed that we knew much about long sparks. One of us had studied spark discharges experimentally for thirty years. The other author had experience in generalizing experimental data available and had made an attempt at a theoretical treatment of sparks in his book 'Gas Discharge Physics'. But, alas, our confidence vanished soon after we began to write this book. When we tried to describe the primary spark element—an ionization wave, or a streamer—we realized that the conventional sequence of material presentation would hardly provide the reader with a consistent picture of the spark discharge, because such a description would look like a multicolor quilt with holes between the patches. Although the amount of experimental discharge data is large, the facts necessary for building a complete picture of the phenomenon are surprisingly scarce. Laboratory research has largely been focused on external spark characteristics, such as discharge current, transported charge, spark velocity, and breakdown voltage, especially for air, which has been studied in detail as the main insulation medium on the Earth. We felt disappointed each time we tried to find data on the plasma parameters, on the electric fields near the ionization wave front, in the streamer and leader channels and even in regions in front of a discharge. At best, we found descriptions of single experiments carried out with many reservations and assumptions.

Some of the theoretical treatments of experimental findings published in periodicals seem questionable. Sometimes, analysis of the same data can lead one to a hypothesis totally opposite to that developed by a particular author, and our own experiments are not an exception to this. So we were faced with a dilemma: either to give up the very hope to write this book or to try to formulate general theoretical concepts of the basic stages in spark development by simplifying the physical models and focusing on the key aspects. In other words, we tried to 'paint a picture with large strokes'. This is not a fast or easy way of writing a book, so it took us longer than we had originally thought.

We had no intention to compete with the well-known book by Meek and Craggs, which is probably the only fundamental publication available on long sparks in English. Neither was there an intention to give the reader a detailed survey of experimental and theoretical work on long sparks. What this book strives for is a consistent, up-to-date description of the spark discharge as we understand it. The bibliography is far from being complete, and references have been given only to those studies that are directly related to the problems under consideration.

Anyone involved in developing a physical model has a strong temptation to check the calculations by direct measurements. We, too, were seeking for such a checkup, trying to find reliable experimental evidence. Good luck, however, was not our frequent guest; more frequent were our doubts concerning the data validity and the adequacy of the techniques used. The accumulated findings, no matter how negative they might be, eventually formed the basis for a whole chapter on measurements. This chapter is aimed at giving some help in data assessment to a theoretician or a specialist working in adjacent areas of physics and engineering. We believe that our analysis of common methodological errors may also come in handy to a beginning experimental reseacher.

We will not speak here in detail about the problems discussed in this book—the reader can find their full list in the contents. The most important chapters are those on streamers and leaders. They contain a fairly large number of theoretical speculations, but we tried to present complicated phenomena in a simple way, without screening the physics with sophisticated mathematical formulas so popular among some theoreticians. Readers with little experience would, at best, admire such formulas without understanding their physical meaning, but more experienced researchers would rather start thinking of a theory of their own. We belong to the latter category.

We hope that the present book will fill a gap in manuals on spark breakdown of gases, because it is intended not only for professional engineers and research physicists but also for undergraduate and graduate students and for those working in adjacent areas.

Much attention is given here to numerical simulation. Although a computer model is primarily expected to produce numerical results, it is more important, in our opinion, to reveal the functional relationships and the role of basic processes, when an analytical theory fails to do so. In this approach, a computer becomes a research instrument rather than a machine for superfast calculations. The combination of physical and computer experiments is the principal gain from computerization of long spark studies. This trend in spark research has not made much progress yet, but we did include most of the available results in this book.

The so-called return stroke has not been discussed deliberately, because this stage in the spark development is of little importance for the breakdown mechanism. The return stroke plays an important role in a lightning discharge, being the principal cause of the destructive action of atmospheric electricity. We plan to consider this problem in detail in a special monograph on lightning, which will be a natural continuation of this book.

The authors would like to thank L. N. Smirnova for the translation of this book. They are also grateful to M. A. Smirnova and L. N. Smirnova for the total preparation of the camera-ready copy.

Authors

Eduard M. Bazelyan was born in Moscow in 1936. He graduated Moscow Power Engineering Institute, where he specialized in high voltage engineering. Since that time, E. M. Bazelyan has been with the Krzhizhanovsky Power Engineering Institute. He learned to do high voltage experiments from I. S. Stekolnikov, a well-known specialist in long sparks. E. M. Bazelyan received his Ph.D. degree in 1964 and his Doctor of Science degree in 1978. His scientific interests are focused on long sparks and lightning. He has studied spark discharges in air and other gases for 30 years, and his hobby is fundamental experiments aimed at clarifying discharge mechanisms. In recent years, he has also been concerned with numerical simulation of spark discharges. His results have made a substantial contribution to both spark theory and engineering. E. M. Bazelyan is an author of over 100 papers and 2 books on long sparks and lightning protection. Today, he is Head of the Physical Modeling Laboratory.

Yuri P. Raizer is Head of the Physical Gasdynamics Department of the Institute for Problems in Mechanics, Russian Academy of Sciences, Moscow (since 1965). He is also a Professor of Physics at Moscow Institute of Physics and Technology (since 1968). Yu. P. Raizer is a Fellow Member of the Russian Academy of Natural Science. He was born in 1927, graduated Leningrad Polytechnical Institute in 1949, and received his Ph.D. degree in 1953. He received his Doctor of Science degree in 1959 from the Chemical Physics Institute, the USSR Academy of Sciences. Yu. P. Raizer has worked in various fields of gasdynamics, explosion physics, gas discharge physics, and interaction of laser radiation with ionized gases. He is an author of over 150 papers, 3 patents, and 5 books, 4 of which were published in English. His 'Physics of Shock Waves and High Temperature Hydrodynamic Phenomena', co-authored by Ya. B. Zel'dovich, Academic Press, 1968, is a world-known handbook for researchers and students. Its impact parameter is very high. His monographs 'Laser-Induced Discharges', Consultants Bureau, 1978 and 'Gas Discharge Physics', Springer-Verlag, 1991 are also quite popular. Professor Raizer is a Lenin Prize winner (the highest scientific award in the former USSR) and was awarded the Penning Prize (in Ionized Gas Physics) in 1993. He is a member of the Editorial Board of the journal, 'Plasma Sources: Science and Technology'.

Contents

1

The Concepts of Spark, Corona, and Breakdown

The spark discharge is one of the most fascinating phenomena in nature. Man had observed spark discharges long before he made his first attempts at a scientific understanding of the world around him. Lightning, which is, in itself, a tremendous spark discharge, even attracted the attention of primitive people, who armed some of their gods with it to emphasize their power.

A scientific study of spark discharges began with the invention of the electrophore, so they soon lost their fairytale halo to become an object of laboratory research. But even in the laboratory, the spark discharge preserved much of its enigmatic spirit. With short crackling noises, fire arrows would jump between metallic balls of the electrophore, discharging the Leyden jar connected to it. A picture of this spectacular experiment can be found in any old textbook on physics.

In the middle of the 18th century, B. Franklin established experimentally that a laboratory spark and lightning had a common nature. With the development of the electrical power industry, the spark discharge was introduced into technology, bringing about a great many problems. An electric spark could suddenly jump across the space between the wires of a high voltage transmission line or between its equipment and some neighboring grounded objects, inducing *short circuiting*. Short circuit current was very dangerous and sometimes could lead to real trouble. When this current was as high as several thousands or dozens of thousands of amperes, it could damage the transmission line equipment and even cut off the power supply. Electrical engineers termed this phenomenon as the *breakdown* of air insulation. *Spark breakdown* and *spark discharge* soon became a subject of interest for fundamental physicists and engineers.

1.1   Lightning and long spark leaders

Spark discharges arise in gaps of a few millimeters or centimeters long, at atmospheric pressure and above it. It is easier to initiate a spark in a gap with a strongly nonuniform distribution of the electric field between the electrodes, because this requires a lower voltage. In very long gaps, a nonuniform field appears in a natural way, since voltage is always applied to electrodes of limited size, making a uniform field practically unfeasible. The gap length has no upper limit, and lightning is justly believed to be the longest spark discharge observable. Lightning develops between the ground and a cloud, or between two oppositely charged clouds. Intercloud sparks have been found to extend to tens of kilometers. As for their appearance, a laboratory spark and lightning have much in common, only differing in the scale. Spark channel diameters are always smaller than their lengths, sometimes by many orders of magnitude, like in lightning, so that a spark always looks like a filament.

If a superhigh-speed movie camera oriented towards the sky is turned on during a thunderstorm, one may be lucky enough to catch a lightning discharge in the vision field of the camera, before it runs out of film. This is one of the approaches used in practice for lightning registration for research purposes. A thin, often branching zigzag filament crosses the dark sky with 'lightning speed' and disappears, before one can hear the peal of thunder. This strong sound is known as *shock wave*. It is produced by the abruptly increased air pressure due to intense Joule heat released in the discharge channel under the action of a strong electric current. The current amplitude in a powerful lightning discharge may exceed 200 kA.

Having developed the film, one can see the discharge channel strike the ground. The speed of lightning has been measured and found to be fairly high, $(2\text{--}3) \times 10^7$ cm s^{-1}, or an order of magnitude higher. The discharge was first time-scanned by B. Schonland in the 1930s [1.1] using a Boys camera with a continuously moving film. He termed the traveling lightning channel as a *leader*. Schonland observed a bright wave of light to travel back up the channel to the cloud after the contact of the leader with the ground, the wave speed being a few tenths of light velocity. This is the *return stroke*, when the electrical charge of the leader, which has transported a high potential to the ground, becomes neutralized. This potential is comparable to that of the cloud. The mere existence of a return stroke indicates a very high conductivity of the leader, which was also pointed out by B. Schonland, who is justly believed to be the pioneer of research into the nature of lightning. For lightning see [1.2, 1.3].

Lightning is not a very convenient object for the registration of the spark

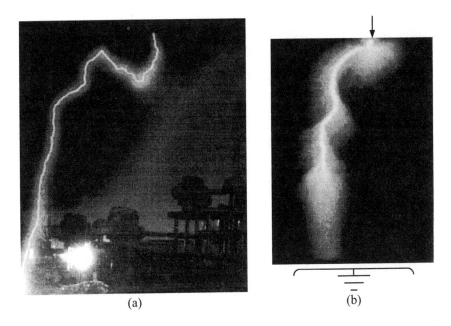

(a)   (b)

FIGURE 1.1
Discharge in a rod-plane gap of 10 m long: an ordinary photograph (a); a
photograph taken by an electron-optical image converter with 300 μs
exposure (b).

fine structure. It cannot be made to appear in the right place or in the right
time. One cannot approach a lightning discharge close enough to exclude
the absorption of its shortwave radiation by the air; so, many short-living
and weak details escape the observer. For research purposes, 'artificial
lightning', or *long sparks*, is produced in laboratories, with lengths as great
as tens or hundreds of meters (Figures 1.1 and 1.2). For this, voltage sources
with an energy storage of 100–1000 kJ and pulse amplitude of 3–5 MV are
used today. A discharge is excited in the laboratory building or, more often,
in the open air, using a relatively small-radius electrode. Sometimes, one
can see long sparks 'jump' off the high voltage end of a capacitor battery of a
pulse generator and watch them run along a winding trajectory right above
the ground or go up, ignoring the natural path towards the ground, which
would seem to be prescribed by the electric field vector in the discharge
gap.

The ability of a long spark to penetrate into the gap areas with a low external field and even to propagate for some time in the direction opposite
to the field surprised the researchers. The field strength was often registered at the ground surface during thunderstorms. Right before a lightning

FIGURE 1.2
Superlong negative discharge to a 110 kV transmission line wire. Voltage pulse
amplitude, 5 MV. Courtesy of A. Gaivoronsky and A. Ovsyannikov, the
Siberian Institute for Power Engineering.

discharge, the strength could be even lower than 100 V cm^{-1} (the field in
the air in the vicinity of a desk lamp cord is stronger), and only within the
few fractions of a second of the leader flight did the field near the ground
abruptly increase, the increase being the greater the closer was the leader
to the observer. It appears that the leader itself carries its strong electric
field. The situation with a laboratory spark is quite similar to this one.

When regular laboratory studies of long sparks were undertaken in the
1930s, T. Allibone in Great Britain [1.4] and I. S. Stekolnikov in Rus-
sia [1.5] independently confirmed the leader nature of spark development
in atmospheric air. Using a Boys camera, they observed the propagation of
a leader channel through a gap of several meters long, between a rod and
a plane. They were able to estimate the leader velocity, which turned out
to be an order of magnitude less than that of lightning.

1.2 Corona and spark discharges

Modern laboratory equipment permits the study of structural details of a corona and a spark discharge. Some of them can be observed visually; in any case, one can see an external manifestation of these forms of discharge. It is convenient to use a high voltage source, for example, a constant voltage generator or a transformer cascade, from which alternating voltage of commercial frequency is applied to a discharge gap. A high voltage electrode can be suspended a few meters above a grounded metallic plane. The shape of the electrode does not matter much, so it may be a rod, a sphere, a wire, etc. What is important is that the electrode radius should be much smaller than the distance to the plane and that the electric field in the gap should be strongly nonuniform.

If the voltage is raised gradually, the visible discharge features begin to manifest themselves as a *corona*. A classical corona can be best observed on a thin polished wire. At a certain voltage, its surface is suddenly covered with a beautiful, blue and violet sparkling envelope of a few fractions of a millimeter thick. If the electrode radius is over a few centimeters, the envelope does not necessarily look continuous, but some of its fragments are always present. One can say that these are the precursors of a leader process. So we think it appropriate here to outline briefly the corona discharge.

1.2.1 Corona discharge

A corona appears only in a nonuniform field at a lower voltage than a spark. If the voltage is raised slowly, it manifests itself as an initially weak glow in the vicinity of the small radius electrode (tip or wire), where the field is considerably stronger. It is in this small area that the gas becomes ionized and glows. The electric current is closed by a flow of charges, of this or that sign, produced in the *ionization region* and extending through the *external region* to the other electrode. The external region, through which the charges drift, does not glow. In an electrically negative gas, such as air, the current from a coronal electrode of any sign is transported across the external region by ions, because electrons become attached to oxygen molecules at the very beginning of the drift process. Ionic current is weak, so it cannot reduce the voltage across the discharge gap, even though the source has a low power. A corona discharge does not possess the spectacular features of a spark: it is a low current discharge in its appearance and characteristics. A corona can only be seen on wires of high voltage transmission lines on a dark night or in bad weather, when the fog or rain locally enhance the field and ionization at the wires. More often, one can hear the characteristic crackling, by which a corona can be easily identified in a transistor radio set. When the voltage approaches a threshold value, a corona may burn for hours and days without inducing short circuiting in the transmission line. Power losses through corona discharges, however, present a serious problem for electrical engineers.

As the voltage is slowly increased further, well above the corona threshold value, the discharge process may take any of the following two paths. In the case of very thin polished wires, one may observe no change in the discharge appearance, except for a higher corona current—its growth is proportional to squared voltage. This is so-called an *ultracorona*. In atmospheric air, a laboratory ultracorona on a wire of a few fractions of a millimeter in diameter can sometimes be sustained at a great *overvoltage*, 10–20 times over that necessary for a corona to be formed. The average gap field rises up to 20–22 $\mathrm{kV\,cm^{-1}}$ (a very large value for average fields in air), when suddenly a dazzling spark appears, immediately bringing the experiment to short circuiting.

Quite different is the corona behavior in a gap with large electrodes. As the voltage rises, the continuously glowing envelope breaks up into separate patches, in which longer filaments, fancifully shaped channels of cold light, appear and disappear again within a few microseconds. These are known as *streamers*, and such a corona, discontinuous in space and time, is called, respectively, a *streamer corona*. A streamer corona produced in a large air gap at voltages above 1 MV looks like fireworks. A streamer may be as long as several meters; numerous bright streamer channels cross the dark space of the laboratory room with such a noise that the experimenter cannot hear his own voice. The pulse current of a propagating streamer only lasts a few fractions of a microsecond and is measured in amperes, even in tens of amperes. But until the streamers reach the other electrode, the current in the external region with no ionization is still transported by an ion flux, whose time-average value has nearly the same order of magnitude as an ultracorona. If the laboratory voltage source permits, a streamer corona can be observed lasting for many hours. It does not lead to short circuiting, because, for this, it is necessary that the streamer channels cross the whole gap space, which normally occurs in atmospheric air at an average field of about 5 $\mathrm{kV\,cm^{-1}}$.

1.2.2 A leader and streamers

Let us turn to the leader process. A leader is formed against the background of a corona discharge, as the voltage is slowly raised further. In its 'pure' form, this phenomenon can be best observed at pulsed voltage. Gas discharge laboratories usually use pulses reaching their highest ampli-

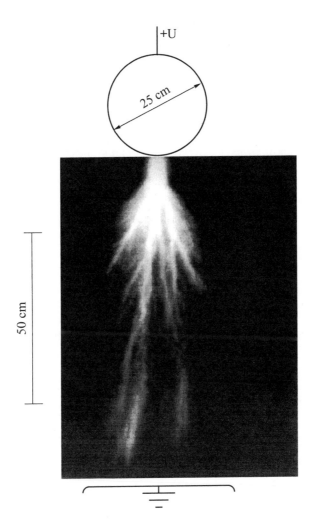

FIGURE 1.3
A still photograph of a corona flash taken from the screen of an electron-optical image converter.

tude within one microsecond or, sometimes, within a few tens or hundreds of microseconds. In this case too, a leader follows a corona, but it usually has a series of short streamer outbursts called *corona flashes*. Streamers start from a common stem at a high voltage electrode. As a rule, streamers branch, so they are often called corona branches. The botanical terminology reflects the fact that coronal outbursts look in photographs like a poorly cared for garden bush (Figure 1.3). Depending on the voltage, streamers can cross a larger or a smaller portion of the gap space, or reach the opposite electrode, but they never cause a gap breakdown because of low conductivity. Even the total current of all streamers is unable to decrease the gap voltage, as it happens in short circuiting (Section 1.5).

The leader channel arises from the stem of a corona flash, where the gas is heated to a maximum value by the total current flowing in from all streamer branches. The leader development is accompanied by repeated streamer outbursts from the channel front end known as *leader head*, or tip. Streamers can move much faster than their leader, their minimum velocity in air being close to 10^7 cm s^{-1}, or even 10^9 cm s^{-1} in a strong electric field. The role of streamers in the leader process was first established experimentally by V. S. Komelkov in 1947–1950 [1.6]: the total streamer current provides the leader channel with power, heats it and maintains the plasma conductivity at a sufficiently high level.

Transporting some of the parent electrode potential, a conducting leader necessarily possesses an appreciable electrical charge distributed along its channel. This follows from general electrostatic laws, since the channel, like any conductor, possesses electrical capacitance. After the leader has crossed the gap to reach the opposite electrode, the leader charge becomes neutralized. This process is similar to the lightning return stroke and is, by analogy, termed as *return stroke of a long spark*. At this stage, the leader channel transforms to a highly conducting spark, which is capable of passing high current typical of short circuiting.

1.2.3   The forms of spark discharge

The further fate of a discharge channel varies with the voltage supply parameters. If the supply represents a capacitor battery, the capacitors rapidly lose their charge, leading to the discharge extinction. Powerful transformer supplies can feed a channel with high current for a very long time. In this case, a cathode spot is formed changing the discharge to an *arc*. Even in a short-living spark, the plasma is generally similar to an arc plasma, so that the final spark stage can be regarded as an *arc flash*. Very long sparks exhibit such an arc-like state within the leader channel long before the leader has reached the other electrode.

What we have described looks very much like a general sketch of a car or a flower, because this kind of picture lacks many important details characteristic for a particular object. In reality, the spark discharge is rich in forms and modifications. By varying the gap length, the gas used, the electric field distribution, the voltage growth rate or duration, one can change the role of each discharge stage, the functions and characteristics of the spark structural elements. One can observe a leader-free breakdown, when an extremely fast streamer overlaps a gap to transform to an arc. Or, on the contrary, one can produce a very slow leader process, when streamers at the channel head become hardly noticeable. Such modifications are feasible owing to an extremely wide range of spark parameters. In order to illustrate this situation, at least partly, we will turn to *average breakdown strength* of long air gaps under normal atmospheric conditions. This value can be obtained by dividing the breakdown voltage by the gap length ($E_b = U_b/d$); it allows us to compare electrical strengths of gas insulation in various conditions.

In a nearly uniform electric field between two large spheres, whose radius is several times larger than the gap length, the average breakdown field should be raised to $E_b = 30\,\mathrm{kV\,cm^{-1}}$. Then, to break a 1 m gap, we need about 3 MV, a value sufficient to produce a spark of 100 m long (Figure 1.2) in a very nonuniform field. The reduction of average breakdown strength by two orders, down to $0.3\,\mathrm{kV\,cm^{-1}}$, is certainly a very impressive fact. If we reduce the gap to 1 m, leaving the electrode geometry the same as for a 100 m gap (the field remains strongly nonuniform), we will not be able to predict the result from the similarity theory. Average breakdown strength does not retain its low value, nor does it return to its maximum characteristic of uniform gaps. In the new conditions, an average field of about $5\,\mathrm{kV\,cm^{-1}}$ is needed for a breakdown to occur, and there is not even a hint of similarity laws being effective, as is typically the case with low pressure discharges.

The conditions for a long spark can be varied further by exciting a *creeping discharge* along thin dielectric plates or glass, polymer and ebonite films. A creeping discharge arises when, say, a rod electrode is attached to a dielectric, while the other electrode is a metallic plate fixed to the other end of the dielectric. If the latter is thin, a long spark may develop at a very low voltage. Spark channels of several meters long have been successfully ignited in air, at an average strength below $50\,\mathrm{V\,cm^{-1}}$, along polymer films of tens of microns thick. Similar behavior is characteristic of a spark creeping along a weakly conducting surface, for example, an electrolyte.

The great diversity of spark manifestations accounts for and, to some extent, justifies the nearly universal passion for phenomenological descriptions of this type of discharge in the past, and, partly, at present. For half a century, vast experimental data have accumulated that are still awaiting to be interpreted. Much less effort has been made to try to analyze the physical nature of the phenomenon and to develop a consistent theory of its fundamental aspects. We have to recognize, however, that the spark is one of the most sophisticated phenomena in gas discharge physics.

1.3   Avalanche and streamer mechanisms of gas breakdown

The studies of lightning and laboratory sparks in air are not the only source of information about the nature of gas breakdown. Another line of research, which initially seemed more logical and better grounded, stemmed from the work on discharges in low pressure gases. Regular experiments with rarefied gases were started at the beginning of the 20th century. They did not use high voltages and could be performed in a laboratory room. Discharge processes in such gases turned out to be less sophisticated than in sparks and more accessible to observation and theoretical interpretation. So the major advances in the early history of discharge physics were associated with low pressure gases. One of the results was the avalanche breakdown theory by Townsend; its basic conceptions are as follows.

Suppose an electron appears, for some reason, in a plane gap of length d in a uniform field. In fact, there are several or many electrons produced in the gas or knocked out of the cathode by cosmic rays, radioactive or ultra-violet radiation. The cathode is often irradiated to accelerate a breakdown process. In the electric field $E = U/d$, where U is the voltage applied to the plane electrodes, an electron gains energy to ionize an atom (or molecule) and thereby loses this energy. As a result, there are two slow electrons: the original electron and the new one. These gain energy from the field and again ionize atoms and molecules, and so on. In this way, an *electron avalanche* develops, in which all electrons quickly drift, due to the field, toward the anode, while heavier, positive ions slowly drift toward the cathode. Having reached the cathode, an ion knocks out, with a certain probability γ, another electron from the cathode. This process is known as cathode or secondary ion-electron emission. A new electron gives rise to a new avalanche, whose ions again strike the cathode to knock out new electrons.

If the number of electrons from the cathode decreases in each subsequent cycle, as it happens in a weak field and low intensity ionization, nothing special occurs in the gas. Just from time to time, charged particles appear and then disappear towards the electrodes under the action of the field. But if the field is strong enough, the number of electrons in each new cycle increases. Then the process of electron multiplication in every avalanche is accompanied by multiplication of avalanches themselves, so that charges, whose density continuously rises for some time, quickly fill up the gap. The gap cross section may be filled by charges, even though the process may have started with one electron, because avalanche electrons diffuse and repel from one another in their own electric field, also traveling in the transverse direction. In addition to electrons, every avalanche produces photons, among them highly energetic ones capable to ionize the gas at some distance from the avalanche region. Photons are uniformly emitted in all directions, including the direction normal to the avalanche motion, so that the avalanche region of the gap expands. When the whole gap between the electrodes becomes ionized to this or that degree, we can speak of *bulk discharge*. This is the way an avalanche gap breakdown occurs.

If the voltage source is a capacitor, it becomes discharged with no consequences. But if it can generate voltage for a long time, a steady-state discharge arises in the gap, which is usually a glow or Townsend-type discharge, when the source is weak. A powerful source can ignite an arc. In this case, the ionized gas cannot fill up the whole interelectrode space, but it persists only within the arc channel, which, as was mentioned above, somewhat resembles a spark filament. Pinching, or contraction, of a plasma channel to a thin filament carrying electric current (the only discharge element with this function) can also be observed in a glow discharge. A contracted glow discharge has much in common with an arc, but they are not identical. This, however, is not directly related to the problem under consideration, and we only mention this to emphasize the fact that a plasma filament is not necessarily an evidence for a spark discharge.

The Townsend mechanism underlying the quantitative theory permitted an adequate description of many essential properties of low pressure gas discharges. It was found, in particular, that discharges exhibited certain similarity laws. For instance, breakdown voltage depends on the product of gap length d and gas pressure p rather than on each individual parameter. This is correct at $pd < 200$ Torr cm, when the Townsend mechanism is completely valid. For large values of pd, the theory gives deviations from experiment, which become fatal at $pd > 1000$ Torr cm. Therefore, at atmospheric pressure in gaps of 1 cm long and more, the theory encounters real difficulties, the major difficulty being the time, for which a breakdown develops. Avalanche multiplication is a slow process. Time is necessary for repeated travels of avalanches through the gap to provide consecutive enhancement of ionization. An avalanche moves at the electron drift velocity (Section 2.1), and, according to the avalanche theory, no breakdown can occur, before the avalanche has crossed the gap at least once. Everything happens much faster in a real process occurring at large pd values, especially in long gaps with a strongly nonuniform field. Here, the discrepancy between experiment and theory is as large as several orders of magnitude.

The avalanche theory was unable to explain another important experimental fact, namely the ability of a spark to run through a gap section with a strongly nonuniform field, where the voltage is too low to induce ionization. By the early 1930s, the need for a new theory based on the concepts of spark discharge became quite acute.

The foundation for a new theory that was to explain spark breakdown, termed at that time as *streamer breakdown*, was laid by the work of Loeb, Meek and Raether [1.7–1.10] in the 1940s. This theory was based on the concept of streamer, a thin ionized channel making its way through the interelectrode space, following the positive trace of a powerful primary avalanche. The trace involves a large number of *secondary avalanches*. They arise near the primary trace from electrons produced under the action of photons, emitted by excited atoms during the propagation of the primary and secondary avalanches. The streamer channel originates in the area of strong external field near the electrode with a small curvature radius. Owing to the fairly high conductivity of the ionized gas in the channel, the initial potential seems to attach itself to the streamer head, thus enhancing the local electric field. This is why a streamer is able to propagate through a weak external field in the gap. Later investigations provided much experimental and theoretical evidence that considerably changed some original estimates and even fundamental concepts, but the basic ideology of this theory remained unshaken. We mean the conception, in which a plasma channel propagates through a gap by ionizing the gas in front of its charged head owing to a strong field induced by the head itself.

1.4   What is a breakdown?

Frankly speaking, the notion of breakdown still remains too general and vague in gas discharge physics. Breakdown is commonly understood as an act of fast formation of a strongly ionized state under the action of applied electric or electromagnetic field, while the consequences of this event are not regarded as particularly important. If the field or the radiation source operates for a long time, a self-sustained discharge, no matter of what kind, is ignited by breakdown (we use this term here in its general meaning). Indeed, at low pressures, there is no need to distinguish between the breakdown and ignition events, so the breakdown voltage is often termed the ignition voltage.

Quite different is the situation at high pressures and in a spark breakdown or discharge, when the electrical strength of the gas insulation becomes an extremely serious engineering problem. In high voltage engineering, *breakdown* is understood as the formation of a highly conductive channel capable of carrying such a strong current that the voltage in the insulation gap sharply drops to produce *short circuiting*. It is this situation that is of interest. Of interest, therefore, are the voltages leading to such a situation. These are *breakdown voltages*. Their estimation and the understanding of their relation to other discharge characteristics are the goal of both experimental and theoretical efforts in spark physics. Our further presentation of the subject will be based on this concept of breakdown.

For short circuiting to occur, the resistance of a channel that has overlapped the gap must become lower than that of the external circuit, including the voltage supply resistance. This, in fact, is a quantitative criterion for breakdown in the above sense. The necessary prerequisite for a breakdown, when an ionized channel has overlapped the gap, is not so much a high conductivity of the channel as an adequate conductivity evolution. It is essential that the channel resistance be reduced rapidly with time and that the external circuit elements, rather than the channel, eventually limit the short circuit current. For this, the channel must be unstable, and the rising current must reduce the voltage necessary to maintain the gas ionization. In other words, an ionized channel must possess a *descending current-voltage characteristic*. This will guarantee short circuiting due to further reduction in the channel resistance. It eventually does not matter why the resistance will decrease—due to a higher degree of plasma ionization or because of new ionized gas layers attaching themselves to the channel side. Whether the resistance per unit channel length will start to decrease during the plasma propagation through the gap or after the contact with the opposite electrode is unimportant either—both events have been observed experimentally. The important thing is that these events should be irreversible.

A self-sustained discharge may not lead to breakdown at high pressures (atmospheric and higher) in strongly nonuniform fields, when at least one electrode has a small curvature radius, much smaller than the gap length. The ionization region may be localized at the electrode surface, where the field is strong, and it may not move farther into the gap space. This happens in the ignition of a self-sustained corona, which does not lead to short circuiting. The total gap resistance is so great, whereas the corona current is so weak, that the gas insulation can be considered as remaining intact.

Even the gap overlap by a discharge does not necessarily produce a breakdown. In laboratory experiments with constant voltage or with alternating commercial frequency voltage, one can observe a sudden appearance of a bluish discharge column consisting of numerous thin filaments between the high voltage electrode and the plane. It can only arise for a moment, but sometimes it allows the researcher to admire it for a long time without leading to short circuiting. The channel conductivities are very low; they do not tend to increase after the overlap, and the weak current in the external circuit closed by such a poor conductor does not resemble short circuit current.

Among the various types of steady-state discharges, the arc is closest to the spark, because it possesses a descending current-voltage characteristic and is capable of passing high currents, up to several thousand kiloamperes.

Therefore, it is reasonable to think that spark breakdown is completed by the formation of an arc channel, if the voltage source is strong enough.

To conclude, the ignition voltage, also called *initial voltage*, of a self-sustained discharge does not coincide with breakdown voltage in high pressure gaps with a strongly nonuniform field. The latter corresponds to a gap overlapped by a channel with a descending current-voltage characteristic, and the breakdown voltage may be considerably larger than the initial voltage.

2

Basic Information on Processes in Ionized Gases

At its various stages and in its different structures, a spark discharge may contain a weakly ionized nonequilibrium glow-type plasma and a relatively strongly ionized quasiequilibrium plasma of the arc type. In the former, the mean electron energy is much greater than the thermal energy of molecules. Electrons mostly collide with neutral particles rather than with one another or with ions. The electron energy spectrum is far from a Maxwellian one and directly depends on the electric field strength. The field also determines directly the rate of ionization by electrons. A strongly ionized plasma is closer to the thermally equilibrium state, which means that an electron gas and a heavy particle gas possess comparable temperatures. The temperature is a key factor determining the ionization degree.

2.1   Drift and diffusion of electrons and ions

We will first consider processes in a weakly ionized plasma with an ionization degree less than 10^{-3}, when electrons mostly collide with neutral molecules but not with one another. The motion of electrons in field $\boldsymbol{E}$ includes their random motion with 'thermal' velocity v and the drift motion along the field with velocity $\boldsymbol{v}_{\mathrm{e}}$. The latter is established by compensation of the momentum, gained from the electric force $-e\boldsymbol{E}$ during the time between collisions, and the drift momentum loss in scattering by elastic collisions with molecules. The electron drift velocity approximately is [2.1]

$$\boldsymbol{v}_{\mathrm{e}} = -\frac{e\boldsymbol{E}}{m\nu_{\mathrm{m}}} = -\mu_{\mathrm{e}}\boldsymbol{E} \qquad (2.1)$$

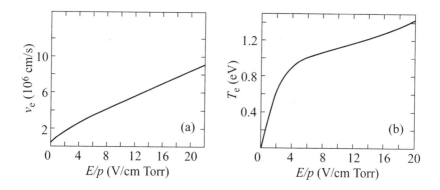

FIGURE 2.1
Electron drift velocity (a) [2.2] and temperature (b) [2.3] in air.

Here $\nu_m = \nu_c \left(1 - \overline{\cos\theta}\right)$ is the collision frequency necessary for the momentum transfer (effective collision frequency), ν_c is the actual frequency of elastic collisions, $\overline{\cos\theta}$ is the mean scattering angle cosine, and μ_e is the electron mobility; $\nu_c = N\overline{v}\sigma_c$, where N is the number of molecules per cm^3, $\overline{v}$ is thermal velocity, and σ_c is the scattering cross section. In the approximation of equation (2.1), the collision frequency ν_m is taken to be independent of velocity v or of electron energy $\varepsilon = mv^2/2$ and is chosen, for calculations, from experimental data available. The drift velocity in moderate fields is small, as compared to the random velocity $v_e \ll \overline{v}$, and depends on the E/N ratio. The latter relationship is a manifestation of the similarity principle inherent in many parameters of a weakly ionized nonequilibrium gas in a field. This principle is also valid, when there is no strict proportionality $v_e \sim E/N$ and the mobility μ_e is variable [Figure 2.1(a)].

For the desired E/N range, the mobility may be defined as an average value of v_e/E for this range. For instance, at $E = 300$ kV cm^{-1} and $p = 1$ atm in air (at room temperature $N = 2.51 \times 10^{19}$ cm^{-3}, $E/N = 0.795 \times 10^{-15}$ V cm^2), we have $v_e = 8.7 \times 10^7$ cm s^{-1} and $\mu_e = 270$ cm^2 V^{-1} s^{-1}. Gas discharge physicists often use pressure $p = NkT$ instead of gas density N. This is convenient, when the gas is cold and its temperature may be taken to be constant. At $T = 293$ K, we obtain E/p [V cm^{-1} Torr^{-1}] $= 1.32\, E/p$ [kV cm^{-1} atm^{-1}] $= 3.3 \times 10^{16}\, E/N$ [V cm^2] $= 0.33\, E/N$ [Td], where one Townsend is: 1 Td $= 10^{-17}$ V cm^2.

Ion mobility is hundreds of times less than electron mobility; therefore, ions generally make a small contribution to electric current, except when electron density n_e is much less than ion density n_+. At $n_e \sim n_+$, the current density and conductivity are

$$j = -e n_e v_e = e n_e \mu_e E = \sigma E \tag{2.2}$$

$$\sigma = e \mu_e n_e = \frac{e^2 n_e}{m \nu_m} = 2.82 \times 10^{-4} \frac{n_e \, [\mathrm{cm}^{-3}]}{\nu_m \, [\mathrm{s}^{-1}]} \, \Omega^{-1} \, \mathrm{cm}^{-1} \tag{2.3}$$

The conductivity of a weakly ionized gas is proportional to the degree of its ionization $x = n_e / N$.

The field does, on an electron, the work $\langle -e\boldsymbol{E}\boldsymbol{v}\rangle = -e\boldsymbol{E}v_e$ per second, where the angular braces indicate the velocity averaging. The energy $e\boldsymbol{E}v_e n_e = \sigma E^2 = jE$ released per 1 cm^3 per second represents the Joule heat of current. Apart from gaining the field energy equal, on the average, to $e\boldsymbol{E}v_e/\nu_m = e^2 E^2 / (m\nu_m^2)$ per effective collision, an electron gives off, to an 'immobile' molecule, the portion $\delta = 2m/M$ of its energy ε (M is the mass of the molecule). So we have

$$\frac{d\varepsilon}{dt} = \left(\frac{e^2 E^2}{m \nu_m^2} - \delta\varepsilon \right) \nu_m \tag{2.4}$$

In molecular gases, electrons can excite molecular vibrations with a high probability. An electron gives off to a molecule an appreciably larger energy than $2m/M$, but this is, on the average, a small portion as well. In nitrogen, for example, this portion at $\varepsilon \sim 1$–3 eV is 2.1×10^{-3} and $2m/M = 4 \times 10^{-5}$. Equation (2.4) remains valid, although here δ should be understood as the actual fraction of the energy transported. At higher energies $\varepsilon \sim 10$–20 eV, electrons excite electronic states and ionize the gas molecules, losing much of their energy. Then $\delta \sim 1$, and equation (2.4) valid for small losses does not make much sense any more.

The mean electron energy, $\bar{\varepsilon}$, is generally established very quickly and can be found from equation (2.4), taking $d\varepsilon/dt = 0$. In moderate fields, its value is much smaller than the potentials of electronic level excitation and ionization. To evaluate the mean energy, it is convenient to assume the path length, rather than ν_m, to be constant. It is then $l = \bar{v}/\nu_m = 1/N\sigma_m$; here, $\sigma_m = \sigma_c \left(1 - \overline{\cos\theta} \right)$ is the momentum transfer cross section known as transport cross section. At $d\varepsilon/dt = 0$, equation (2.4) yields a convenient expression

$$\bar{\varepsilon} = \frac{3}{2} k T_e = \frac{\sqrt{3\pi}}{4} \frac{eEl}{\sqrt{\delta}} \approx 0.8 \frac{eEl}{\sqrt{\delta}} \sim \frac{E}{N} \tag{2.5}$$

The electron 'temperature' T_e has been introduced conventionally. The calculations have taken $l = \bar{v}/\nu_m$ and $m\overline{v^2} = (16/3\pi)\left(m\overline{v^2}/2\right)$, as if we dealt with a Maxwellian electron velocity distribution. On the assumption of $l = \mathrm{const}$, the mean energy is proportional to E/N and is much larger than the energy gained from the field per unit path length eEl. The similarity principle $\bar{\varepsilon} = f(E/N)$ is valid in a rigorous theory, too. From equations (2.1), (2.5) and $l = \text{const}$, the drift velocity is

$$v_e = \left(\frac{3\pi\delta}{16}\right)^{1/4} \left(\frac{eEl}{m}\right)^{1/2}$$

Together with equation (2.5), it yields the relation

$$\frac{v_e}{v} = \frac{\sqrt{3\pi\delta}}{4}$$

which shows the drift velocity to be appreciably smaller than the mean random velocity. This justifies the calculation of collision frequency $\nu_m = N\bar{v}\sigma_m$ from random velocity. One can see from equation (2.4) that the mean electron energy corresponding to a particular field strength is established for the time τ_u equal to

$$\tau_u \equiv \nu_u^{-1} = \frac{\nu_m^{-1}}{\delta} \qquad \nu_u = \nu_m\delta \qquad (2.6)$$

The quantity ν_u is referred to as the energy loss frequency. The relaxation time for the mean electron energy is very small, for example, $\tau_u \sim 10^{-10}$ s at atmospheric pressure in air ($\nu_m \approx 3 \times 10^{12}$ s^{-1}, $l \approx 4 \times 10^{-5}$ cm).

The mobility μ_e and the electron diffusion coefficient $D_e = \overline{v^2}/3\nu_m \approx l\bar{v}/3$ are interrelated as

$$\frac{D_e}{\mu_e} = \frac{mv^2}{3e} = \frac{2}{3}\frac{\bar{\varepsilon}}{e} = \frac{kT_e}{e}$$

This relationship is a particular case of Einstein's general relation

$$\frac{D}{\mu} = \frac{kT}{e} \qquad (2.7)$$

which has a thermodynamic nature and is valid for any particles and diffusion mechanisms or mobility. The experimental D_e/μ_e ratio describes the mean electron energy, which is difficult to measure experimentally [Figure 2.1(b)].

The ion drift velocity is generally found from Equation (2.1) with the electron mass substituted by the reduced mass M' of a colliding ion-molecule pair and ν_m by the ion collision frequency. In moderate fields, ions performing an effective energy exchange with molecules acquire their temperature T, so that $T_+ \approx T$ in an ionic gas, in contrast with an electron gas with $T_e \gg T$. The ion diffusion coefficient is related to the mobility through equation (2.7). For instance, at room temperature and atmospheric pressure in nitrogen, we have $\mu_+ \approx 2$ cm^2 V^{-1} s^{-1} and $D_+ \approx 5.2 \times 10^{-2}$ cm^2 s^{-1}. Of course, these are somewhat arbitrary data, because ions tend to form complexes, while the transport coefficient of, say, N_2^+ and N_4^+ ions may differ by a factor of 1.5–2.

2.2 Molecular ionization and excitation

2.2.1 Ionization

The principal mechanism of charge production in gas discharges is ionization of unexcited atoms or molecules by electron impact. The ionization rate, that is the number of ionization events per 1 cm^3 per second, is

$$\left(\frac{dn_e}{dt}\right)_i = \int_I^\infty N v \sigma_i(\varepsilon) \, n_e(\varepsilon) \, d\varepsilon = k_i N n_e = \nu_i n_e \qquad (2.8)$$

where σ_i is the cross section of ionization by electrons with energy ε; $n_e(\varepsilon)$ is their energy distribution function; I is the ionization potential; the factor k_i with the densities is called the ionization rate constant; $\nu_i = k_i N$ is the ionization frequency, or the average number of ionization events per electron per second. The latter parameter is the principal characteristic of the rate of the process; it is proportional to the gas density and is defined by the electron energy distribution, which, in turn, varies with the E/N ratio. At constant ionization frequency and in the absence of electron losses, the number of electrons grows with time in an avalanche manner as $n_e = n_0 \exp(\nu_i t)$, where n_0 is the initial electron density.

An electron avalanche develops in an electric field not only in time but also in space. All the electrons produced travel together with the same drift velocity, which is established very quickly, approximately over the time of one collision event. For this reason, it is convenient to describe the ionization rate with the ionization coefficient α, which denotes the number of ionization events per electron per 1 cm path along field E. Evidently, we have

$$\alpha = \nu_i/v_e \qquad \nu_i = \alpha v_e \qquad (2.9)$$

In a constant field, an electron avalanche grows in the direction of motion x as $n_e = n_0 \exp(\alpha x)$. Although in experiment one usually measures α and v_e, the primary parameter is the ionization frequency calculated rigorously with equation (2.8). Townsend's semi-empirical formula,

$$\alpha = A p \exp(-Bp/E) \qquad (2.10)$$

where A and B are constants chosen from experimental data (Figure 2.2), is often used for theoretical estimations: for air $A = 15$ cm^{-1} Torr^{-1},

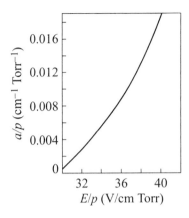 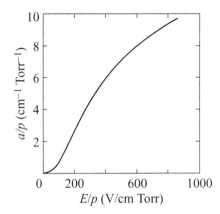

FIGURE 2.2
The Townsend ionization coefficient for air (from data of [2.4, 2.5]).

$B = 365\ \text{V cm}^{-1}\,\text{Torr}^{-1}$ at $100 < E/p < 800\ \text{V cm}^{-1}\,\text{Torr}^{-1}$. For the E/p ranges characteristic of some spark structures in air, other interpolation formulas may appear to be more convenient:

$$
\begin{aligned}
\alpha/p &= 1.17 \times 10^{-4}\,(E/p - 32.2)^2 & 44 < E/p < 176 \\
\alpha/p &= (0.21 E/p)^{1/2} - 3.65 & 200 < E/p < 1000 \\
\alpha/p &= 5.9\,(E/p)^{3/2} & 110 < E/p < 530
\end{aligned}
\qquad (2.11)
$$

where E/p is measured in $\text{V cm}^{-1}\,\text{Torr}^{-1}$ and α/p in $\text{cm}^{-1}\,\text{Torr}^{-1}$. Especially convenient is the third formula, because one can take $\alpha/p \sim (E/p)^{3/2}$ and $\nu_\text{i}/p \sim (E/p)^{5/2}$ at elevated values of E/p.

Among other ionization processes occurring in spark discharges, of interest are photoionization and associative ionization. For example, the photoeffect cross sections σ_ν at the threshold $h\nu \approx I$ are $2.6 \times 10^{-17}\ \text{cm}^2$ for N_2, $10^{-18}\ \text{cm}^2$ for O_2, and $3.5 \times 10^{-17}\ \text{cm}^2$ for Ar. Associative ionization represents a process in which two atoms or excited molecules combine to form a molecular ion and a free electron; for example,

$$
N + O \to NO^+ + e \qquad N_2^* + N_2^* \to N_4^+ + e \qquad (2.12)
$$

The molecular binding energy released is added to the energy of the atoms or excited molecules and is spent for extracting an electron. The rate constant of the first of reactions (2.12) is

$$
k = 5 \times 10^{-11} T^{-1/2} \exp\left(\frac{-32500}{T}\right) \qquad (2.13)
$$

where T is the gas temperature measured in Kelvin degrees [2.6, 2.7].

2.2.2 Inelastic energy losses

Crossing a potential difference of 1 V in a uniform field, an electron produces α/E electrons (ion pairs). In order to produce one ion pair, it must gain an average energy $w = eE/\alpha$ from the field. This value depends on E/p and has a minimum. In the approximation of equation (2.10), $w = \bar{e}eB/A$, where $\bar{e}$ is the natural logarithmic base; the minimum is reached at $E/p = B$. Even under these conditions most favorable for ionization, at $E/p = 365$ V cm^{-1} Torr^{-1} in air, an electron expends, for one ion pair, the energy w_{min} (the Stoletov constant), which is equal to 66 eV for air. This value is several times over the ionization potentials of N_2 and O_2 molecules (15.6 and 12.2 eV, respectively), indicating great energy losses for the electronic excitation of molecules. The electronic levels are largely excited at high electron energies $\varepsilon > 10-15$ eV. An ionization event is always accompanied by an excitation event, and the number of excited molecules produced is generally larger than that of ions. This fact is very essential for spark breakdown, since some of the excited molecules and atoms emit photons with a high probability. Photons are involved in the production of primary electrons, which trigger off an avalanche-like ionization.

At electron energies of tens of electron volts, inelastic energy losses by electrons considerably exceed elastic losses. Here, elastic collisions are generally of minor importance, and electrons are primarily scattered forward

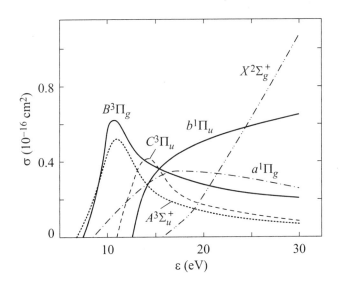

FIGURE 2.3
Excitation cross sections for various electronic states and ionization in nitrogen.

in such collisions. Under these conditions, the term for the energy loss by an electron in equation (2.4) should be changed. Then we go from time to the coordinate in accordance with the expression $dx = v_e dt$ and rewrite the energy equation approximately as

$$\frac{d\varepsilon}{dx} = eE(x) - L(\varepsilon) \qquad L = N\sigma_i I + N \sum_k \sigma_k^* E_k^*$$

Here, $L(\varepsilon)$ is the function of inelastic losses for the ionization and excitation of various molecular levels (σ_k^* are the respective cross sections and E_k^* are the excitation potentials). The character of the $L(\varepsilon)$ function can be easily understood, if one mentally sums up the $\sigma(\varepsilon)$ cross sections in Figure 2.3. Like all cross sections, $L(\varepsilon)$ has a peak at electron energy $\varepsilon \sim 10-20$ eV, which means that an electron traveling in a field $E > E_r = e^{-1}L_{\max}$ will be steadily accelerated in spite of the inelastic losses. This effect is known as electron runaway. For nitrogen, for example, we have $(E/p)_r = 365$ V cm^{-1} Torr^{-1} = 277 kV cm^{-1} atm^{-1}. In a longer field exceeding E_r, an electron can be accelerated to energies over 1 keV in a dense gas (not in vacuum!).

2.2.3   Excitation and relaxation of molecular vibrations

Low energy electrons can actively excite molecular vibrations. In nitrogen, this usually happens at $1.8 < \varepsilon < 3.3$ eV [2.1]. Electrons can also excite rotational levels, but this process is of lesser importance than the excitation of vibrations. At moderate values of $E/p \sim 3-30$ V cm^{-1} Torr^{-1} available in some structural elements of a long spark, electrons spend up to 90–95% of the gained energy for the excitation of molecular vibrations in air and nitrogen. At higher E/p values, the vibrational losses give way to inelastic losses for the excitation of electronic levels and ionization.

The significance of vibrational excitation of molecules in discharges is enhanced, because the relaxation in cold air and nitrogen is very slow, such that the vibrational energy does not for a long time transform to the translational energy (or 'heat'), thereby making no contribution to the gas temperature. This circumstance is very essential for the streamer-leader process. At room temperature, the transformation frequency of the vibrational energy of an excited nitrogen molecule to the translational energy (VT relaxation frequency) approximately is

$$\nu_{VT} = \left(23x_{N_2} + 2.1 \times 10^2 x_{O_2} + 1.3 \times 10^5 x_{H_2O}\right) p \,[\text{atm}] \text{ s}^{-1}$$

where x_k are molar fractions of the air components. Water molecules can considerably deactivate vibrations. In dry air at 1 atm, $\tau_{VT} = \nu_{VT}^{-1} \approx 1.7 \times 10^{-2}$ s; with air humidity of 0.8×10^{-5} g cm^{-3} close to normal, $x_{H_2O} \approx 1\%$,

$\tau_{VT} \approx 7 \times 10^{-4}$ s. In hot humid air $\tau_{VT} \approx 8 \times 10^{-5}$ s at $T = 1000$ K and $\tau_{VT} \approx 1 \times 10^{-5}$ s at 2300 K. On heating, the relaxation process exhibits a rather complicated behavior through the VV exchange, or the vibrational energy transfer to higher vibrational levels, which relax much faster than lower levels. In some cases, an instability of the nonequilibrium state with strongly excited vibrations is observed: the VT relaxation becomes 'self-accelerated'. The relaxation leads to a higher temperature, which, in turn, accelerates the relaxation, and so on, until a common translation-vibration temperature is established. It is not inconceivable that such an instability may affect the gas heating in the leader head.

2.3 Electron loss and liberation

2.3.1 Recombination

In recombination of electrons with positive ions unaccompanied by other processes, for instance by ionization, the electron density in an electrically neutral plasma with $n_e = n_+$ decreases in time as

$$\left(\frac{dn_e}{dt} \right)_r = -\beta n_e^2 \qquad n_e = \frac{n_0}{1 + \beta n_0 t} \qquad n_0 = n_e(0) \qquad (2.14)$$

where β is an electron-ion recombination coefficient. The characteristic time τ_r is equal to reciprocal recombination frequency $\nu_r \equiv \tau_r^{-1} = \beta n_0$.

Among the various recombination mechanisms, the fastest one is dissociative recombination, which looks like this:

$$A_2^+ + e \rightarrow A + A^* \qquad (2.15)$$

The dissociative recombination coefficients β_{dis} are about 10^{-7} cm^3 s^{-1} and decrease with electron temperature as $\beta_{dis} \sim T_e^{-1/2}$ at T_e lower than several thousand degrees, while at higher temperatures they decrease as $T_e^{-3/2}$. Recombination generally occurs in this way even in weakly ionized monatomic inert gases, such as argon. Molecular ions are produced by the primary, atomic ions during a conversion reaction of the type:

$$A^+ + A + A \rightarrow A_2^+ + A \qquad (2.16)$$

But because of its two-step structure, the recombination process in monatomic gases is one or two orders of magnitude slower than in molecular gases. In fast processes characteristic of spark discharges, reactions of the above type (2.16), involving heavy particles only, do not have enough time to develop; as a result, the monatomic plasma slowly decays. But even in molecular gases, the recombination is too slow for electrons to be lost to the discharge structures with very fast processes. For instance, the typical recombination time $\tau_{\rm r} = (\beta n_{\rm e})^{-1} = 10^{-6}$ s is relatively large at $\beta = 10^{-7}$ cm^3 s^{-1} and $n_{\rm e} = 10^{13}$ cm^{-3}. But recombination is of little significance in slower processes lasting over one microsecond, say, in the decay of a long streamer or leader plasma. The recombination rate of complex ions like O_4^+ is an order of magnitude higher than in the case of simple ions like O_2^+:

$$O_4^+ + e \rightarrow O_2 + O_2 \qquad \beta_{\rm dis} = 1.4 \times 10^{-6} \left(\frac{300}{T_{\rm e}\,[{\rm K}]}\right)^{1/2} {\rm cm}^3\,{\rm s}^{-1}$$

Complex ions are produced during a conversion reaction of the type:

$$O_2^+ + O_2 + O_2 \rightarrow O_4^+ + O_2$$

The rate constants of the forward (conversion) reaction, K_1, and of the reverse (decomposition of O_4^+) reaction, K_2, vary with the gas temperature:

$$K_1 = 2.4 \times 10^{-30} \left(\frac{300}{T\,[{\rm K}]}\right)^3 {\rm cm}^6\,{\rm s}^{-1}$$

$$K_2 = 3.3 \times 10^{-6} \left(\frac{300}{T\,[{\rm K}]}\right)^4 \exp\left(-\frac{5030}{T}\right) {\rm cm}^3\,{\rm s}^{-1}$$

In electrically negative gases at low temperatures, when the majority of electrons quickly change to negative.ions, charges may be lost due to the ion-ion recombination. Its coefficient is $\beta_{\rm ii} \sim 10^{-7}$–$10^{-6}$ cm^3 s^{-1}, the largest coefficient for air (about 2.2×10^{-6} cm^3 s^{-1} for simple ions) being achieved at pressures close to atmospheric pressure. The rate constants of electron-ion-molecular processes are given in [2.8–2.10] for air and in [2.10–2.14] for heated air. For reactions involving complex ions, see [2.15–2.17].

2.3.2 Electron attachment

Attachment is a major process of electron loss in negative gases. In cold air at zero field, electrons become attached to O_2 in triple collisions

$$e + O_2 + M \rightarrow O_2^- + M \tag{2.17}$$

where M stands for O_2, N_2, and H_2O molecules. The binding energy of the electron in an O_2^- ion is about 0.5 eV. The electron loss is described by the kinetic equation

$$\left(\frac{{\rm d}n_{\rm e}}{{\rm d}t}\right)_{\rm a} = -n_{\rm e} \sum k_{\rm M} N_{O_2} N_{\rm M} = -\nu_{\rm a} n_{\rm e} \qquad \nu_{\rm a} = \sum k_{\rm M} N_{O_2} N_{\rm M} \tag{2.18}$$

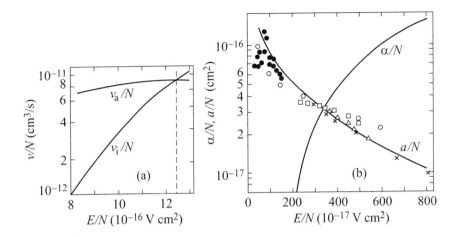

FIGURE 2.4
Coefficients of ionization α and of attachment a: for air (a); for SF_6 (b) [2.18].

$$n_e = n_0 \exp\left(-\nu_a t\right) \tag{2.19}$$

where k_M are the reaction rate constants and ν_a is the attachment frequency. Thermalized electrons in cold air of atmospheric pressure ($T_e = T = 300$ K) possess $k_{O_2} = 2.5 \times 10^{-30}$ cm^6 s^{-1}, $k_{N_2} = 0.16 \times 10^{-30}$ cm^6 s^{-1}, $k_{H_2O} = 14 \times 10^{-30}$ cm^6 s^{-1}, and $\nu_a = 0.9 \times 10^8$ s^{-1}; that is, the lifetime of an electron is $\tau_a = \nu_a^{-1} = 1.1 \times 10^{-8}$ s.

After the electrons have gained an energy of several electron volts from the field, dissociative attachment reactions occur, which, in contrast to reactions (2.17), require energy for molecular dissociation:

$$e + O_2 + 3.6\,eV \rightarrow O + O^- \tag{2.20}$$

$$e + H_2O + 4.25\,eV \rightarrow OH + H^-$$

$$e + H_2O + 3.6\,eV \rightarrow H_2 + O^- \tag{2.21}$$

$$e + H_2O + 3.2\,eV \rightarrow H + OH^-$$

The major process occurring at low humidity is the attachment to oxygen. The cross section for reaction (2.20) becomes larger at a higher gas temperature, whereas the reaction threshold becomes lower than 3.6 eV. This is accounted for by the involvement of vibrationally excited molecules, whose energy is also expended for the reaction. The cross sections for reactions (2.20) and (2.21) are presented in [2.1].

Like ionization, the attachment of electrons in a constant field is accompanied by their drift. The attachment coefficient $a = \nu_a/v_e$ similar to α

from equation (2.20) denotes the number of attachment events per electron per 1 cm path length along the field. Dissociative attachment is described by the same similarity law $a/N = f(E/N)$ as α. Avalanche multiplication of electrons is described by the equation $dn_e/dx = (\alpha - a) n_e$ and defined by the resultant coefficient $\alpha_{\text{eff}} = \alpha - a$. The ionization coefficient α for air shows a stronger dependence on E/N than the attachment coefficient a, since the ionization requires several times the energy for the dissociative attachment. This is why the curves $\alpha(E)$ and $a(E)$ in Figure 2.4 intercept. When calculated with the kinetic equation, they intercept at $(E/p)_i \approx 41$ V cm^{-1} Torr^{-1} ≈ 31 kV cm^{-1} atm^{-1}. At smaller E/p values, $\alpha < a$, so no electron avalanche can develop. It is not at all accidental that the breakdown threshold for medium-size gaps with a uniform field is close to this value. For another gas important for high voltage engineering, SF$_6$ (elegas), the α and a curves intercept at $(E/p)_i \approx 117.5$ V cm^{-1} Torr^{-1} $=$ 89 kV cm^{-1} atm^{-1} [Figure 2.4(b)]. Very high is the respective breakdown threshold, which is the principal reason, along with its other acceptable properties, why elegas is used as a gas of high electrical insulation.

2.3.3 Electron detachment

Negative O$_2^-$ ions are decomposed by collisions with molecules possessing an energy high enough to detach an electron. This process follows the scheme:

$$\left(\frac{dn_e}{dt}\right)_{\text{d}} = -\left(\frac{dn_-}{dt}\right)_{\text{d}} = \nu_{\text{d}} n_- = k_{\text{d}} N n_- \qquad (2.22)$$

where ν_{d} and k_{d} are the detachment frequency and rate constant. The rate constants are of the order of 10^{-10}–10^{-9} cm^3 s^{-1} per active molecule. Especially effective in this respect are excited nitrogen molecules: for N$_2$(A), $k_{\text{d}} = 2.1 \times 10^{-9}$ cm^3 s^{-1} and for N$_2$(B), $k_{\text{d}} = 2.5 \times 10^{-9}$ cm^3 s^{-1} (A and B are the lowest excited electronic states with the excitation energies $E^* \approx 6$ eV and 7.3 eV). Weakly excited oxygen molecules are less active: $k_{\text{d}} = 2 \times 10^{-10}$ cm^3 s^{-1} for O$_2$(a) and $k_{\text{d}} = 3.6 \times 10^{-10}$ cm^3 s^{-1} for O$_2$(b). Their number is however much larger due to a low excitation energy, so they make a larger contribution than nytrogen. The detachment rate constants calculated for any molecule sharply increase with temperature, since the proportion of excited molecules grows. For the reaction O$_2^-$ + O$_2 \to$ O$_2$ + O$_2$ + e the calculated rate constant per molecule is equal to

$$k_{\text{dO}_2} = 8.6 \times 10^{-10} \exp(-6030/T) [1 - \exp(-1570/T)] \text{ cm}^3 \text{ s}^{-1} \qquad (2.23)$$

At room temperature and zero field, this is only $k_{\text{dO}_2} \sim 10^{-18}$ cm^3 s^{-1}. At $T = 1500$ K, we have $k_{\text{dO}_2} \approx 10^{-11}$ cm^3 s^{-1}; negative ions decompose in about 10^{-8} s in dry air of atmospheric pressure. The detachment process in humid air is somewhat slower, because it involves the formation of hydrated ions $O_2^-[H_2O]_n$ ($n = 1, 2, 3 \ldots$). The electron binding energy in such a cluster grows with the number n, while the coupling of H_2O molecules in it becomes weaker with increasing n. Clusters disintegrate through a successive splitting off of H_2O molecules during collisions followed by detachment. Disintegration of hydrated ions at $T \approx 1500-2000$ K and 1 atm takes $10^{-6}-10^{-5}$ s.

The presence of field and free electrons favors the production of excited molecules, which decompose negative ions [2.19]. In general, indirect evidence (no direct measurements are available) provides $k_d \sim 10^{-14}$ cm^3 s^{-1} per any molecule. The value of $k_d \sim 10^{-10}$ cm^3 s^{-1} per active molecule corresponds to the concentration of active molecules of about 10^{-4}.

2.4 Electron and ion gases in electric fields

2.4.1 The hydrodynamic approximation

At any appreciable density of electrons and ions, when the average distance between the charged particles is much less than that characteristic of gas discharges (for instance, the spark radius), the particle community can be regarded as a continuous gas medium. The particle behavior is then described by equations of the gasdynamic type, also known as the hydrodynamic approximation. The number balances of each particle species are represented as continuity equations, whose right-hand sides contain charged particle sources. The hydrodynamic equations for a non-electronegative gas have the form:

$$\frac{\partial n_e}{\partial t} + \text{div}_e \boldsymbol{\Gamma} = \nu_i n_e - \beta n_e n_+ \qquad \boldsymbol{\Gamma}_e = -n_e \mu_e \boldsymbol{E} - D_e \, \text{grad} \, n_e$$

$$\frac{\partial n_+}{\partial t} + \text{div}_e \boldsymbol{\Gamma} = \nu_i n_e - \beta n_e n_i \qquad \boldsymbol{\Gamma}_+ = n_+ \mu_+ \boldsymbol{E} - D_+ \, \text{grad} \, n_+ \quad (2.24)$$

Here $\boldsymbol{\Gamma}$ is the particle flux density. The fluxes are of the drift and diffusion nature, thermodiffusion being of minor importance. Physically clear expressions for current densities can be derived from the equations of motion, in which the inertia terms, unessential at high pressure, have been omitted. Drift velocities are established over the time between collisions, and this fact allows particle acceleration to be neglected. For electronegative gases, we should write three instead of two equations (2.24): for n_e, n_+ and n_-, taking into account attachment, detachment and ion-ion recombination. These equations will contain nothing principally new, as compared to equation (2.24).

The diffusion coefficients D_e and D_+ are related to the mobilities μ_e and μ_+ through Einstein's ratios [equation (2.7)] involving electron and ion temperatures. The ion temperature is commonly the same as the gas temperature T. The latter is not needed for finding D_+, since ion diffusion is often of no importance at all. The temperature T is, however, necessary to find the molecular density $N = p/kT$, which determines the ionization frequency and mobility. If one needs to know T, it is necessary to include an equation for the gas energy balance with allowance for the Joule heat release and heat conduction. As for the electron temperature T_e, it may be of use for finding the coefficients of diffusion D_e and of recombination β, as well as the ionization frequency ν_i, if it is given as a function of T_e. In that case, equation (2.24) is supplemented by the hydrodynamic equation of electron energy balance, which is a generalization of the simplified electron energy equation (2.4) including a number of additional factors: electron heat conduction, electron pressure, etc. Since this equation is too sophisticated [2.1], workers dealing with weakly ionized plasmas often restrict themselves to equation (2.24), choosing reasonable values of T_e to find D_e and β. The ionization frequency and Townsend's ionization coefficient [equation (2.9)] can be regarded as a direct function of field strength by omitting the intermediate step—the direct dependence of ν_i and α on the electron energy distribution. Note that the use of equation (2.9) allows one to represent conveniently the ionization term $\nu_i n_e$ as $\alpha |\boldsymbol{\Gamma}|$.

2.4.2   The electrostatic field equation

Equation (2.24) should be supplemented with the equation for electric field strength. In the majority of cases, the induction effects are weak and the electric field can be taken to be quasipotential. Then, we have

$$\operatorname{div}\boldsymbol{E} = \frac{\rho}{\varepsilon_0} = \frac{e}{\varepsilon_0}\left(n_+ - n_e\right) \tag{2.25}$$

$$\boldsymbol{E} = -\operatorname{grad}\varphi \qquad \triangle\varphi = -\frac{\rho}{\varepsilon_0} \tag{2.26}$$

where ρ is space charge density and $\varepsilon_0 = 1/\left(36\pi\right) \times 10^{-9}\ \mathrm{F\,m^{-1}} = 8.85 \times 10^{-14}\ \mathrm{F\,cm^{-1}}$ is vacuum dielectric permittivity; φ in equation (2.26), which is Poisson's equation, is an electric potential. The space charge density in the presence of negative ions is $\rho = e\left(n_+ - n_e - n_-\right)$. Equations (2.24) and (2.25) with the chosen T, T_e and ν_i form a closed set of equations for charged particles n_e, n_-, and n_+ and field $\boldsymbol{E}$.

2.4.3 Space charge conservation and relaxation

By multiplying all equations (2.24) by electron charge e and subtracting the continuity equations from each other, we obtain the formulas

$$\frac{\partial \rho}{\partial t} + \operatorname{div} \boldsymbol{j} = 0 \qquad \rho = e\,(n_+ - n_e) \qquad \boldsymbol{j} = e\,(\boldsymbol{\Gamma}_+ - \boldsymbol{\Gamma}_e) \qquad (2.27)$$

which express the electric charge conservation law. In the expression for current density $\boldsymbol{j}$, it is sufficient to leave only one term—the electron drift current, which usually dominates over the other currents. Then, we obtain for $\boldsymbol{j}$ the familiar expression (2.2).

Substituting ρ from equation (2.25) into equation (2.27), we have

$$\operatorname{div}\left(\boldsymbol{j} + \varepsilon_0 \frac{\partial \boldsymbol{E}}{\partial t}\right) = 0 \qquad (2.28)$$

Expression (2.28) states that the vector under the divergence sign has no sources, and this allows us to regard this expression as the conservation law for the 'total current': the conductivity current $\boldsymbol{j}$ plus the displacement current $\varepsilon_0 \partial \boldsymbol{E}/\partial t$. The latter is not a current in the strict sense and is only associated with the field variation due to charge motion.

Substitute the formula for Ohm's law $\boldsymbol{j} = \sigma \boldsymbol{E}$ into equation (2.27), assuming the conductivity σ to be nearly uniform in space and expressing the field through ρ from equation (2.25). The resulting expressions are

$$\frac{\partial \rho}{\partial t} = -\frac{\sigma \rho}{\varepsilon_0} \qquad \rho = \rho\,(0) \exp\left(-\frac{t}{\tau_{\mathrm{M}}}\right) \qquad \tau_{\mathrm{M}} = \frac{\varepsilon_0}{\sigma} \qquad (2.29)$$

If there is space charge in a plasma at initial time $t = 0$ and the plasma is unaffected by an external field capable of producing or supporting space charge, the latter is dissipated with characteristic time $\tau_{\mathrm{M}} = \varepsilon_0/\sigma$, known as Maxwellian time. This characteristic time is also sufficient for the plasma field to vanish, after the external field is turned off, as is seen from equation (2.28) after substituting $\boldsymbol{j} = \sigma \boldsymbol{E}$ into it. We would like to emphasize that space charge and field will eventually turn to zero in any unaffected, neutral conducting medium—this is only a matter of time. For example, at $p = 1$ atm and $n_e = 10^{13}$ cm^{-3} in air, the plasma conductivity is $\sigma = 4.2 \times 10^{-5}\ \Omega^{-1}\,\mathrm{cm}^{-1}$ and $\tau_{\mathrm{M}} = 2.1 \times 10^{-9}$ s.

2.4.4 Ambipolar diffusion and the Debye radius

At moderate charge densities n_e and n_+, even a slight disturbance of the plasma neutrality caused by factors other than the external field gives rise to great attractive Coulomb forces (polarization fields) that prevent further charge separation. If the plasma has n_e and n_+ gradients, charges diffuse towards the lower density region, the lighter and more mobile electrons tending to run ahead of the ions. The induced polarization field tends to decelerate the electrons, but at the same time, it pull ions towards them, so that charges of opposite signs diffuse together. This phenomenon is known as *ambipolar diffusion*.

Consider diffusion either at zero external field or in the direction normal to it, say, in the radial direction in the case of a tube with flowing current. Then the field in expression (2.25) is related exclusively to the plasma polarization. Suppose the disturbance of the electrical neutrality is slight, that is $(n_+ - n_e) \ll n_e \approx n_+ = n$. To prevent its appreciable growth, the charge fluxes normal to the current should not differ much. Let us project equalities (2.24) for $\boldsymbol{\Gamma}_e$ and $\boldsymbol{\Gamma}_+$ onto the x-axis aligned with the diffusion flux and normal to the current. Eliminate the polarization field E_x from the expressions

$$\Gamma_x \approx -n\mu_e E_x - D_e \frac{\partial n}{\partial x}$$

$$\Gamma_x \approx n\mu_+ E_x - D_+ \frac{\partial n}{\partial x}$$

by dividing the first one by μ_e and the second by μ_+ and then summing them up. We will eventually find that the total flux Γ_x has a conventional diffusion form

$$\Gamma_x = -D_a \frac{\partial n}{\partial x} \qquad D_a = \frac{D_+ \mu_e + D_e \mu_+}{\mu_e + \mu_+}$$

with a common ambipolar diffusion coefficient D_a.

Suppose the plasma is in a tube of radius R. When reaching the walls, electrons and ions become neutralized, so that the plasma density at the walls turns out to be much lower. Assume the bulk recombination to be unessential: the densities n_e and n_+ are small or the pressure is low, so that the recombination losses rank below the diffusional escape to the walls. By solving the ambipolar diffusion equation derived from equations (2.24) with the boundary condition $n = 0$ at $r = R$, we can find the radial distribution of the plasma density in the tube: $n \sim J_0(2.4r/R)$, where J_0 is the Bessel function. In a steady state, plasma production by ionization is compensated by its diffusional escape to the walls and further neutralization. The diffusion loss rate $\nu_{dif} n$, equal to the ionization rate $\nu_i n$, is characterized by the diffusion loss frequency ν_{dif}. This is a value equal to the reciprocal time of ambipolar plasma diffusion from the tube bulk to the walls. The solution to the diffusion equation yields

$$\nu_{dif} = \tau_{dif}^{-1} = \frac{D_a}{\Lambda^2} \qquad \Lambda = \frac{R}{2.4}$$

where Λ is the so-called characteristic diffusion length, in our case for the cylindrical geometry.

If a plasma channel is not limited by walls, as is the case with a spark discharge, the diffusional expansion of the channel in time is also described by a law of the type $R^2 \approx 4D_a t$, where R stands for the effective radius of the plasma column.

The plasma polarization field is established such that the relatively large drift and diffusion electron fluxes would compensate one another with an accuracy of a relatively small value of the ambipolar flux. With the equality (2.24) for $\mathbf{\Gamma}_e$ and expression (2.7) for a radial polarization field (for cylindrical geometry), we can find

$$E_r \approx \left(\frac{D_e}{\mu_e}\right)\left(\frac{1}{n}\right)\frac{\partial n}{\partial r} \approx \frac{kT_e}{eR}$$

However, the charge separation in a plasma is caused by diffusion, that is, by thermal motion of electrons. Therefore, the field-induced potential difference $\delta\varphi \approx E_r R$ cannot exceed a value of the order of kT_e/e. The polarization field is induced by space charge $\rho = e\delta n = e(n_+ - n_e)$ which, according to expressions (2.25) and (2.26), is described by an approximate equality

$$\frac{e\delta n}{\varepsilon_0} \approx \frac{E_r}{R} \approx \frac{\delta\varphi}{R^2} \qquad \delta\varphi \approx \frac{kT_e}{e}$$

Hence, we can estimate the extent of plasma neutrality disturbance as

$$\frac{\delta n}{n} \approx \frac{\varepsilon_0 kT_e}{e^2 n R^2} = \left(\frac{d_D}{R}\right)^2 \qquad d_D = \left(\frac{\varepsilon_0 kT_e}{e^2 n}\right)^{1/2}$$

where d_D is the Debye radius. The necessary condition for plasma neutrality is a small Debye radius, as compared with its minimum characteristic dimension, for example, with a channel radius. For sparks, this condition is usually met with a considerable reserve. For example, at $T_e = 1$ eV and $n = 10^{13}$ cm^{-3}, the Debye radius is $d_D = 1.7 \times 10^{-4}$ cm, whereas $R \sim 10^{-2}$–10^{-1} cm.

2.5 Breakdown by avalanche multiplication

Let us consider in some detail the breakdown mechanism for a small plane gap d with a uniform field $E = U/d$, as was mentioned in Section 1.3. Suppose an accidental electron has left the cathode. As a result of multiplication, $\exp(\alpha d)$ electrons will reach the anode; in other words, one primary electron will produce $\exp(\alpha d) - 1$ new electrons and the same number of positive ions. Having reached the cathode, the ions will knock out of it $\mu = \gamma[\exp(\alpha d) - 1]$ secondary electrons, which will, in turn, produce new avalanches, and so on. This process will be enhanced with time, causing a breakdown, if the number of secondary electrons in each cycle exceeds that of primary electrons, $(\mu > 1)$. The breakdown will develop faster the larger the difference is between μ and unity. The value of μ dramatically depends on E, like the exponent in expression (2.10). Therefore, the condition $(\mu = 1)$ characterizes, with sufficient accuracy, the breakdown threshold field E_b or the breakdown voltage $U_b = E_b d$. This condition

$$\gamma \left[\exp\left(\alpha d\right) - 1\right] = 1$$

$$\alpha d = \ln\left(1 + \frac{1}{\gamma}\right)$$

is referred to as the Townsend criterion for gap breakdown, or as the criterion for self-sustained discharge ignition.

Using Townsend's formula (2.13), one can easily obtain explicit expressions for the breakdown voltage and the breakdown field strength:

$$U_b = \frac{B\,(pd)}{C + \ln\,(pd)} \qquad \frac{E_b}{p} = \frac{B}{C + \ln\,(pd)}$$

$$C = \frac{A}{\ln\,(1 + 1/\gamma)}$$

With an adequate choice of the constants A and B, the last formula describes fairly well the experimental dependencies $U_b(pd)$, or so-called Paschen curves; the poorly known quantity γ appears in the formulas only under the logarithmic sign. The breakdown mechanism concerned is characterized by a breakdown voltage minimum. With $A = 15$ cm^{-1} Torr^{-1}, $B = 365$ V cm^{-1} Torr^{-1} and $\gamma = 10^{-2}$, we have for air $C = 1.18$ and $U_{min} = \bar{e}B/A\ln(1 + 1/\gamma) = 300$ V, $(E/p)_{min} = 365$ V cm^{-1} Torr^{-1}, $(pd)_{min} = 0.83$ Torr cm. The minimum voltage is, of course, negligible compared to typical breakdown voltages for air sparks at atmospheric pressure, while the field or, more exactly, the E/p ratio is very large. Such E/p values are only obtained at a streamer head.

At the boundary of the pd range, for which the breakdown mechanism due to avalanche multiplication is still operative ($pd \sim 10^3$ cm Torr in air), the ratio $E_b/p \approx 45$ V cm^{-1} Torr^{-1} $= 34$ kV cm^{-1} atm^{-1} drops to a value slightly exceeding the critical value for the ionization $E_i/p \approx 30$ kV cm^{-1} atm^{-1} with $\alpha \approx a$.

2.6 Induced charge and circuit current

When a single charge q, or a group of charges, travels through a noncon-
ductive gas gap, an electric current is induced in the closed external circuit, although there is no conduction current between the charge and the
electrodes. This is due to the metallic electrodes being polarized by the
charge field, so that charges opposite in sign to charge q are induced on
the electrode surfaces: q_c on the cathode and q_a on the anode (Figure 2.5).
The induced charge is the larger, the closer is charge q to the electrode
surface; it grows, as charge q approaches the respective electrode, simultaneously decreasing on the other electrode. Therefore, the charge flows
from one electrode to the other, inducing current i in the external circuit.
Let us calculate this current. When charge q covers a distant $d\boldsymbol{r} = \boldsymbol{v}dt$ at
velocity $\boldsymbol{v}$, the field $\boldsymbol{E}$ created by the voltage U applied to the gap does
the work $q\boldsymbol{E}d\boldsymbol{r}$ over this charge. The work is always positive, irrespective
of the charge sign, because a negative charge travels opposite to the field.
This work is eventually done by the voltage source for time dt and is $iU dt$.
Hence, we can find the current in the external circuit

$$i = \frac{q\boldsymbol{v}\boldsymbol{E}}{U} \tag{2.30}$$

where $\boldsymbol{E}$ and $\boldsymbol{v}$ are the vectors of the electric field and charge velocity at
the point r and moment t. The $\boldsymbol{E}/U$ factor depends only on the system
geometry but not on the voltage; it remains the same even with the limit
$U \to 0$. This means that expression (2.30), as well as those for induced
charges, remains valid, when a charge moves in a zero field for some other
reason. For instance, an electron escapes an electrode at finite initial velocity due to the photoeffect.

When a charge is driven by the gap field, current i is always positive,
irrespective of the charge sign (if $q > 0$, $\boldsymbol{v}\boldsymbol{E} > 0$; at $q < 0$, $\boldsymbol{v}\boldsymbol{E} < 0$) and
is directed in the external circuit as it should—from the cathode to the

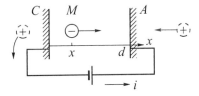

FIGURE 2.5
Schematic diagram of charges induced on the electrodes and of current through
the circuit during the charge movement in the gap.

anode. In a plane gap of length d, the geometry factor is $E/U = 1/d$, so that the current through the circuit is defined as

$$i = qv/d \qquad (2.31)$$

where v is the algebraic velocity value, which is positive, if the vectors $\boldsymbol{v}$ and $\boldsymbol{E}$ coincide. There is a displacement current in the space between the electrodes, but outside charge q if it has a finite size. This statement means that the field induced by charge q and by charges q_c and q_a is added to the external field and varies with time at any point of space. This situation obeys the general equation (2.28). If several charges move through the gap, the current of each charge is calculated with the other charges ignored, and the currents obtained are just summed up. The superposition principle is met because of the linearity of the electrostatic and motion equations.

We now find q_a and q_c charges induced on the electrode surfaces. They are accumulated there in accordance with the equations $\dot{q}_a = -\dot{q}_c = i$, which can be integrated using equation (2.30). But we must first set the initial conditions. When charge q crosses all space between the two electrodes, a charge exactly equal to q flows through the external circuit. Indeed, by integrating equation (2.30) over the whole time t_d of the charge flight through the gap, we find

$$\int_0^{t_d} i dt = \frac{q}{U} \int_0^{t_d} \boldsymbol{E} v dt = \frac{q}{U} \int_0^d \boldsymbol{E} d\boldsymbol{r} = q \qquad \int_0^d \boldsymbol{E} d\boldsymbol{r} = U$$

because the integral of field strength along the whole gap is, by definition, equal to the electrode potential difference U. Incidentally, one can see from the calculations that for the time the charge crosses part of the gap, a charge smaller than q will flow through the circuit; it is proportional to the potential difference covered by q.

Thus, the travel of charge q between the electrodes is accompanied by a flow of exactly the same charge from one electrode to the other. Therefore, when charge q comes in contact with the electrode, a maximum possible charge equal to $-q$ is induced on it. The other electrode has no charge. With this initial condition in mind, we will now integrate the equation $\dot{q}_a = i$, assuming that the charge starts its travel from the cathode. On the anode, we then have

$$q_a(t) = \int_0^t i dt = \frac{q}{U} \int_{r_c}^r \boldsymbol{E} d\boldsymbol{r} = -\frac{q\varphi(r)}{U} \qquad (2.32)$$

where r_c and r are the initial charge coordinate on the cathode and the one at time t; $\varphi(r)$ is the potential at the point r for the electrode potential difference U and for the cathode potential taken to be zero (without losing the generality); the minus is due to the opposite vectors $\boldsymbol{E}$ and $d\boldsymbol{r} = v dt$.

The induced cathode charge at this moment is $q_c(t) = -q - q_a(t)$, or

$$q_c = -q - q_a(t) = -q - \frac{q}{U}\int_{r_c}^{r} \boldsymbol{E}\mathrm{d}\boldsymbol{r} = \frac{q}{U}\int_{r}^{d} \boldsymbol{E}\mathrm{d}\boldsymbol{r} = -\frac{q\varphi'(r)}{U'} \qquad (2.33)$$

where $\varphi'(r)$ and $U' = -U$ are the potentials at the point r and on the cathode counted off from the anode, whose potential is now taken to be zero. Expressions (2.32) and (2.33), with the intermediate calculations omitted, represent the well-known Shockley-Rameau theorem for an arbitrary electrode geometry: the charge q_k induced by a charge q on the k-th conductor is equal to $q_k = -q\varphi/U$, where φ is the potential at the point q calculated on the assumption that the potential of the k-th conductor is U and that of the others is zero. Induced charges can also be calculated from the superposition principle. The net induced charge can be defined as the sum of all charges induced by each individual charge present in the space.

We will illustrate this with reference to a plane gap of length d, the charge coordinate being counted from the cathode. Then $\varphi'(x)/U' = (d-x)/d$, $\varphi(x)/U = x/d$, $q_c = -q(d-x)/d$, $q_a = -qx/d$, $q_a + q_c = -q$. Consider another example. Let us find the charge induced on an isolated metallic sphere of radius r_a (which is grounded or connected to the ground through a voltage source U) by charge q located on the radius $r > r_a$. If the sphere possesses potential U, the potential at the point r is $\varphi(r) = U_a r_a/r$, so that the geometry factor is $\varphi/U = r_a/r$. The induced charge $q_i = -q r_a/r$ is smaller than q. The other charge $-q - q_i$ is 'induced' in infinity, to which the field lines of force go.

2.7 Strongly ionized equilibrium plasma

A plasma will be termed as strongly ionized, when its ionization degree is so high that collisions of electrons with ions and one another become as essential as those with neutral molecules, although the absolute ionization degree may be as low as 10^{-4}–10^{-3}. Generally, the plasma conductivity can be described by equation (2.2) with the only difference that the electron-ion collision frequency is added to ν_m. In strong ionization, the plasma often exhibits a thermodynamic equilibrium of the ionization degree corresponding to the electron temperature. Then, the ionization degree and n_e are described by Sach's equation and are proportional to $\exp(-I/2kT_e)$.

The conductivity of equilibrium argon, nitrogen, and air plasmas (Figure 2.6) in the temperature ranges typical of arc and spark discharges can

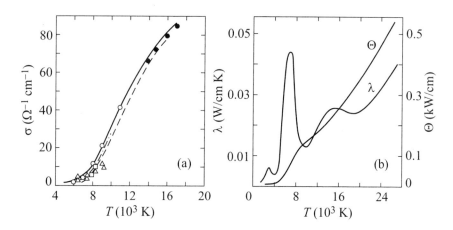

FIGURE 2.6
Electrical conductivity σ (a), heat conductivity λ, and heat flow potential Θ (b) in heated equilibrium air.

be described approximately by an interpolation formula of the type

$$\sigma \approx b \exp\left(-\frac{I_{\text{eff}}}{2kT}\right) \tag{2.34}$$

where $b = 830\ \Omega^{-1}\,\text{cm}^{-1}$, $I_{\text{eff}}/k = 72$ kK, and $8 < T < 14$ kK.

Since the behavior of an equilibrium plasma is related to its temperature, the knowledge of the temperature is important for the understanding of processes involving such a plasma, for instance, those in spark discharges. The plasma temperature is determined by its energy balance, that is, by the competition between the Joule heat released by currents and heat removal. The latter often occurs through heat conduction. Thermal conductivity depends on temperature in a complicated, nonmonotonic manner [Figure 2.6(b)], because the transport of potential energy of dissociation and ionization of atoms and molecules (reactive conduction) is added to that of kinetic energy. Molecules in a hotter plasma region dissociate to produce atoms, while atoms transported by diffusion to a cooler region recombine to release the binding energy, which was earlier expended for dissociation in the hotter region.

The ionization process in a strongly ionized plasma differs much from that in a weakly ionized plasma, where molecules are ionized by electrons that have acquired their energy directly from the electric field. For this reason, the parameters ν_i and α for weakly ionized plasmas are often taken to be direct functions of the field, omitting such 'intermediate' and unessential parameters as the mean energy or the electron 'temperature'. On the contrary, the field effect in a strongly ionized plasma seems to be smeared out: the field supplies its energy to the electron gas as a whole. Electrons become thermalized by collisions with one another, eventually acquiring a Maxwellian distribution. The gas is ionized by electrons, which have gained sufficient energy through the energy exchange with other particles rather than from the field. The thermal ionization frequency is defined by the same integral as in equation (2.8) but with a Maxwellian distribution; therefore, it depends on the electron temperature in terms of the Boltzmann law $\nu_i \sim \exp\left(-I/kT_e\right)$. How the plasma gains its energy does not really matter, but if it does gain it from the field, then T_e varies with the field strength.

It is important to understand why a weakly ionized plasma is nonequilibrium, while a strongly ionized plasma is commonly equilibrium and is easier to sustain. A weakly ionized plasma sustained by an electric field is nonequilibrium in two respects. First, there is a considerable difference between the electron temperature T_e and the gas temperature T (meaning by temperature the measure of mean electron energy, which is not entirely unacceptable). Second, the ionization degree of such a plasma is much lower than in the thermodynamic equilibrium corresponding to the actual temperature (mean energy) of electrons.

The reason for T_e being much higher than T is that, in weak ionization, the Joule heat of current proportional to $\sigma \sim n_e/N = x$ is insufficient to heat a gas of large heat capacity. The electron temperature must be high, about 1 eV = 11600 K, otherwise the number of energetic electrons would be too small to ionize the atoms, that is, there would be no conditions necessary for the production of plasma itself. A high electron temperature is achievable via increasing the electric field strength. The electron temperature in a strongly ionized plasma is approximately the same as in a weakly ionized plasma, 10^4 K, because the necessity to ionize the atoms does remain. But owing to the high ionization degree, high conductivity and great energy release, the latter turns out to be sufficient to heat the gas up to the electron temperature.

Thermodynamically equilibrium ionization is feasible due to the compensation of forward and strictly reverse processes, which follows from the detailed balancing principle. In a cold steady-state plasma, the ionization and charge loss are well balanced, but the charge loss is not a reverse process with respect to the basic mechanism of atomic ionization:

$$A + e \to A^+ + e + e \qquad (2.35)$$

Dissociative recombination or diffusion toward the walls, as in a low pressure glow discharge, are much faster processes than recombination in three-body collisions, which is a reverse process for reaction (2.35). Fast charge losses by a 'foreign' mechanism do not allow the ionization degree to reach thermodynamic equilibrium with available T_e. Due to the high gas temperature, a strongly ionized plasma does not practically contain molecular ions. Moreover, such a plasma is commonly produced at high pressures (densities), when the diffusion is very slow. So the actual mechanism of charge loss is the recombination in three-body collisions following a pattern reverse to reaction (2.35). As a result, the ionization degree can rise to the equilibrium level.

There is another feature, which makes a strongly ionized equilibrium plasma different from a weakly ionized nonequilibrium plasma and which is important for streamer-leader processes in a spark discharge. We will show below that the plasma in a leader and, obviously, in a spark channel is closer to an equilibrium plasma, while a streamer plasma is essentially nonequilibrium. It is much easier to sustain an equilibrium plasma in the sense that this requires a much lower field. To illustrate this statement, we will present some data on other discharges. The equilibrium plasma in a nitrogen arc burning at atmospheric pressure in a cooled tube of radius $r = 1.5$ cm at current $i = 10$ A is sustained by the field $E = 10$ V cm^{-1}. The other parameters are: $T = 8000$ K, the gas density $N = 10^{18}$ cm^{-3}, $n_e = 2 \times 10^{15}$ cm^{-3}, $x = n_e/N \approx 2 \times 10^{-3}$ and $E/N \approx 10^{-17}$ V cm^2. In a nonequilibrium glow discharge plasma, when the gas (nitrogen) is weakly heated, and the density and tube radius are approximately the same as above, we have $E/N \approx 2 \times 10^{-16}$ V cm^2 and $E \approx 200$ V cm^{-1}, which is an order of magnitude higher ($n_e \sim 10^9$–10^{10} cm^{-3}, $x \sim 10^{-9}$–10^{-8}). The appreciable difference is due to the different nature of the ionization process mentioned above. In both situations, electrons gain energy from the field in small portions during collisions [equation (2.4)]. But the field in a weakly ionized plasma must be high to provide an electron with energy $I \sim 13$–15 eV necessary for the ionization. The field in a strongly ionized equilibrium plasma can provide an electron only with $kT \approx 1$ eV. Electrons supply their energy to atoms, and during the energy exchange in collisions, all particles 'club together' to give off their energy to some electrons, which induce the ionization.

2.8   Equilibrium plasma arc with current

We have pointed out that leader and spark plasmas have much in common with an arc plasma. We have also noted (Section 1.4) that for a spark breakdown to develop, the plasma channel must possess a descending current-voltage characteristic (CVC). It is exactly this type of CVC that an arc possesses. It is appropriate here to give an outline of arc plasma properties

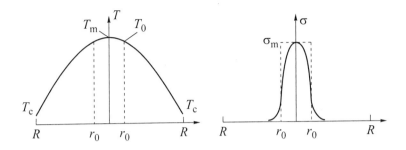

FIGURE 2.7
Schematic distributions of temperature T and conductivity σ along the arc column radius. The dashed line is an approximation in the arc model.

before discussing this subject further.

Consider a one-dimensional plasma column, whose steady state is maintained via the compensation of the Joule heat release of current ($W_1 = iE$ per unit length) by the heat removal to a cooled coaxial tube of radius R. Generally, an arc can burn in a free atmosphere as well. A spark channel is never surrounded by 'walls'. In the case of an arc, the heat can be removed by an air flow, while a leader channel just expands slowly due to the energy spread in the radial direction owing to heat conduction. But this does not principally affect the characteristics of a highly conductive channel, so the model of an arc in a cooled tube seems quite acceptable. This is especially so, because the basic parameters of a current column will be shown to depend only logarithmically on the tube radius.

Since the conductivity σ of an equilibrium plasma is the definite function of temperature, as in equation (2.34), the problem reduces to solving the energy balance equation (with λ as the diffusion constant)

$$-\frac{1}{r}\frac{\mathrm{d}}{\mathrm{d}r}rJ + \sigma\left(T\right)E^2 = 0 \qquad J = -\lambda\frac{\mathrm{d}T}{\mathrm{d}r} \qquad (2.36)$$

known as the Elenbaas-Heller equation. It has the boundary conditions: $\mathrm{d}T/\mathrm{d}r = 0$ at $r = 0$ and $T = T_c$ at $r = R$. The tube wall temperature T_c, small compared to that on the channel axis T_m, can be assumed to be zero. The field E in equation (2.36) can be considered as a given parameter. As a rule, the actually controllable parameter is current rather than voltage, so equation (2.36) should be supplemented with the equality

$$i = E\int_0^R \sigma 2\pi r\mathrm{d}r \qquad (2.37)$$

expressing Ohm's law. The set of equations (2.36) and (2.37) is nonlinear and can only be solved by approximate or numerical methods.

Consider the physical properties of a discharge, making use of the so called arc channel model. Since the conductivity [equation (2.34)] drops with decreasing temperature, the current can actually flow only inside a thin channel of radius $r_0 \ll R$. It is natural to assume $T \approx \text{const} = T_m$ and $\sigma \approx \sigma_m \equiv \sigma(T_m)$ in the channel and $\sigma = 0$ outside it (Figure 2.7). Hence, Ohm's law [equation (2.37)] takes a very simple form:

$$i = \pi r_0^2 \sigma_m E \tag{2.38}$$

Integration of equation (2.36) over the current-free region $r_0 < r < R$ yields

$$\Theta \equiv \int_0^T \lambda(T)\, dT = \frac{W_1}{2\pi} \ln \frac{R}{r_0} \qquad W_1 = \frac{i^2}{\pi r^2 \sigma_m} \tag{2.39}$$

where $\Theta(T)$ is the so-called heat flow potential (Figure 2.6). In accordance with the above approximation, we do not distinguish the temperature T_0 at the channel boundary from the temperature T_m on its axis.

It does not, however, suffice to know how the heat is removed from the channel surface through a current-free region to the tube walls. Indeed, equations (2.38) and (2.39) contain three unknown terms: T_m, r_0, and E. It is necessary to supplement these equations with another one to describe how the heat is removed from the channel bulk to its surface. With account of the small difference $T_m - T_0$ in the channel, we obtain $W_1 \approx 2\pi r_0 \lambda_m (T_m - T_0)/r_0$ in order of magnitude; here $\lambda_m = \lambda(T_m)$. This evaluation can be made rigorous by integrating equation (2.36) with respect to the range $0 < r < r_0$ on the above assumption of uniform energy sources σE^2. We obtain

$$4\pi \lambda_m (T_m - T_0) = W_1 \tag{2.40}$$

Equality (2.40) does not close the set of equations (2.38) and (2.39), because it has introduced a new unknown term—the channel boundary temperature T_0. To find this value is the same as to give a quantitative description of the rather arbitrary classification of plasma into conducting and nonconducting. When we say 'channel', we do not provide a tool for the calculation of its radius r_0. So it is natural to call a 'channel' a region, through which most of the total current flows. In this case, we deliberately 'sacrifice' the other portion of current. On the other hand, we should not include in the channel an excessively large portion of current, because, in that case, all the content of the tube will become 'a channel'. Therefore, we will assume the channel boundary to be limited by the radius r_0, at which the conductivity σ drops by a factor of e, as compared to its value σ_m on the axis. In other words, the temperature at the channel boundary will be taken to be $\sigma(T_0) = \sigma(T_m)/e$. Having accepted the relation (2.34) for $\sigma(T)$ and keeping in mind that $I_{\text{eff}}/kT_m \gg 1$, we obtain

$$T_m - T_0 \approx 2\frac{kT_m^2}{I} \tag{2.41}$$

By equating the expressions for $T_m - T_0$ from equations (2.40) and (2.41), we relate the maximum plasma temperature to the energy release power

$$T_m = \left[\frac{I}{8\pi\lambda_m k}W_1\right]^{1/2} \tag{2.42}$$

which closes the set of equations for T_e, r_0, and E. The temperature rises with W_1 slowly, even slower than with $W_1^{1/2}$, since the thermal conductivity λ_m grows with temperature. Because of a strong temperature dependence of conductivity, the former is confined within a more or less narrow range. To increase the temperature, one must greatly increase the power. This tendency is more remarkable at $T > 11000$–12000 K, when the energy balance involves radiation losses neglected in the calculation above.

The equalities obtained define all arc parameters. By giving W_1, we can find T_m from equation (2.42), r_0 from the first ratio of equation (2.39) and i from the second one; finally, we find $E = W_1/i$. Equalities (2.39) and (2.42) actually provide the arc CVC, but in the parametric form only. Let us calculate, for example, the plasma temperature for an arc in atmospheric air. At current $i = 200$ A in the channel, the measured temperature was found to be $T_m \approx 10000$–11000 K [2.20]; the experimental CVC gives the field $E \approx 2.5$ V cm^{-1}. Hence, the linear (per unit length) power is $W_1 = 500$ W cm^{-1}. With $\lambda_m \approx 1.5 \times 10^{-2}$ W cm^{-1} K^{-1} (Figure 2.6) and $I_{\text{eff}} = 6.2$ eV, as in equation (2.34), we find from equation (2.42) $T_m = 9800$ K, a value reasonably consistent with measurements. At a relatively low current $i = 10$ A, the experimental CVC gives $E = 20$ V cm^{-1} so that $W_1 = 200$ W cm^{-1}. Keeping in mind that the plasma temperature and the ionization degree are now lower, we take $\lambda_m = 2 \times 10^{-2}$ W cm^{-1} K^{-1} and $I_{\text{eff}} = 10$ eV. With the latter value, the experimental data on $\sigma(T)$ are better approximated at lower temperatures, when I_{eff} is closer to the actual ionization potential. Equation (2.42) yields $T_m \approx 7000$ K. The arc CVC for nitrogen at atmospheric pressure, plotted according to equations (2.39) and (2.42), shows a fairly good agreement with the measured CVC.

For a better understanding of the cause-effect relationships in an arc, it is desirable to find explicitly the relationships between its parameters and the controllable value of discharge current. We can come very close to obtaining them on the assumption of constant thermal conductivity, $\lambda(T) = \text{const}$. This approximation, no doubt, introduces a quantitative error, but it reflects correctly the qualitative character of the results.

Having omitted the algebraic manipulations, we will write down the appropriate formulas derived from equations (2.34), (2.39), and (2.42):

$$T_{\mathrm{m}} = \frac{I/2k}{\ln\left(8\pi^2 \lambda b k T_{\mathrm{m}}^2/I\right) - \ln\left(i/R\right)} \approx \frac{\mathrm{const}}{\mathrm{const} - \ln\left(i/R\right)}$$

$$\sigma_{\mathrm{m}} = \left(\frac{Ib}{8\pi^2 \lambda k T_{\mathrm{m}}^2}\right)^{1/2} \frac{i}{R} \approx \frac{i}{R} \frac{\mathrm{const} - \ln\left(i/R\right)}{\mathrm{const}}$$

$$W_1 = \frac{8\pi \lambda k T_{\mathrm{m}}^2}{I} \approx \frac{\mathrm{const}}{\left[\mathrm{const} - \ln\left(i/R\right)\right]^2}$$

$$E = \frac{8\pi \lambda k T_{\mathrm{m}}^2}{Ii} \approx \frac{\mathrm{const}}{i\left[\mathrm{const} - \ln\left(i/R\right)\right]^2}$$

$$r_0 = R\left(\frac{\sigma_{\mathrm{m}}}{b}\right)^{1/2} = \left(\frac{I}{8\pi^2 \lambda k T_{\mathrm{m}}^2 b}\right)^{1/4} (iR)^{1/2} \qquad (2.43)$$

Assuming these relationships to be explicit, we make a slight error, for we have to take the logarithm of the quantity with T_{m} constant. This, however, can be justified by the fact that the temperature varies within a narrow range, and the logarithm is only slightly sensitive to this variation.

Expressions (2.43) show that the current rise is accompanied by a nearly proportional increase in the plasma conductivity, allowing it to pass a higher current. Due to the strong dependence of σ on T in equation (2.34), the temperature rises much slower. The power, which is closely related to the temperature, is limited by the energy transport to the walls through heat conduction; it also grows slowly with current. This is why a weaker field, $E = W_1/i$, is sufficient to sustain a plasma, if the current increases. As a result, we have a descending CVC. The channel radius changes in such a way $(r^2 \sim i)$ that the current increase is accompanied by an increase in the channel volume rather than in the current density, which grows as slowly as the temperature: $j \sim i/r_0^2 \sim T_{\mathrm{m}}$. At this current, all the parameters only depend logarithmically on the tube radius R, except for the channel radius which shows a somewhat stronger dependence, $r_0 \sim R^{1/2}$.

3

Basic Processes in Spark Breakdown

3.1 Streamer as an ionization wave

3.1.1 Physical background

The basic phenomenon of any discharge, including a spark, is the ionization of gas atoms and molecules by electron impact. The ionization rate depends dramatically on the electric field strength, so it is only in sufficiently strong fields that the rate may become high enough to produce a plasma. The quantitative criterion for 'sufficiently strong fields' varies with the gas and other conditions. In negative gases, such as atmospheric air or halogen-containing mixtures (for example, elegas SF_6) and under the nonstationary conditions of a fast spark discharge, the gas can become ionized only if the rate of ionization exceeds that of electron attachment. In air, this occurs at $E > E_i \approx 30$ kV cm^{-1} ($E_i/p \approx 40$ V cm^{-1} Torr^{-1}) and in elegas at $E > E_i \approx 90$ kV cm^{-1} ($E_i/p \approx 120$ V cm^{-1} Torr^{-1}). But even in a positive gas like nitrogen, in which there is no attachment in the absence of oxygen, the ionization in a field less than 30 kV cm^{-1} is so slow at atmospheric pressure that no fast discharge can develop, even if all other electron losses (diffusional and recombinational) are neglected.

Generators available at modern high voltage laboratories can provide voltage U as high as millions of volts. Sparks produced at such voltages in gaps with strongly nonuniform fields may be tens or hundreds of meters long. Figure 1.1 illustrates a spark ignited in air at $U = 2$ MV. It has started from a spherical anode of radius $r_a = 10$ cm and struck a grounded plane cathode located at distance $d = 10$ m. The field of 200 kV cm^{-1} at the anode surface decreases with distance approximately as $E = Ur_a/r^2$ and drops below the critical ionization value of 30 kV cm^{-1} for a 25 cm radius, that is, at 15 cm from the anode. Then the spark travels through external field $E < E_i$, clearly too low for the ionization to go on, and eventually reaches the cathode in a very weak field, $E \approx 2Ur_{\mathrm{a}}/d^2 \approx 0.2\,\mathrm{kV\,cm^{-1}}$ (the factor 2 is due to the anode image charge in the metallic plane, see Section 3.2.1). Such situations are also characteristic for other gases and for other nonuniform field geometries. A common feature is that a spark ignited at the smaller electrode, where the field is highest, is free to travel through anomalously low fields known to be unable to sustain ionization.

This fact clearly indicates that a plasma channel capable of propagating in a weak external field induces at its head, or tip, a field strong enough to maintain the ionization. The region of intensive ionization, moving together with a strong field, transforms the gas to a plasma. Available observations show that the radius of a traveling plasma channel is very small, thousands of times smaller than its length, and that the channel looks like a filament. A hardly noticeable, often unobservable, lateral expansion of the filament has been found to occur much slower than its elongation, while the head radius seems to be independent of the filament length at all. This means that the channel head, where the gas is transformed to a plasma, does not practically vary with time, propagating forward with the channel speed. In other words, the head structure remains stationary in the related coordinate system, or, more exactly, quasistationary if the channel velocity shows a relatively slow time variation. It is then appropriate to speak of a strong field wave and an ionization wave, the only difference from a one-dimensional plane wave (shock wave, combustion wave, etc.) being its extremely small transverse dimension comparable with its front width.

It is clear that a channel travels due to the production of new plasma regions, when the gas is ionized by strong head field E_{m}. This process is initiated by free electrons always present in the air because of cosmic radiation or the Earth's radioactive background. The excited molecules produced emit photons, giving rise to primary electrons in front of the head. During the ionization, the electron density n_{e} increases with time in an avalanche-like manner, starting from the initial background value n_0, namely,

$$\frac{\mathrm{d}n_{\mathrm{e}}}{\mathrm{d}t} = \nu_{\mathrm{i}}n_{\mathrm{e}} \qquad n_{\mathrm{e}}(t) = \exp\left(\int_0^t \nu_{\mathrm{i}}\mathrm{d}t\right) \qquad (3.1)$$

where ν_{i} is the frequency of gas ionization by electron impact, strongly dependent on the field strength E. The main source of field is the head space charge. The appearance of a strong field is, in a sense, due to the high head potential U, which, owing to the plasma conductivity, can greatly exceed the external potential at the head induced by the voltage applied to the electrode. Let us assume, for the sake of definiteness but without sacrificing the presentation generality, that the voltage is applied to the anode (Figure 3.1). In the limiting case of the perfectly conducting channel, the

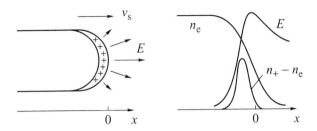

FIGURE 3.1
Schematic diagram of a cathode-directed streamer and qualitative distributions
of field E, electron density n_e, and 'space charge' density $(n_+ - n_e)$ along the
axis in the streamer head region.

head potential U, as well as the total channel potential, just coincides with
the anode potential. In the real case of finite, sometimes very low, plasma
conductivity, the head potential is lower than the electrode potential but is
still sufficient for the head charge to induce a strong electric field. Let r_m
denote a conventional head radius (radius of the tip 'curvature'), then the
field at the head top, which is the frontmost point of the elongating channel,
equals, in order of magnitude, to $E_m \approx U/r_m$. This is strictly valid only for
a separate, uniformly charged sphere. But, say, for a metallic hemisphere
adjacent to a metallic cylinder of the same radius r_m, the field at the hemisphere top is about half that of a sphere with the same U and r_m. This
conclusion follows from electrostatics, and we will comment on it later, in
Section 3.2. Here, our qualitative considerations and evaluations ignore the
factor of 2. Therefore, use will be made of all electrostatic relations valid
for a separate sphere. In this approximation, the head charge Q, the head
capacitance C, and the field E_m, maximal at the head, have the form

$$Q = 4\pi\varepsilon_0 U r_m \qquad C = \frac{Q}{U} = 4\pi\varepsilon_0 r_m \qquad E_m = \frac{Q}{4\pi\varepsilon_0 r_m^2} \qquad (3.2)$$

Although the near-head field is largely induced by the head charge, an
additional source of a high head potential may be the charge distributed
along the conducting channel. In case of a perfect conductor, the contributions of both components are identical (Section 3.2.1).

The head charge Q is partly built up by electron escape from the head
into the channel under the action of its electric field; these electrons eventually reach the anode. In the limiting case of perfect conduction, the channel behaves like a metallic rod, whose conductivity is 'infinite' and the field
is zero. Their product, or the current density, is, however, finite, because,

as the rod becomes longer, it needs more positive charge to keep its potential constant and equal to the anode potential. The latter condition is imposed by electrostatics. In the same limit of a perfectly conducting head, its charge is concentrated on its surface, very much like the channel charge is concentrated on its cylindrical surface. In reality, a plasma channel has a finite conductivity and a finite longitudinal field (potential gradient). This potential gradient causes current to flow from the anode into the channel, pumping positive charge into it. In both the channel and head, the charge is now distributed in space and becomes space charge.

Turn now to the process of streamer propagation through a gas. The streamer velocity v_s and the charge density behind the head, n_c, are established by themselves via a self-regulated process, which depends on the 'external' parameters: maximum field at the head, E_m, or its potential U, and radius r_m. The values of v_s and n_c are established such that for the time of flight through a strong field 'attached' to the charged head (the time of ionization development), the local plasma could gain a sufficiently high conductivity to push out or reduce the longitudinal field. By the latter, we mean the axial field component, which drops to a value too small to maintain the same ionization rate as at the head. The requirement of the field being pushed out of this site is equivalent to that of the head charge leaving the head to occupy a new site, while the head is displaced with velocity v_s at a distance Δr of the order of its size, $\Delta t \approx \Delta r / v_s$. The charge is displaced by current, actually by electrons that drift in the head field towards the anode, in the direction opposite to the streamer elongation. The electrons flow over the ionic charge of the older head, partly exposing the ions of the new one (Figure 3.1). This is the way a streamer develops.

3.1.2 The problem formulation

We will now formulate the problem of a streamer head as an ionization wave. The qualitative picture of streamer propagation will be approached approximately. This can provide knowledge of the causative relationships among ionization wave parameters and give their estimates. Clearly, a comprehensive description of a developing streamer involves two more or less independent problems. One problem is that of ionization wave propagation, its velocity, the degree of ionization, or the initial channel conductivity. Since the wave is quasistationary in the coordinate system related to its front, its characteristics largely depend on processes occurring within the front (the head) and, naturally, on some ambient parameters. These are maximum field E_m or head potential U, interrelated through the head radius. A special problem is that of the head radius r_m, which coincides with the initial channel radius. The mechanisms by which a plasma conductor acquires a definite and fairly small effective radius (effective because a plasma conductor does not have such a sharp boundary as a metallic one) seem to go far beyond the steady state processes of ionization, charge movement, and electrostatic field formation in a mature ionization region. We should recognize that these mechanisms are not quite clear at present. This important issue will be discussed in Section 3.5.4, but for the time being, the wave radius r_m will be taken to be a known quantity. This is justifiable because the radius size does not follow directly from the wave behavior. One may suggest from some indirect data available that it is $r_m \approx 10^{-2}-10^{-1}$ cm for atmospheric air.

The second problem is that of the streamer structure and the processes occurring in them. These processes are slow relative to the streamer velocity and take a characteristic time larger than the time necessary for a streamer to cover a distance of about the head size. It is because of the low rate of kinetic, electrical, and plasma processes in a long streamer (or because of the low rate of charge variation in the plasma bulk) that an ionization wave can be considered as being quasistationary. Indeed, when a long streamer increases its length by several head radii, the field induced by the channel charges at its head remains practically unchanged. Here, we will focus on the first problem (Figure 3.1).

According to the generally accepted, mostly speculative, views and scarce experimental data, initial electrons arise in front of an ionized streamer head due to the absorption of energetic photons emitted by the plasma. Photoelectrons give rise to elementary electron avalanches, which move towards the streamer head in its strong positive field. If the number of initial electrons is large enough, the avalanches will overlap one another, before they come in contact with the head. So there is a certain initial electron density n_0 in a relatively weak field in front of the ionization wave, which is many orders of magnitude less than the final density n_c behind the wave.

The actual density of initial electrons at the wave front decreases with distance r from the head center as $n_0 \sim r^{-2} \exp\left(-r/\lambda\right)$, where λ is a distance of about the path length for photon absorption (initial electrons are thought to be produced by photons). Moving together with the wave is the head halo of weak primary ionization extending at distance λ. At atmospheric pressure, λ is limited to about 10^{-1} cm and cannot be much larger than r_m.

The processes of ionization, electron drift, and space charge formation, as well as the field build-up, are the principal events in an ionization wave, which can be described by the continuity equations for the densities of electrons n_e and ions n_+ [equations (2.24)] and by the equation for the electric field [equation (2.25)]. In the former, we can neglect charge diffusion, which is slow compared to the drift, and the slow drift of ions; these do not show displacement over the time of the head displacement, $r_m/v_s \approx 10^{-10}-10^{-9}$ s (at $v_s \approx 10^7-10^9$ cm s^{-1} and $r_m \approx 0.01-0.1$ cm). Having an axial symmetry, this process is two-dimensional (Figure 3.1) and quite sophisticated mathematically. This difficulty can be overcome by using approximate one-dimensional models that do not distort the qualitative picture but only introduce a quantitative error. In the treatment of field distribution, the head front and the region before it will be taken to be 'spherical'. The radial field in the cylindrical channel adjacent to the head is appreciably lower than the maximum field at the head. Otherwise, the wave front would move not only forward but also in the transverse direction, rapidly expanding the channel; this, however, contradicts the initial assumptions. The axial field having a maximum at the head front decreases from the head to the cylindrical channel and drops to the longitudinal field value in the channel.

For a 'spherical' head top and 'spherical' space before the wave front, equation (2.25) has the form

$$\frac{1}{r^2}\frac{\mathrm{d}}{\mathrm{d}r}r^2 E = \frac{e}{\varepsilon_0}\left(n_+ - n_e\right) \qquad \frac{\mathrm{d}E}{\mathrm{d}r} = \frac{e\Delta n}{\varepsilon_0} - \frac{2E}{r} \qquad (3.3)$$

with $\Delta n = n_+ - n_e$. The field at this moment has a maximum, $E = E_m$, at distance r_m from the center, where $\mathrm{d}E/\mathrm{d}r = 0$. With equation (3.3), the maximum field is related to r_m and the space charge density at this point, $\rho = e\Delta n_m$, by the ratio

$$E_m \approx \frac{e\Delta n_m r_m}{2\varepsilon_0} \qquad (3.4)$$

The radius r_m, at which the field strength is maximum, should be considered as an effective head radius. At $r > r_m$, the ionization degree before the head and the space charge at a small distance from it become small. The field decreases radially and along the extension of the x-axis as $E \approx E_m\left(r_m/r\right)^2$, as for a separate charged sphere. Behind the maximum in the axial direction, the effect of geometry gives way to that of space charge. Closer to the sphere center, where the hemispherical head changes to the cylindrical channel, the axial field decreases under the action of space charge together with the density of that charge. The lines of force become straight, because the field in the channel bulk is mostly longitudinal, although there is a weak radial field induced by the charge distributed along the channel. Figure 3.1 shows the distribution of longitudinal field E, electron density n_e, and space charge $e\Delta n$ along the axis in the head vicinity.

In calculating the axial distributions of E, n_e, and Δn, we will neglect the insignificant transverse electron motion and use the balance equations (2.24) for electrons and ions for the plane case. The distribution patterns in Figure 3.1 move along the axis at velocity v_s without distortion. In such a wave, $E = E(x - v_s t)$, $n_e = n_e(x - v_s t)$, that is, the distributions remain steady in the coordinate system moving together with the streamer head. With the neglect of charge diffusion and ion drift, equations (2.24) become

$$-\frac{d}{dx} n_e (v_s + v_e) = \nu_i n_e \qquad -v_s \frac{dn_+}{dx} = \nu_i n_e \qquad (3.5)$$

where $v_e = \mu_e E$ is the electron drift velocity module. The first of equations (3.5) differs from the simple ionization kinetic equation (3.1) by having an unimportant term, $n_e dv_e/dx$. Indeed, the time derivative in the latter describes electron multiplication in an avalanche moving at drift velocity: $dn_e/dt = \partial n_e/\partial t - v_e \partial n_e/\partial x$. As one goes to the ionization wave coordinates, equation (3.1) changes to equation (3.5) without the above term associated with field variation along the x-axis. This term can be ignored for strong waves moving much faster than drifting electrons, $v_s \gg v_e$.

Equations (3.5), multiplied by e and subtracted from one another, yield

$$\frac{d}{dx}(en_e v_e - e\Delta n v_s) = 0 \qquad \Delta n = n_+ - n_e$$

which is equivalent to equation (2.28) and expresses the total current conservation law. At a fixed point of space, the field induced by head charges varies, as the head moves on, creating displacement current. In the head coordinates, the gas flows into the head at velocity v_s, and if space charge ρ exists in the medium, it flows into the head together with the gas. When one goes from equation (2.28) to equation (2.25) for the moving coordinates and plane geometry, the displacement current turns to the convective current of charge $-\rho v_s = -en_e v_s$.

Let us integrate the equation written above. There are neither electrons nor space charge in front of the ionization wave. Hence, the integration constant equal to the total current density vanishes in the given approximation:

$$j_t = en_e v_e - e\Delta n v_s = 0 \qquad \Delta n v_s = n_e v_e \qquad (3.6)$$

Electron drift current flowing in the direction opposite to space charge convective current compensates it, so that there is no current anywhere in the wave coordinate system. This is because the field far ahead of the wave, where there are no charges at all, shows no variation in the framework of the problem formulation that the wave is steady within its coordinates. This accounts for zero displacement current in front of the wave. But by virtue of the total current conservation, there is no current behind the wave either, where $n_e \neq 0$. Therefore, there is neither space charge nor field behind the wave: $\Delta n \to 0$, $E \to 0$. The plasma becomes electrically neutral.[1] The

---

[1]The actual presence of current j_c and field E_c behind the head corresponds to the

ionization wave problem can be solved using equation (3.3) with $r = x$, the first of equations (3.5), equation (3.6) together with the ratio (3.4), if the function $\nu_i (E)$ and initial electron density in front of the head are given.

3.1.3 Ionization wave parameters

We will take advantage of the fact that the space charge density inside a sphere of radius r_m is much higher than outside it. (Otherwise there would be no maximum, and the field would have to keep on rising with increasing radius.) The space charge effect in front of the head will be neglected, and the field variation law there, $E \sim r^{-2}$, will be extrapolated as far as the head surface, $r = r_m$: $E = E_m (r_m/r)^2$. For convenience, the ionization frequency increasing with the field will be approximated by the power function $\nu_i = \nu_m (E/E_m)^k$. Consider, as an illustration, strong waves only, with $v_s \gg v_{em} \equiv v_e (E_m)$. The first of equations (3.5) will be integrated from point r_0 with $n = n_0$ to the head surface with $n_e (r_m) \equiv n_m$:

$$\ln \frac{n_m}{n_0} \approx \int_{r_m}^{r_0} \frac{\nu_i}{v_s} dr = \frac{\nu_m r_m}{(2k - 1) v_s} \tag{3.7}$$

$$v_s \approx \frac{\nu_m r_m}{(2k - 1) \ln (n_m/n_0)} \tag{3.8}$$

On the other hand, with equations (3.6), (3.4) and with $v_e = \mu_e E$, we have

$$v_s = \frac{e \mu_e n_m r_m}{2\varepsilon_0} \tag{3.9}$$

By equating (3.9) and (3.8), we find the electron density on the head surface near the axis, considered as the ionization wave front:

$$n_m = \frac{2\varepsilon_0 \nu_m}{(2k - 1) e \mu_e \ln (n_m/n_0)} \tag{3.10}$$

The density at the wave front, n_m, exceeds, by many orders of magnitude, the value of n_0, since most electron generations are produced before the wave front; otherwise no charge could be formed in the head behind it. However, the last few generations, which will be shown to be responsible for an order of magnitude higher plasma density, n_c, are produced in the head itself. To calculate n_c, we will neglect the wave curvature, assuming it to be plane. This assumption can be justified by the fact that the wave thickness is small relative to radius r_m. Then, according to equations (3.3)

---

presence of displacement current in the un-ionized medium before the head even in its own coordinates, but these quantities cannot be found without analyzing the processes in the channel itself (Section 3.2).

and (3.6), we have behind the wave front

$$\frac{dE}{dx} = \frac{e\Delta n}{\varepsilon_0} = \frac{e\mu_e E n_e}{\varepsilon_0 v_s}$$

This equation is divided by the first of equations (3.5) without the term containing v_e. We obtain a differential equation for $n_e(E)$ and a quadrature for $x[E, n_e(E)]$. By integrating the equation for $n_e(E)$ over the whole wave thickness from its front, with $E = E_m$ and $n_e = n_m$, to the region behind the wave, with $E \approx 0$ and $n_e = n_c$, we find the final plasma density

$$n_c = n_m + \frac{\nu_m \varepsilon_0}{e\mu_e k} \approx \frac{\nu_m \varepsilon_0}{e\mu_e k} \qquad (3.11)$$

The last approximate term is due to the condition $n_c \gg n_m$. Indeed, from equations (3.11) and (3.10), we have

$$\frac{n_c - n_m}{n_m} = \frac{2k-1}{2k} \ln\left(\frac{n_m}{n_0}\right) \approx \ln\left(\frac{n_m}{n_0}\right) \gg 1 \qquad (3.12)$$

The effective thickness of the ionization region inside the head, estimated from the above quadrature, has the scale:

$$\Delta x \sim \frac{r_m n_m}{n_c} \sim \frac{r_m}{\ln(n_m/n_0)} \qquad (3.13)$$

It is proportional to the head radius but is much smaller, which justifies the assumption of a 'plane' wave. Expressions (3.9) and (3.11) relate the wave velocity to the final plasma density in the channel behind the wave:

$$v_s = \frac{e\mu_e n_c r_m}{2\varepsilon_0 \ln(n_m/n_0)} \qquad (3.14)$$

Before analyzing the physical meaning of relationships among wave parameters, we will give an example of numerical calculation for a fast streamer in atmospheric air. The Townsend coefficient in the field range $E \approx 80-300$ kV cm^{-1} ($E/p \approx 100-400$ V cm^{-1} Torr^{-1}) with an acceptable error of 10–20% is approximated by the power function

$$\alpha \approx 4500 \left(\frac{E\left[\text{kV cm}^{-1}\right]}{300}\right)^{3/2} \text{cm}^{-1}$$

and corresponds to $\nu_i = \alpha v_e \sim E^{5/2}$, $k = 2.5$.

It has been mentioned that the lowest accuracy is exhibited by the parameter r_m which, according to indirect data, varies from 10^{-2} to 10^{-1} cm. Here r_m is taken to be 10^{-1} cm and $E_m = 300$ kV cm^{-1}. For an isolated sphere, this field value is obtained at potential $E_m r_m = 30$ kV. It will be shown below that the potential is to be twice as large, up to 60 kV, for a

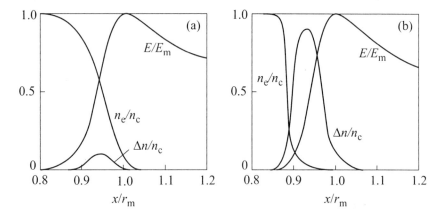

FIGURE 3.2
Calculated distributions of E, n_{e}, and $\Delta n = n_{+} - n_{\mathrm{e}}$ along the axis of an ionization wave in atmospheric air for the streamer head radius $r_{\mathrm{m}} = 0.1$ cm: a strong wave, $E_{\mathrm{m}} = 300$ kV cm^{-1}, $n_{\mathrm{c}} = 4.7 \times 10^{14}$ cm^{-3}, $v_{\mathrm{s}} = 4.9 \times 10^8$ cm s^{-1} (a); a weak wave, $E_{\mathrm{m}} = 80$ kV cm^{-1}, $n_{\mathrm{c}} = 2 \times 10^{13}$ cm^{-3}, $v_{\mathrm{s}} = 5.8 \times 10^5$ cm s^{-1} $\ll v_{\mathrm{e\,m}} = 2.2 \times 10^7$ cm s^{-1} (b).

hemispherical head of the same radius, which corresponds to many experimental conditions (voltages in air may be higher).

At $E_{\mathrm{m}}/p \approx 400$ V cm^{-1} Torr^{-1}, the drift velocity is $v_{\mathrm{e\,m}} \approx 8 \times 10^7$ cm s^{-1} and the estimated mobility is $\mu_{\mathrm{e}} = v_{\mathrm{e}}/E_{\mathrm{m}} \approx 270$ cm^2 V^{-1}s^{-1}. The Townsend coefficient at the wave front is $\alpha_{\mathrm{m}} = 4500$ cm^{-1} and the ionization frequency is $\nu_{\mathrm{m}} = 3.6 \times 10^{11}$ s^{-1}. With $n_0 = 10^6$ cm^{-3}, this value will be shown to be acceptable, especially if we recall that the other parameters depend on it logarithmically only.

Expression (3.10) yields $n_{\mathrm{m}} = 2.2 \times 10^{13}$ cm^{-3} and $\ln(n_{\mathrm{m}}/n_0) = 17$, which means that $K_{\mathrm{m}} = \ln(n_{\mathrm{m}}/n_0)/\ln 2 = 24$ electron generations are produced before the wave front. It follows from formula (3.11) that the ionization wave leaves behind it a plasma of density $n_{\mathrm{c}} = 3.2 \times 10^{14}$ cm^{-3}. This is the plasma density in the streamer channel behind the head; the plasma conductivity is $\sigma = e\mu_{\mathrm{e}} n_{\mathrm{c}} = 1.4 \times 10^{-2}$ Ω^{-1} cm^{-1}. Approximately four electron generations are produced behind the wave front. The streamer velocity estimated with expressions (3.8) or (3.9) is $v_{\mathrm{s}} = 5.3 \times 10^8$ cm s^{-1}, which is 6.6 times higher than $v_{\mathrm{e\,m}}$; it is a really strong wave. Such velocities have been registered experimentally under the conditions described. The wave phase velocity v_{s} greatly exceeds electron drift velocities, $v_{\mathrm{e}} \leq v_{\mathrm{e\,m}}$; therefore, electrons cannot considerably expose the ion charge, $\Delta n_{\mathrm{m}}/n_{\mathrm{m}} \approx$ 15%. The head charge is $Q \approx 4\pi r_{\mathrm{m}}^2 \varepsilon_0 E_{\mathrm{m}} \approx 3 \times 10^{-9}$ C.

The streamer parameters calculated in terms of a simple theory are quite

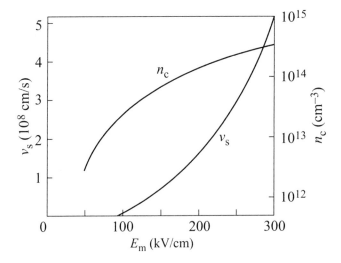

FIGURE 3.3
Calculated dependence of the streamer velocity and plasma density behind the
head on maximum field at the head.

reasonable and agree with both measurements and calculations from more
sophisticated problem formulations (Chapter 5). Figure 3.2 presents the
axial distributions of wave parameters calculated without simplifying equations (3.3) and (3.5) for the same conditions. The dependence of v_s and n_c
on E_m is shown in Figure 3.3.

3.1.4 Relationships among wave parameters

The physical sense of equation (3.8) for strong wave velocity is so evident
that this parameter can be derived from a simple reasoning to an accuracy
of a numerical factor. The field before the head and the ionization rate drop
at a distance of the head radius r_m. During the period of time $\Delta t \sim r_m/v_s$,
while the head is covering this distance, the number of electron generations
produced is $K_m = \ln(n_m/n_0)/\ln 2$. It takes the time of $\nu_m^{-1}\ln 2$ for one
generation to be produced at maximum field and by a factor of K_m longer
time for K_m generations. Since the ionization occurs in a lower field, the
actual time necessary for this is longer by a factor of $(2k - 1)$. This is
exactly what equation (3.8) shows. Evidently, the rate is largely defined
by the ionization frequency at a maximum field, or by the field itself.

Expressions (3.14) and (3.9) relating the wave velocity to the plasma
densities behind the wave and at its front also have a clear physical mean-

ing. The subsurface layer of the head front has charge

$$q_1 = \int e\Delta n dx = \varepsilon_0 E_{\mathrm{m}}$$

per unit front area. For the time the head is displaced at a distance of the layer thickness, Δx, in the laboratory coordinates, this charge, according to expression (3.13), is transported by conductivity current $j \sim en_{\mathrm{c}}\mu_{\mathrm{e}}E_{\mathrm{m}}$ to a new position. Therefore, this charge is $q_1 \sim j\Delta t \sim j\Delta x/v_{\mathrm{s}}$, leading to expression (3.14).

This problem can be approached somewhat differently. Space charge in the plasma behind the wave front is dissipated for a Maxwellian time $\tau_{\mathrm{M}} = \varepsilon_0/\sigma = \varepsilon_0/en_{\mathrm{c}}\mu_{\mathrm{e}}$ (Section 2.4). The charge must move on together with the head to take a new position for the time $\Delta t \sim \Delta x/v_{\mathrm{s}}$; hence, $\tau_{\mathrm{M}} \sim \Delta t$, giving the same expression (3.14). Similar reasoning leads one to equation (3.9) for v_{s}, but here we deal with the charge before the wave front rather than with the main space charge of the wave. This layer thickness is about r_{m}, and the Maxwellian time corresponds to a lower n_{m} at the front. The formula $\tau'_{\mathrm{M}} = \varepsilon_0/en_{\mathrm{m}}\mu_{\mathrm{e}} \sim \Delta t' \sim r_{\mathrm{m}}/v_{\mathrm{s}}$ leads to equation (3.9).

It follows from expressions (3.10) and (3.11) that the plasma densities at and behind the ionization front are not directly related to the head radius but are only defined by the field in it, E_{m}. The streamer velocity is, on the contrary, proportional to this radius. The wave thickness is also proportional to the radius, while the wave front displacement at this distance takes the Maxwellian time τ_{M}, defined only by the plasma density; hence, $v_{\mathrm{s}} \sim r_{\mathrm{m}}/\tau_{\mathrm{M}}$.

This reasoning can be extended to interpret expression (3.11) for the final plasma density in the ionization wave. Space charge largely arises in the head bulk due to a high degree of gas ionization in front of the head. Few electron generations are produced in the head, before it takes a new position. Therefore, the displacement time for most space charge, or the Maxwellian time for its dissipation in the previous position, $\tau_{\mathrm{M}} \approx \varepsilon_0/en_{\mathrm{c}}\mu_{\mathrm{e}}$, is equal, in order of magnitude, to the time necessary to produce an electron, ν_{m}^{-1}. As a result, we have expression (3.11).

According to expression (3.11), the wave velocity is proportional to the product $n_{\mathrm{c}}\mu_{\mathrm{e}}$. The lower is the electron mobility, the slower is their displacement relative to ions, and the smaller is $\Delta n/n_{\mathrm{e}}$, so the gas must be ionized to a much greater degree to provide the head charge $Q \sim e\Delta n r_{\mathrm{m}}^3$ and field $E_{\mathrm{m}} \sim Q/r_{\mathrm{m}}^2$ (both proportional to Δn) to achieve a particular ionization frequency ν_{m}. Since v_{s} is phase velocity, it would seem to have no limits; but in reality, its upper limit is light velocity c, because initial electrons are produced by photons emitted from the plasma channel. Hence, formula (3.14) with $v_{\mathrm{s}} = c$ yields the upper limit for a possible plasma density in the ionization wave; for air, $(n_c)_{max} \approx 10^{16}$ cm^{-3}. This, of course, does not exclude further ionization far behind the streamer head, as its temperature rises.

The wave velocity weakly, logarithmically depends on the density of initial electrons, while the final plasma density is independent of it, according to equation (3.12). What actually matters is that initial electrons must appear from this or that source. No doubt, if the rate of their production were too low, the question would arise concerning the statistical time delay of the onset of the streamer movement, its possible discontinuity and branching.

The problem of initial electrons has always been a subject of interest to those concerned with the theoretical treatment of ionization waves. The general belief has been that it is the initial electron concentration that determines many of the wave properties. It follows from the foregoing, however, that an electron concentration sufficient to initiate ionization without a time delay at any point on the wave path will have little effect on the streamer process.

The velocity of a fast wave ($v_s \gg v_e$) depends only on the ionization frequency. From equation (3.8), the velocity is, naturally, proportional to the frequency and is independent of electron mobility. From the second formula of (3.6) and from equation (3.8) with constant E_m and r_m, the ratio $\Delta n / n_e$ characterizing the displacement of electrons relative to ions is proportional to mobility μ_e. But since the value of space charge at given E_m and r_m is also constant, the plasma density is $n_e \sim 1/\mu_e$; it is inversely proportional to electron mobility at v_s corresponding to the given E_m.

We can imagine a hypothetical gas, in which electrons would immediately attach themselves to molecules to form negative ions; so negative charges possess a mobility μ_e/μ_+, which is 300 times less than that of electrons. Even under these exotic conditions, the wave velocity would remain more or less constant, but the gas in the wave would become ionized to a degree $\mu_e/\mu_+ \approx 300$ times greater (150 times, if we take into account that positive and negative ions drift with nearly the same velocity in opposite directions). In reality, the effect of mobility on the ionization wave is better revealed, when the gas density, rather than its chemical composition, is changed.

The ionization wave behavior has been examined in several publications [3.1–3.6], where qualitative formulas of the type (3.8) and (3.11) for v_s and n_c were given in this or that form. We have tried to present a consistent but simplified idea of an ionization wave and to derive working expressions for parameter estimation, retaining, where possible, all numerical factors.

3.2 Electrical properties of a plasma channel

3.2.1 Capacitance and charge

Although a streamer has a finite, often moderate conductivity, it is still useful to estimate its parameters on the assumption of perfect plasma conduction. This assumption would permit many calculations to be made analytically in order to reveal the physics of the streamer plasma and, in some cases, even to obtain reasonable quantitative results. An adequate accuracy of electrostatic characteristics can be obtained only by numerical methods (Chapter 5).

Assume a streamer to be a straight cylindrical rod with a hemispherical head of the same radius. Owing to ideal conductivity, all points of the rod have the same potential U. Let us define the electrical capacitance of the rod. Obviously, analytical calculations can only be made within simple models ignoring the two-dimensional geometry of an electrostatic field of a finite rod (neglecting end effects). Two situations are possible here. If the distance h between the charged rod and the grounded object nearest to its lateral surface is less than the rod length l, one can consider an infinite rod of radius r_m above a parallel grounded metallic plane at distance h. The problem for $h \gg r_m$ can be easily solved by the image method. For this, it suffices to replace the thin rod and its image in the plane by charged axes with linear (per unit length) charge densities $+\tau$ and $-\tau$, respectively, and then to write, with the Gaussian theorem, the field strength at the radial distance r from the rod as a function of r: $E(r) = \tau (2\pi\varepsilon_0)^{-1} [1/r + 1/(2h - r)]$. Integration of $E(r)$ over the range from r_m to h yields the voltage between the rod and the plane, $U \approx \tau (2\pi\varepsilon_0)^{-1} \ln (2h/r_m)$. The linear capacitance of the rod parallel to the plane is

$$C_1 = \frac{\tau}{U} = \frac{2\pi\varepsilon_0}{\ln (2h/r_m)} \approx \frac{0.555}{\ln (2h/r_m)} \, \text{pF} \, \text{cm}^{-1} \qquad (3.15)$$

The capacitance only logarithmically depends on the approximate value of h, which characterizes the distance between the rod and grounded objects; hence, equation (3.15) permits reasonable estimations to be made.

This formula will be used in further analysis, but the conditions of interest involve grounded objects located at a distance $h > l$, when the electrostatic characteristics depend not so much on h as on the rod length l. Although the actual charge distribution $\tau (x)$ along a perfectly conducting rod is not quite uniform, the following approach may be useful for evaluations. Suppose the linear charge is uniform, then we can find the corresponding potential distribution $U(x)$, neglecting the anode charge effect. We will see that the potential variation along the rod will be insignificant, so we can conventionally define linear capacitance as the ratio $C_1(x) = \tau/U(x)$. Suppose further that this ratio would be valid for the 'true' linear capacitance, for which we would have to solve an analytically intractable reverse problem of $\tau(x)$ leading to $U(x) = \text{const}$. This assumption is meaningful, since the nonuniformity of $C_1(x)$ has been found to be small, and numerical calculations justify it.

With $\tau = \text{const}$, $U(x)$ at the point x on the rod is defined by an easily calculable integral, summing up the potentials of all elementary charges τdx:

$$U(x) = \frac{\tau}{4\pi\varepsilon_0}\left[\int_0^x \frac{dz}{(z^2 + r_m^2)^{1/2}} + \int_0^{l-x} \frac{dz}{(z^2 + r_m^2)^{1/2}}\right] \tag{3.16}$$

$$= \frac{\tau}{4\pi\varepsilon_0}\ln\left(\frac{\left[x + (x^2 + r_m^2)^{1/2}\right]\left\{x - l + \left[(x-l)^2 + r_m^2\right]^{1/2}\right\}}{r_m^2}\right)$$

The potential drops less than by half from the rod center towards its ends:

$$U(l/2) = \frac{\tau\ln(l/r_m)}{2\pi\varepsilon_0} \qquad U(0) = U(l) = \frac{\tau\ln(2l/r_m)}{4\pi\varepsilon_0} \tag{3.17}$$

This is clear qualitatively: the basic contribution to the potential is made by charges located close to the particular point. At the center, they are localized on both sides, while at the ends only on one side. Remote charges make only a logarithmically increasing contribution, so we have $2l$ and l under the sign of logarithm in equation (3.17). Calculation with equation (3.16) shows $U(x)$ to be closer to its value at the rod center for most of its length. Obviously, charges act on both sides of this point, and a rod section comparable to its radius r_m is sufficient to make their effect noticeable. Therefore, it is natural to take the linear capacitance of a long rod, $l \gg r_m$, to be independent of x and equal to its central value with the total rod capacitance $C = C_1 l$:

$$C_1 = \frac{2\pi\varepsilon_0}{\ln(l/r_m)} \qquad C = \frac{2\pi\varepsilon_0 l}{\ln(l/r_m)} \tag{3.18}$$

Consider now the effect of the head on the system charges and potential. The charge Q of an isolated metallic sphere is uniformly distributed over its surface and gives the sphere the potential $U = Q/4\pi\varepsilon_0 r_m$ (its capacitance $C = 4\pi\varepsilon_0 r_m$). If a charged head is in contact with a charged channel, the head potential is due to both charges, their contributions depending on the charge distribution along the channel. In experiments, the head potential is sometimes determined by the channel charge rather than by its own charge. Consider this problem with reference to a metallic rod with a hemispherical head. The potential, U, at the rod center is primarily due to the neighboring charges and is defined by the first of equations (3.17), where τ is the average linear charge. The potential at the head center is the sum of the potential caused by the channel charges and given by the second of equations (3.17) and the potential $\Delta U = Q/4\pi\varepsilon_0 r_\mathrm{m}$ due to charge Q distributed over the head hemisphere. The sum represents the same potential U as at the channel center. In other words, the head potential is added to $U(l)$ calculated from formula (3.16), $U(l/2) = U$. Hence, we have $\Delta U \approx U/2$ with an accuracy to the small difference in the logarithms of equation (3.17). The head charge Q at the given potential U and its capacitance $C_\mathrm{t} = Q/U$ are half the values for an isolated sphere:

$$Q \approx 2\pi\varepsilon_0 r_\mathrm{m} U \qquad C_\mathrm{t} \approx 2\pi\varepsilon_0 r_\mathrm{m} \tag{3.19}$$

From formulas (3.18) and (3.19), the head charge makes a negligible contribution to the total streamer charge CU at $l \gg r_\mathrm{m} \ln(l/r_\mathrm{m})$.

The ionization frequency and wave velocity are directly and strongly affected by the maximum field near the head surface. It is this field that is of special interest to us. The field at the head top is created mainly by the head charge. Indeed, the potential of a charged unit volume decreases with distance as $1/r$ and the field strength as $1/r^2$. This means that the channel charges will make a much smaller contribution to the field than to the potential. This is confirmed by a direct calculation similar to that with formula (3.16). The longitudinal field at the hemisphere top, induced by charges on the lateral surface of the adjacent cylinder, is

$$E' = \frac{\tau}{4\pi\varepsilon_0} \int_{r_\mathrm{m}}^{l} \frac{x\,\mathrm{d}x}{\left(r_\mathrm{m}^2 + x^2\right)^{3/2}} = \frac{\tau}{4\sqrt{2}\pi\varepsilon_0 r_\mathrm{m}} = \frac{U}{2\sqrt{2}r_\mathrm{m} \ln(l/r_\mathrm{m})} \tag{3.20}$$

Here $\tau = C_1 U$ has been substituted and C_1 taken from formula (3.18). On the other hand, the field created by Q may be evaluated as $E_\mathrm{m} \approx Q/4\pi\varepsilon_0 r_\mathrm{m}^2$. But according to expressions (3.19), the charge Q is approximately half that of an isolated sphere with the same potential. So the field at the hemisphere top will also be half the value for a sphere:

$$E_\mathrm{m} \approx \frac{U}{2r_\mathrm{m}} \qquad U \approx 2r_\mathrm{m} E_\mathrm{m} \tag{3.21}$$

A qualitative explanation may be as follows. The field at the top is the potential gradient along the channel axis near its head. But since this potential is due not only to the head charge but also to distant channel charges, it decreases along the axis slower than in the absence of distant charges. Comparison of expressions (3.21) and (3.20) shows that the contribution of the head charge to the field at the head top is by a factor of

$\sqrt{2}\ln\left(l/r_{\mathrm{m}}\right)\gg 1$ larger than that of the channel charges, making it possible to neglect the latter in estimations. Thus, the maximum field at the head top, which determines the wave velocity and ionization degree behind its front, will require twice as large potential as for the case of electrostatic channel effect being ignored. Rigorous calculations of electrostatic characteristics of a rod with a hemispherical head support this qualitative result.

Note that the effect of the hemispherical head of a rod is, in a sense, analogous to that of the rod end, which we neglected in the derivation of formula (3.17). The end charge may be said to equalize the potential by raising the end potential to the value at the rod center. The other end of the rod, attached to the anode, is affected in a qualitatively similar way by charges on the anode surface. This effect becomes excessive when the anode is large and has much charge on it. As a result, the linear charge density at a distance, from the anode, of the size of its radius turns out to be smaller than the average value for the streamer channel. Expressions (3.18) for linear and total capacitances of a metallic rod agree better with reality than what follows from their derivation.

3.2.2 Streamer and external currents

A streamer propagates by adding new portions of plasma to its head. These become charged, acquiring a potential $U\left(l\right)$, which generally depends on the streamer length l. In the model of a perfectly conducting streamer, $U\left(l\right)$ is equal to the anode potential U_0. The streamer elongates at a distance v_{s} per unit time, with a new charge $\tau v_{\mathrm{s}} = C_1 U\left(l\right) v_{\mathrm{s}}$ appearing here. Strictly, this charge raises the potential not by $U\left(l\right)$ but only by $U\left(l\right) - U_0\left(l\right)$, where $U_0\left(l\right)$ is the potential due to the anode charge in the absence of channel. The field of a small anode is known to decrease very rapidly, and $U\left(l\right) \gg U_0\left(l\right)$ if the channel is long enough and has sufficient conductivity. For this reason, we will neglect $U_0\left(l\right)$. Charge τv_{s} flows into the new portion of the channel (in reality, it is transported by electrons from the head), so that the current behind the head is:

$$i_{\mathrm{c}} = \tau v_{\mathrm{s}} = \frac{2\pi\varepsilon_0 U v_{\mathrm{s}}}{\ln\left(l/r_{\mathrm{m}}\right)} \qquad U = U\left(l\right) \qquad (3.22)$$

where the linear capacitance for the channel center from formula (3.18) has been taken to be a constant value. If a streamer is a perfect conductor, it is always under a potential equal to the anode potential, $U = U_0$. If the latter is constant, the linear charge varies little with time (only because of the logarithmic dependence of the capacitance on l); as a result, a nearly constant current flows through the channel. In this case, current i_0 close to i_{c} of expression (3.22) flows into the channel base from the anode. Passing through the channel, it pumps the charge into its new portion.

A current i_{ex}, which is smaller than i_0, flows into the external circuit through the voltage source. Currents i_{ex} and i_0 differ by the value of $\dot{q}_i$, where q_i is the charge induced on the anode by the positive charge of the elongating channel, $i_{ex} = i_0 + \dot{q}_i$. The induced charge can be found with the Shockley-Rameau theorem. The charge dq_i, induced on an isolated spherical anode of radius r_a by unit charge dq located at distance r from the center, equals $dq_i = -dq r_a / r$ (see the example in Section 2.6). The total charge, uniformly distributed along a straight streamer of length l with linear density τ, and the respective current are:

$$q_i = -\int_{r_a}^{l+r_a} \frac{\tau r_a}{r} dr = -\tau r_a \ln \frac{l+r_a}{r_a} \qquad \dot{q}_i = -\tau v_s \frac{r_a}{l+r_a} \qquad (3.23)$$

The induced current of a long channel is $|\dot{q}_i| \ll \tau v_s = i_c$. For a perfectly conducting rod, we have $i_c = i_0$, so that the external current $i_{ex} \approx i_0$ is close to the base current. But at the beginning, at $l < r_a$, the external current $i_{ex} = i_0 + \dot{q}_i = \tau v_s l / (r_a + l)$ may be much lower than current $i_0 = \tau v_s$ flowing into the channel. In this case, the charge deficit is covered by the anode surface charge. In reality, not all electrons coming from the channel to the anode escape through the external circuit; some accumulate on the anode, being attracted to it by positive channel charges.

The current along a certain section behind the head, or along the whole length of a perfectly conducting channel, is constant and is approximately defined by equation (3.22), whereas the head current rises abruptly. Indeed, the total head charge Q distributed, in average, within a volume of size r_m along the x-axis takes the new position for a time Δt of the order of r_m / v_s. During this time, the head covers a distance equal to its size, and the average current through the head cross section normal to the x-axis is

$$\overline{i_t} \approx \frac{Q}{\Delta t} \approx 2\pi\varepsilon_0 U v_s = \overline{\tau_t} v_s \qquad \frac{\overline{i_t}}{i_c} = \frac{\overline{\tau_t}}{\tau} = \ln\left(\frac{l}{r_m}\right) \qquad (3.24)$$

The average head current is stronger than the channel current, because the average linear charge in the head, $\overline{\tau_t} = Q/r_m$, is by a factor of $\ln(l/r_m)$ larger than in the channel, $\tau = C_1 U$ [formula (3.18)]. A strong current at a given point x of the head exists for a very short time, $\Delta t \approx r_m/v_s$ ($10^{-10}-10^{-9}$ s at $r_m \sim 10^{-1}$ cm and $v_s \sim 10^8-10^9$ cm s^{-1}), and the current pulse position changes with velocity v_s together with the head.

The distribution of Q within the head is extremely nonuniform. Most of it is concentrated along the x-axis to form a layer, whose thickness is by a factor of $\ln(n_m/n_0)$ less than the radius [equation (3.13), Figure 3.2]. The time variation of the current in this cross section is the same as the distribution of linear charge $\tau(x)$, which travels 'as a whole' along the axis with velocity v_s. Hence, the maximum head current is stronger than the average current $\overline{i_t}$ by a factor of $\ln\left(n_m/n_0\right)$. The presence of a current spike at the head at a fixed moment of time should not be misleading. The difference between the channel and head currents corresponds to an algebraic charge accumulation: charge accumulates in the head front before the current spike but disappears after it, because it flows over to a new site together with the head.

Until the channel has reached the cathode, the charge current between them is not closed, but this should not confuse us either. There is a displacement current between the cathode and the channel head, which means that the electric field here, as well as on the cathode, increases with time, as the channel elongates. To give the reader an idea of the respective orders of magnitude, we will make estimations for the examples in Section 3.1.4. For $U = 60$ kV, $v_s = 5.3 \times 10^9$ cm s^{-1}, $r_m = 0.1$ cm, $l = 50$ cm, $\ln\left(l/r_m\right) = 6.2$, and $C_1 = 0.09$ pF cm^{-1} in air, the average linear charge is $\tau = 5.4 \times 10^{-9}$ C cm^{-1} and the current behind the head is $i_c = 2.8$ A.

3.2.3 Longitudinal field behind the head

In the perfect conduction approximation, the whole channel is under the same potential and the longitudinal electric field in it is zero. This is certainly not so in reality. We have shown that the gas in an ionization wave becomes ionized only to a certain degree, which largely depends on the maximum head field defining, in turn, the ionization frequency and wave velocity. Suppose there is no additional ionization in the channel. This assumption has a sense for a certain section behind the head, since the channel field is much lower than the maximum head field, while the ionization rate is, probably, by orders of magnitude smaller. Let us examine the situation with a positive gas, or with a negative gas in a short section behind the head, before the number of electrons has been reduced by attachment. This section has a length $l_a \sim 10-100$ cm in the example for air at the streamer velocity $v_s \sim 10^9$ cm s^{-1} and attachment frequency $\nu_a \sim 10^7-10^8$ s^{-1}. Recombination occurring at the wave plasma densities is of little significance during the short periods of channel elongation. In the same case for air with the electron-ion recombination coefficient $\beta \sim 10^{-7}$ cm^3 s^{-1} and electron density $n_c \approx 3 \times 10^{14}$ cm^{-3}, the recombination frequency $\nu_r = \beta n_c \approx 3 \times 10^7$ s^{-1} is comparable with the attachment frequency. The recombination occurs as far from the head as the attachment.

With the above assumptions and $\mu_e = $ const, the linear resistivity is $R_1 = \left(\pi r_m^2 e \mu_e n_c\right)^{-1}$, remaining constant and equal to the value in the head. Note that the diffusional plasma expansion or the channel radius increase due to an insignificant thermal expansion does not much affect the linear resistivity (if they do, this is only because of a different mobility in the lower density gas). But the number of electrons per unit channel length, $\pi r_{\mathrm{m}}^2 n_{\mathrm{c}}$, remains constant. An appreciable change in R_1 would require a great change in the ionization degree, that is, in the production or loss of electrons. If the resistivity remains at the level behind the head, the longitudinal field can be evaluated in the next approximation relative to that for currents, in which the field behind the head was assumed to drop to zero.

In the first approximation, channel current i_{c} is defined by formula (3.22). In the next approximation, the channel conductivity is taken to be finite, and a longitudinal field $E_{\mathrm{c}} = i_{\mathrm{c}} R_1$ is necessary for the current to flow. By substituting here $R_1 = \left(\pi r_{\mathrm{m}}^2 e \mu_{\mathrm{e}} n_{\mathrm{c}}\right)^{-1}$, i_{c} from expression (3.22), v_{s} from (3.14), and U from (3.21), we obtain the relation

$$\frac{E_{\mathrm{c}}}{E_{\mathrm{m}}} = \frac{2}{\ln\left(l/r_{\mathrm{m}}\right)\ln\left(n_{\mathrm{m}}/n_0\right)} \approx \frac{i_{\mathrm{c}}}{i_{t\ \mathrm{max}}} \tag{3.25}$$

where the current ratio has taken into account equation (3.24) and what follows it. In the approximation concerned, the channel-to-head field ratio practically depends on nothing but the weak logarithmic dependence on the channel length and radius and on the plasma ionization degree. The longitudinal field in the channel is by two orders of magnitude lower than in the head, greatly reducing the ionization rate or, perhaps, entirely excluding ionization, which requires at least $30\ \mathrm{kV\,cm^{-1}}$ in normal air. But the absolute longitudinal field may turn out to be so high as to result in great voltage losses in a long streamer. For instance, in the example discussed in Section 3.1.4, the field was found to be $E_{\mathrm{c}} \approx 0.02 E_{\mathrm{m}} \approx 6\ \mathrm{kV\,cm^{-1}}$ and the voltage drop over a channel of length $l = 50\ \mathrm{cm}$ was $E_{\mathrm{c}} l = 300\ \mathrm{kV}$.

Formula (3.25) permits an important conclusion to be made. The field and voltage losses behind the head cannot be reduced directly by increasing the plasma conductivity in the ionization wave front. The reason for this is the nature of discharge current: until the wave has reached the cathode, the current is determined by the wave velocity v_{s} and the channel potential U rather than by the conductivity of the channel it has created, as follows from formula (3.22). Increase in electron density n_{c} necessary for reducing linear resistivity $R_1 = \left(\pi r_{\mathrm{m}}^2 e \mu_{\mathrm{e}} n_{\mathrm{c}}\right)^{-1}$ is accompanied by a proportional wave velocity increase v_{s} from equation (3.14) and, hence, of current i_{c}. It is this factor that rules out any dependence of channel field E_{c} on electron density. The field is only affected by channel potential U, which determines linear charge τ, as well as the maximum field in the ionization zone, $E_{\mathrm{m}} = U/2r_{\mathrm{m}}$, defining the ionization frequency.

3.2.4 Streamer expansion and channel radius

We have already mentioned a possible expansion of the streamer channel. Its diffusional expansion is a relatively slow process, considering the ambipolar character of plasma diffusion. Somewhat faster is a principally possible thermal or gasdynamic expansion. With a slight excess pressure in the channel, Δp, over the ambipolar pressure p, the thermal expansion rate makes up a small fraction $\Delta p/p$ of the velocity of sound, which covers distance $r_{\mathrm{m}} \sim 10^{-1}$ cm in 3 μs. This indicates that thermal expansion is not a fast process. The most effective mechanism of streamer expansion is that of ionization. With the high potentials and large linear charges actually present in a streamer channel, the radial field on its surface may prove to be high enough to initiate the gas ionization near its lateral surface. This adds new plasma to the channel, increasing its radius and decreasing its linear resistivity. The process of channel expansion is, in a sense, similar to radial ionization wave propagation, so it must cease as soon as the radial field becomes too low to maintain fast ionization. Let us make some evaluations.

From the Gaussian theorem, the radial field at the surface of a charged cylinder of radius r must be

$$E_{\mathrm{r\,m}} \approx \frac{\tau}{2\pi\varepsilon_0 r} \approx \frac{U(x)}{r \ln(r/l)} \tag{3.26}$$

where $U(x)$ is potential in a given cross section x. In a perfectly conducting rod of the same radius as the head, $U(x) = U_{\mathrm{t}} = U_{\mathrm{a}}$ and $r = r_{\mathrm{m}}$, the maximum radial field derived from formula (3.26) is much larger than the longitudinal field from formula (3.25) but less than the maximum field at the head. In the above example with air at $U_{\mathrm{t}} = 60$ kV, $r_{\mathrm{m}} = 0.1$ cm, and $l = 50$ cm, the radial field is $E_{\mathrm{r\,m}} = 96$ kV cm^{-1}. Taking into account that the actual channel potential in the anode vicinity may greatly exceed that of the head, $E_{\mathrm{r\,m}}$ may prove to be much stronger in the same geometry.

To give an idea of the radius limit, we will make use of Peak's empirical formula for the critical radial field E_{ig} near a wire of radius r displaying a self-sustained corona. The appearance of a corona indicates the onset of ionization. For air of relative density $\delta = \gamma/\gamma_0$, the field is

$$E_{\mathrm{ig}} = 31\delta \left(1 + \frac{0.308}{\sqrt{\delta r \, [\mathrm{cm}]}}\right) \text{ kV cm}^{-1} \tag{3.27}$$

where γ_0 is normal atmospheric density. For instance, at $U = 60$ kV and $\delta = 1$, the radial field from formula (3.26) will drop to the ionization 'threshold' defined by expression (3.27) at $r = 0.23$ cm and $E_{\mathrm{r\,m}} = 51$ kV cm^{-1}. This may be said to be the maximum radius, above which the field will drop below a value corresponding to the empirical ionization threshold.

A more rigorous evaluation of the critical radial field and maximum expansion radius can be made on the condition that the necessary number of electron generations $K = \ln(n_c/n_0)/\ln 2 \approx 18-20$ is produced by a field close to the critical field over the time $t \sim l/v_s \sim 10^{-7}$ s, for which the channel can acquire length l. Such estimations have yielded results close to those just presented. The expansion radius grows with discharge voltage, but even at a moderate voltage of 60 kV, the estimated value $r = 0.23$ cm is not quite consistent with the generally accepted concept of a thin channel with $r \sim 10^{-3}-10^{-2}$ cm. We have just seen that general considerations make us question the validity of this concept, which cannot be subjected to a reliable experimental test at present. It should be stressed again that a strong radial field around a very thin channel produces a cylindrically expanding ionization wave. It inevitably expands the channel to an ultimate radius, which is not as small as is often believed.

3.2.5 Spark as a line with distributed parameters

The series of interrelated dependences discussed above determines all parameters of a spark of finite conductivity—its potential, linear charge, current, etc.—since they are dependent on time and on the x-coordinate along the channel axis. Indeed, the field in a channel with electrical resistance is finite; otherwise, there would be no current in it. The voltage in a finite field falls from the anode towards the head. The situation with the time variation of potential at a fixed point in the channel is more ambiguous, because it may grow or fall depending on whether or not a wave propagates through it. In the first case, the charge at this point grows, which means that the anode current responsible for this growth decreases towards the head.

These relationships can be estimated using the concept of spark as a long line with distributed parameters. We actually did this before, when dealing with linear capacitance C_1 and charge τ. In addition, the line must be characterized by linear resistivity R_1. For the general case, we will also need linear inductance L_1, but this quantity can be neglected in the treatment of ionization waves. Equations for a long line express the charge conservation law and, if we ignore the inductance, also Ohm's law. If the amount of charge $\tau(x,t)$ between points x and $x + \mathrm{d}x$ grows with time, this means that more charge flows through the x cross section than through the $x + \mathrm{d}x$ cross section. Hence, the first equation is

$$\frac{\partial \tau}{\partial t} + \frac{\partial i}{\partial x} = 0 \qquad (3.28)$$

Equation (3.28) follows directly from formula (2.27) integrated over the whole cross section. From Ohm's law, the voltage fall per unit channel length is

$$-\frac{\partial U}{\partial x} = iR_1 \qquad (3.29)$$

What remains to be done is to relate τ to U through capacitance. It was pointed out above that when the capacitance at a given point is filled with charge, the potential rises to $U(x,t)$ not from zero, but from 'external' potential $U_0(x,t)$; so we have

$$\tau(x,t) = C_1\left[U(x,t) - U_0(x,t)\right] \qquad (3.30)$$

We will restrict ourselves to cases with constant voltage U_0. Then the substitution of equation (3.30) into equation (3.28) yields

$$C_1\frac{\partial U}{\partial t} = -\frac{\partial i}{\partial x} \qquad (3.31)$$

which, together with equation (3.29), forms a set of second-order equations for U and i. One boundary condition for this set of equations is $U(0,t) = U_0$. The other condition reflects the fact that the conductivity current at the head is not closed and coincides with the kinetic charge flow

$$i(l,t) = \tau(l,t)\, v_s(t) = C_1\left[U(l,t) - U_0(l)\right] v_s \qquad (3.32)$$

Only now does the total charge conservation law become valid:

$$i_0 = \frac{dq}{dt} \qquad q = \int_0^l \tau \, dx \qquad (3.33)$$

This follows from equation (3.28) integrated over the whole channel length. Here, q is the total channel charge and i_0 is the current flowing into its base, the head charge being negligible relative to the total charge. All this is valid only until the channel has bridged the gap.

The channel velocity in equality (3.32) should be taken as a known function of head potential $U_t \equiv U(l)$. It is found by solving the ionization wave problem, say, with formulas (3.8), (3.10), and (3.21). Thus, the problems of an ionization wave and a channel become interrelated.

Let us see in what cases electrical processes in a channel should be treated in terms of the above equations and when simple evaluations of current i_c from formula (3.22) and of voltage drop $U_0 - U_t \approx i_c R_1 l \approx E_c l$ from formula (3.25) are sufficient. Such evaluations may lead to serious errors, if the base current i_0 considerably exceeds the head current $i_c \equiv i_1$, making the field $E = iR_1$ variable with x. Evaluate the relative excess of i_0 over i_1 by integrating equation (3.31) over the channel length and representing the distributed potential as a linear function:

$$i_0 - i_1 = C_1 \int_0^l \frac{\partial U}{\partial t} \, dx \qquad U \approx U_0 - \frac{x}{l}(U_0 - U_1) \qquad U_1 \equiv U(l)$$

By differentiating the channel length l, but not the slowly varying head potential U_1, with respect to t (at fixed x) and assigning the derivative $\partial U/\partial t$ to the channel center $x = l/2$, we will have

$$\frac{i_0 - i_1}{i_1} \approx C_1 v_\text{s} \frac{U_0 - U_1}{2l} \frac{l}{i_1} \approx C_1 \frac{i_1 R v_\text{s}}{2l} \frac{l}{i_1} = \frac{RC v_\text{s}}{2l} \tag{3.34}$$

where $U_0 - U_1 \sim i_1 R$; $R = R_1 l$; and $C = C_1 l$ are the total channel resistance and capacitance. One can notice that the current changes but little along the channel, if the characteristic time RC for charging capacitance C through resistance R is less than the time $t = l/v_\text{s}$ necessary for the channel to change its length, or its capacitance. This is natural because the capacitance can become nearly completely charged at every moment of time, and the charges do not 'get stuck' but move towards the head. At $t \ll RC$, however, the capacitance of the older channel parts is charged slowly, so the current along the channel becomes lower.

If we substitute expressions (3.18) for C, (3.14) for v_s, and the expression $R = \left(\pi r_\text{m}^2 e \mu_\text{e} n_\text{c}\right)^{-1} l$ into equation (3.34), we will obtain

$$\frac{i_0 - i_1}{i_1} \approx \frac{l}{\pi r_\text{m}^2 e \mu_\text{e} n_\text{c}} \frac{2\pi \varepsilon_0 l}{\ln\left(l/r_\text{m}\right)} \frac{e \mu_\text{e} n_\text{c} r_\text{m}}{2\varepsilon_0 2l \ln\left(n_\text{m}/n_0\right)}$$

$$= \frac{l}{2 r_\text{m} \ln\left(l/r_\text{m}\right) \ln\left(n_\text{m}/n_0\right)} \tag{3.35}$$

Thus, even at $l \gtrsim 1$ cm (for $r_\text{m} \sim 0.1$ cm and $\ln\left(l/r_\text{m}\right) \sim 5$), the current along the channel changes so much that the need for the equations of a long line becomes acute. Note that the problem of a long line is similar to that of temperature distribution in an elongating rod with constant temperature maintained at one of its ends. Indeed, taking R_1 and C_1 to be constant, differentiating equation (3.29) with respect to x and substituting the result into equation (3.31), we will find

$$\frac{\partial U}{\partial t} = \varkappa \frac{\partial^2 U}{\partial x^2} \qquad i = \lambda \frac{\partial U}{\partial x} \tag{3.36}$$

Here, U plays the role of temperature T, i of heat flow, $\lambda = 1/R_1$ of heat conductivity, $\varkappa = 1/R_1 C_1$ of thermal diffusivity, and τ of energy for heating unit volume of the rod from the initial temperature $T_0\left(x\right)$, whose role is played by the external potential $U_0\left(x\right)$. The linear capacity C_1 acts as the heat capacity per unit volume. The condition of (3.32) for the end of an elongating channel with an open-circuit current is equivalent to that of thermal insulation of the rod end. Temperature $T_0\left(0\right)$ is given at the rod base, which is analogous to the given potential U_0. Naturally, the law for the rod elongation must also be given. This analogy provides a better understanding and treatment of all parameter distributions.

Finally, let us evaluate the induction. With allowance for self-induction, equation (3.29) takes the form:

$$-\frac{\partial U}{\partial x} = L_1 \frac{\partial i}{\partial t} + R_1 i \tag{3.37}$$

The linear inductance of a long thin isolated conductor can be expressed through its linear capacitance as

$$L_1 = \frac{\varepsilon_0 \mu_0}{C_1} \approx \frac{\mu_0}{2\pi} \ln\left(\frac{l}{r_m}\right) \tag{3.38}$$

with $\varepsilon_0 \mu_0 = 1/c^2$, where μ_0 is vacuum magnetic permeability. The account of self-induction may become important for three reasons: (i) a fast increase in the electrode voltage; (ii) a fast current rise from zero, as the ionization wave acquires an increasingly higher velocity until the establishment of a 'quasistationary' mode; and (iii) when the perturbation responsible for regular current variation (ionization wave) propagates very fast. We will omit the first two factors producing a short-lived effect at the beginning of the process but will focus on the regular effect of channel elongation.

The relative contribution of the self-induction emf to the voltage drop is equal, in order of magnitude, to:

$$\frac{L_1 \partial i/\partial t}{\partial U/\partial x} \sim \frac{L_1 il}{tU} \sim \frac{L_1 l C_1 U v_s}{tU} \sim L_1 C_1 v_s^2 = \frac{v_s^2}{c^2} \tag{3.39}$$

where $t \sim l/v_s$ is the time scale of current variation. Self-induction is unessential, if the wave velocity is much less than light velocity, as is usually the case.

Owing to self-induction, the equations for a long line (3.31) and (3.37) reduce to the wave equation. It is sometimes used for treating spark discharges, in particular lightning, since they have an important stage of powerful perturbation propagating with a velocity comparable to light velocity.

3.3 Gas temperature and energy balance

The problem of gas temperature in a developing spark is of great importance for the understanding of spark breakdown, especially in air gaps, which have always been of primary interest to researchers. The point is that the lifetime of electrons in cold air and moderate field is very short due to their attachment to oxygen. At a typical attachment frequency $\nu_a \sim 10^7 \text{ s}^{-1}$ and ionization wave velocity $v_s \sim 10^8 - 10^9 \text{ cm s}^{-1}$, the plasma conductivity must drop sharply at a distance $v_s/\nu_a \sim 10 - 100$ cm from the wave front. This is what is actually observed at moderate voltages, when the channel is far from being a 'perfect' conductor transporting the anode potential to the head. No breakdown is possible even if such a spark has bridged the gap. The conductivity of a spark that would seem to have connected the electrodes is so small and the current so weak that it has very little effect on the circuit, and there is no short circuiting of a true breakdown.

Nevertheless, spark breakdown of multimeter air gaps is quite feasible and has been observed experimentally. Another example is lightning, whose highly conducting channel covers several kilometers. Breakdown can occur only if the gas temperature is as high as several thousand degrees, in which case electron detachment processes do not allow the conductivity to drop rapidly. At temperatures of about 10^4 K, the situation becomes still more favorable, because thermal ionization of the gas enters the scene, contributing to the plasma conductivity.

3.3.1   Evaluation of the upper temperature

Let us find the temperature to be expected in the channel behind the front of a traveling ionization wave. Its upper limit can be estimated from general energy considerations, irrespective of any assumptions about other channel or plasma characteristics. Suppose voltage U_0 is applied to the anode, and charge q has been incorporated into the channel by the time it acquires length l. For this, the power source has expended the energy $W_0 = U_0 q$. The charge becomes distributed along the channel with a linear density $\tau = C_1 U(x)$, where C_1 is the linear channel capacitance, which can generally be taken to be dependent on x. Then, we have

$$W_0 = U_0 \int_0^l C_1(x) U(x) \, \mathrm{d}x \tag{3.40}$$

The electrical energy stored by the channel capacitance is

$$W_e = \frac{1}{2} \int_0^l C_1(x) U^2(x) \, \mathrm{d}x \tag{3.41}$$

We have used here the expression for the capacitor energy $W_e = CU^2/2$ well known from electricity theory. The energy to be spent for the gas heating (dissipation) cannot be larger than $W_0 - W_e$. Hence, the upper limit of the energy dissipated in the channel is expressed as

$$W < W_0 - W_e = \frac{1}{2} \int_0^l C_1 \left(2U_0 U - U^2 \right) \mathrm{d}x \leq \frac{CU_0^2}{2} \tag{3.42}$$

where $C = \int_0^l C \, \mathrm{d}x$ is the total channel capacitance, and the identity and inequality $2U_0 U - U^2 \equiv U_0^2 - (U - U_0)^2 \leq U_0^2$ have been used.

Clearly, the energy dissipated by this moment does not exceed the electrical energy of a capacitor with a capacitance equal to the total capacitance of a channel charged to voltage U_0. This conclusion is consistent with electricity theory. As the capacitor with constant capacitance C is being charged to the voltage U_0 of the power supply, the amount of dissipated energy remains equal to that of electrical energy stored in the capacitor, irrespective of the circuit resistance R, which may have any value and even vary in time. It is only the characteristic time of charging, of the order of RC, that depends on the resistance. One can easily see this by solving the equation for the RC-circuit, $iR + q/C = U_0$ with $i = \dot{q}$, and by integrating the power $i^2 R$ with respect to the total time of the process. This general property of capacitors has allowed the evaluation of maximum Joule heat from the known channel capacitance without knowing its resistance. It will be shown below that in the general case the inequality in expression (3.42) might as well be replaced (as to the orders of magnitude) by the equality. Therefore, the upper temperature evaluation is informative enough to give valid estimates.

With expressions (3.42) and (3.18), the average temperature rise in the channel can be found as

$$\Delta T \leq \frac{W_1 \left(1 - \varepsilon_{\mathrm{V}}\right)}{\pi r_{\mathrm{c}}^2 c_{\mathrm{v}}} \approx \frac{C_1 U_0^2 \left(1 - \varepsilon_{\mathrm{V}}\right)}{2\pi r_{\mathrm{c}}^2 c_{\mathrm{v}}} \approx \frac{\varepsilon_0 U_0^2 \left(1 - \varepsilon_{\mathrm{V}}\right)}{r_{\mathrm{c}}^2 c_{\mathrm{v}} \ln \left(l/r_{\mathrm{c}}\right)} \tag{3.43}$$

where W_1 is the average linear energy release, r_{c} is the channel average radius, c_{v} is the heat capacity per unit volume, and ε_{V} is the energy transferred to the slowly relaxing vibrational degrees of freedom of molecules, if any. For example, for argon, a monatomic gas with the lowest heat capacity and $\varepsilon_{\mathrm{V}} = 0$, we obtain $W_1 = 1.1 \times 10^{-4}$ J cm^{-1} and $\Delta T \leq 7$ K at $l = 50$ cm, $r_{\mathrm{c}} = 0.1$ cm, $\ln \left(l/r_{\mathrm{c}}\right) = 6.2$, and a characteristic voltage $U_0 = 50$ kV. In air, where 90–95% of the energy gained by electrons from the field is used to excite slowly relaxing vibrations of nitrogen molecules and where the heat capacity is 5/3 times higher, the heating does not exceed a hundred degrees at $U_0 = 1$ MV, a value characteristic for experiments in air. The heating estimates are sensitive to voltage (the values above are typical in experiments) and to the radius of the energy release region. We have even simplified the conditions for heating in our evaluations by ignoring the ionization expansion radius. Note that linear capacitance is not very sensitive to this radius.

Thus, a single ionization wave (for collective waves, see Section 6.7.1) leaves behind it a weakly ionized channel of relatively cold gas consisting of heavy particles. We emphasize again that the gas remains cold irrespective of the ionization degree, and the plasma conductivity was simply ignored in the calculations above. A channel created by such an ionization wave is commonly referred to as a *streamer*. A streamer in air rapidly loses its conductivity due to electron attachment.

3.3.2   Energy balance and gas heating

Our evaluation above has established the upper limit for the channel heating, but it does not help us to find the lower heating threshold. This problem is of interest, because one sometimes deals with a gradual streamer heating and has to explain the related effects. Evaluation of the actual energy release and the heating range at a given linear resistivity is expected to provide information on the conditions for raising the channel temperature to a value high enough to suppress electron attachment in air or to attain high conductivity, for instance, in argon.

Consider the energy balance in a gap with a streamer channel of length l (at $U_0(t) = \text{const}$) starting from the anode. Equation (3.29) is multiplied by i and integrated over the whole channel length. By taking the integral by parts in the left-hand side of this equation and substituting $\partial i/\partial x$ from formula (3.31) and i_1 from formula (3.32), we obtain the relation

$$U_0 i_0 = \int_0^l i^2 R_1 \mathrm{d}x + \int_0^l \frac{\partial}{\partial t}\left(\frac{C_1 U^2}{2}\right)\mathrm{d}x + C_1 U_1\left[U_1 - U_0(l)\right]v_\mathrm{s} \qquad (3.44)$$

which expresses the energy conservation law. With the equality

$$\frac{\mathrm{d}}{\mathrm{d}t}\int_0^l \frac{C_1 U^2}{2}\mathrm{d}x = \int_0^l \frac{\partial}{\partial t}\left(\frac{C_1 U^2}{2}\right)\mathrm{d}x + \frac{C_1 U_1^2}{2}v_\mathrm{s} \qquad (3.45)$$

equation (3.41) can be conveniently transformed to

$$U_0 i_0 = \int_0^l i^2 R_1 \mathrm{d}x + \frac{\mathrm{d}}{\mathrm{d}t}\int_0^l \frac{C_1 U^2}{2}\mathrm{d}x + \left[\frac{C_1 U_1^2}{2} - C_1 U_1 U_0(l)\right]v_\mathrm{s} \qquad (3.46)$$

The power $U_0 i_0$ released in the gap is dissipated in the channel bulk [the first terms in the right-hand sides of equations (3.46) and (3.44)], contributes to the electric field energy [the second term in the right-hand side of equation (3.46)] and is spent for the formation of new capacitance [the last term in equation (3.46)].[2] The power for filling the new capacitance by electrical energy is defined by the last term in equation (3.42), while the sum of the last terms in equations (3.45) and (3.46) gives the total

---

[2]The fact that the capacitance variation requires some energy can be illustrated as follows. Suppose the capacitance of a charged capacitor increases, because the plates are free to approach each other under the action of attractive forces. Accelerated by electrostatic attraction, the plates will acquire kinetic energy from the initial electrical energy. If they are forced to stop, their kinetic energy will remain in the material in the form of thermal energy or elastic strain, but it will not transform back to electrical energy.

energy expenditures for the new capacitance, described by the last term of equation (3.44). If the external potential at the channel end is small, $U_0(l) \ll U_1$, as is the case when the channel length is not much smaller than the anode radius, the expenditures for the new capacitance and filling are nearly the same.

Let us define the energy balance at the time t, when the channel has reached length l. Integration of equation (3.46) with respect to time, starting with the moment $t = 0$ of the channel inception to t, yields the formula:

$$U_0 q = W + \int_0^l \frac{C_1 U^2}{2} dx + \left\langle \frac{C_1 U_1^2}{2} - C_1 U_1 U_0(l) \right\rangle_t l \qquad (3.47)$$

The angular braces indicate that this is the time-average value between 0 and t. When the channel resistance becomes high enough to cause a considerable voltage drop, which is what usually happens in reality, one can use the inequality $\overline{U_1^2} \ll \overline{U^2}$, where the bars stand for the averaging over the channel length. Therefore, the last term in equation (3.47) can be neglected. For the average linear energy release in the gas, $W_1 = W/l$, we can write an approximate equality and inequalities limiting W_1 on both sides:

$$W_1 \approx \left(U_0 - \frac{\overline{U}}{2} \right) \overline{\tau} \qquad U_0 \overline{\tau} > W_1 > \frac{U_0 \overline{\tau}}{2} \qquad \overline{\tau} = \frac{q}{l} \qquad (3.48)$$

where $\overline{\tau} \approx C_1 \overline{U}$ is the average linear charge. The dissipated energy is greater than the energy stored by the electric field, $\overline{U} q/2$, but it is naturally smaller than the expended energy (the energy dissipated in charging constant capacitance is equal to that stored by the capacitor).

It is clear that the range of energy release is quite narrow, which is a good enough argument to conclude that equation (3.43) provides not only the upper limit but also a correct order of the heating magnitude. Higher voltage and effective capacitance (see Section 6.9) make it feasible to heat even molecular gases to 1000–2000 K. The perspectives seem quite good, because a fast release of vibrational energy of gas molecules will create an additional powerful heat input into a discharge.

3.4 Initial ionization mechanisms

3.4.1 Production of seed electrons

Photoionization of an ideal gas is due only to photons emitted by atoms
or molecules excited to energies higher than ionization potential I. The
probability of such 'superexcitation' is unlikely to be very large, but we
cannot support this assertion quantitatively. Photons with $h\nu > I$ can
also be emitted in photocapture of an electron by an ion to the ground
level of the atom. The process cross section is $\sigma_p \sim 10^{-21}$ cm^2. At
$n_e \approx n_+ \approx 10^{13}$ cm^{-3}, the photocapture frequency is $\nu_p \sim n_+ \bar{v}\sigma_p \sim 1$ s^{-1};
the electronic excitation frequency of the 'common' atomic levels lying be-
low the ionization potential is of the same order of magnitude as the ionization frequency and is $\nu^* \sim 10^{10}$ s^{-1} for argon (Section 3.1.3). Therefore,
with a reasonably small probability of superexcitation, photorecombination
radiation is negligible even relative to the radiation by superexcited atoms.
However, the contribution of photons with $h\nu > I$ is greatly decreased by
active absorption in the gas. The absorption coefficient of this radiation is
$\varkappa_\nu \approx 10^3$ cm^{-1} for argon at atmospheric pressure (Section 2.2). At a distance of the streamer head radius $r_m = 0.1$ cm, the radiation intensity falls
as $\exp(\varkappa_\nu r_m) = 10^{43}$; it is absorbed by a very thin layer in front of the head.

These considerations make us discard photoionization of the gas atoms.
It is possible that seed electrons are produced by photoionization of organic impurity molecules having a low ionization potential or by knocking
electrons out of dust particles. This process is due to photons emitted by
excited atoms of the basic gas, and this suggestion is supported by experiments, in which initial electrons were detected in any gas at distances of
several centimeters from the radiation source.

As far as air is concerned, photoionization of oxygen molecules with
$I = 12.2$ eV by radiation of nitrogen excited to energies less than its ionization potential of 15.6 eV has been discussed since the studies by Loeb.
A detailed analysis of this mechanism was made in [3.7]. The principal
conclusions of this work are as follows. Oxygen is ionized by nitrogen
molecules emitting in the Berdge-Hopfield bands in the wavelength range
980–1025 Å. The respective nitrogen levels are excited by electrons in a
wide range of E/N values with a frequency comparable with that of ionization. About 4% of the excited molecules emit photons, the others being quenched by collisions. The emission in the above range is weakly
absorbed by nitrogen and stronger by oxygen. The absorption coefficient
varies greatly and irregularly within this range from $\varkappa_1 = 26.6$ cm^{-1} atm^{-1}
to $\varkappa_2 = 1520$ cm^{-1} atm^{-1} with a 90% photoionization probability at lower
frequencies. Since the absorption in oxygen is quite strong, only weakly

absorbed quanta can cover large distances. With the rapid spectral depletion of actively absorbed quanta, the production rate of photoelectrons at distance r from a point photon source with the power of F photons per second can be written as

$$q_e = \frac{F \exp(-\varkappa_1 p_0 r)}{4\pi r^3 \ln(\varkappa_2/\varkappa_1)} \qquad (3.49)$$

where p_0 is partial oxygen pressure; q_e is measured in electrons per cubic centimeter per second. But at $p_0 r > 0.13$ cm atm (in atmospheric air at $r > 0.6$ cm), this radiation is so actively absorbed by oxygen that impurity ionization by a more penetrating radiation with the effective absorption coefficient $\varkappa \approx 1.9$ cm^{-1} at 1 atm begins to dominate. The role of impurity in 'pure' nitrogen is usually played by oxygen.

The mechanism of resonance radiation diffusion, which transports the excitation far beyond its primary source and is accompanied by dissociative ionization via reaction (2.12), has also been discussed. This mechanism can hardly be considered as competing with direct ionization of oxygen or impurities by the source photons in air, especially if we recall that there is no evidence for resonance radiation diffusion in molecular gases. The diffusion does, of course, exist in argon and other inert gases, but the presence of a readily ionizable impurity is probably more important. The molecular reaction (2.12), like the diffusion, takes much time, more than a fast streamer can normally provide. This problem cannot be considered as being solved.

3.4.2 Initiation of an ionization wave

Generally, the origin of primary ionization, being interesting in itself, is of less importance for the streamer process than is commonly believed. The evaluations to be made will show that there is no need to worry about the lack of initial (seed) electrons. This question can be put in a different way: how high should the density of these electrons be and where should they be located for the ionization wave to behave as was described in Section 3.1?

Let us find the coordinate r_0 of the point on the streamer axis indicating the origin of avalanche ionization. We will place this point formally at infinity. The effective radius r_0 is defined by the condition that only one generation of electrons can be produced along the whole distance from r_0 to infinity. For strong waves ($v_s \gg v_e$), which will be analyzed for simplicity, this condition, according to equation (3.7), has the form

$$\int_{r_0}^{\infty} \frac{\nu_i}{v_s} dr = \frac{\nu_m r_m}{v_s(2k-1)} \left(\frac{r_m}{r_0}\right)^{2k-1} = \ln 2 \qquad (3.50)$$

involving the relations $\nu_i \sim E^k \sim r^{-2k}$. With equations (3.50) and (3.7), the avalanche ionization starts at the radius

$$r_0 = r_m K_m^{1/(2k-1)} \qquad K_m = \frac{\ln(n_m/n_0)}{\ln 2} \tag{3.51}$$

where the field is $E_0 = E_m (r_m/r_0)^2 = E_m K_m^{-2/(2k-1)}$. By substituting numerical values (see below), we can find that E_0 markedly exceeds the critical field for air, $E_i \approx 30 \,\mathrm{kV\,cm^{-1}}$, below which the extrapolation $\nu_i \sim E^k$ is meaningless; hence, our estimation has a 'sense.

For an avalanche ionization to start at point r_0 without a time delay to produce K_m generations of electrons, which would all enter the head cross section πr_m^2, it is absolutely necessary that the initial electrons be present in a cylinder of radius r_m and length δ, the latter producing one electron generation. We mean a cylinder aligned with the streamer axis with its far end located at distance r_0 from the head center. The length δ is defined by a condition similar to that of equation (3.50):

$$\int_{r_0-\delta}^{r_0} \frac{\nu_i}{v_s}\mathrm{d}r = \frac{\nu_m r_m}{v_s(2k-1)}\left[\left(\frac{r_m}{r_0-\delta}\right)^{2k-1} - \left(\frac{r_m}{r_0}\right)^{2k-1}\right] = \ln 2$$

With equation (3.50) and assuming $\delta \ll r_0$, we can find $\delta \approx r_0/(2k-1)$.

Generally, an avalanche may start even with a single electron present in such a cylinder. But if the electron is located at the cylinder end nearest to the streamer head, the number of generations produced will be smaller by one. Moreover, if the average number of electrons in the cylinder is small, the probability of their entire absence may be high, which will result in statistical breaks in the streamer propagation. In order to reduce the probability of such breaks to, say, 1%, we should assume the average number of electrons in the cylinder to be 100. This sets the following requirements on the minimum electron density at distance r_0 from the head center:

$$n_{0\,\mathrm{min}} > \frac{100}{\pi r_m^2 \delta} = \frac{100(2k-1)}{\pi r_m^3 K_m^{1/(2k-1)}} \tag{3.52}$$

If this condition is met, we can use the treatment of an ionization wave as a 'continuous medium', without focusing on individual avalanches.

Numerical calculations with equations (3.51) and (3.52) will be made for the conditions in air discussed in Section 3.1.4, where K_m was found to be 24 at $r_m = 0.1$ cm, $k = 2.5$ and $E_m = 300 \,\mathrm{kV\,cm^{-1}}$. We have found $r_0 \approx 2r_m = 0.2$ cm, $\delta = 0.05$ cm and $n_{0\,\mathrm{min}} = 5 \times 10^4 \,\mathrm{cm^{-3}}$. The field at point r_0 is $E_0 = 75 \,\mathrm{kV\,cm^{-1}}$; it is indeed appreciably stronger than $E_i \approx 30 \,\mathrm{kV\,cm^{-1}}$. Therefore, photoionization must provide an electron density of, at least, 10^4–$10^5 \,\mathrm{cm^{-3}}$ at distance $r_0 \approx 0.2$ cm. Let us calculate the actual density, using equation (3.49), which is quite suitable, because a

0.2 cm radius is much smaller than the validity limit in this formula (about 0.6 cm). Photoelectrons are accumulated at the point of interest during the whole period of time of the ionization wave propagation. The count should be made, when the streamer head center approaches this point at distance r_0:

$$n_0 \sim \int_r^\infty q\left(r\right) \frac{\exp\left[-\nu_a\left(r - r_0\right)/v_s\right]}{v_s} \mathrm{d}r$$

$$\sim \frac{F \exp\left(-\varkappa r_0\right) \exp\left(-\nu_a/\varkappa v_s\right)}{4\pi r_0^3 \varkappa v_s \ln \varkappa_2/\varkappa_1} \tag{3.53}$$

We have found $\varkappa = \varkappa p_0 = 5.6$ cm^{-1} in air at $p = 1$ atm. The cofactor $\exp\left[-\nu_a\left(r - r_0\right)/v_s\right]$ allows for the electron losses due to attachment, whose frequency was assumed to be constant and corresponding to field E_0 at point r_0. This gives overestimated losses. Let ω be the number of ionizing photons, emitted in avalanche ionization per electron produced, and τ^* be the average lifetime of an excited nitrogen molecule with respect to the necessary quantum emitted. In accordance with [3.7], the number of photons is $\omega \sim 10^{-2}$–10^{-1} and the lifetime is $\tau^* \sim 10^{-9}$–10^{-8} s. For time τ^*, the head is displaced forward from the excited molecule at distance $x^* = v_s \tau^* \sim 1$–10 cm $\approx (10$–$100)\, r_m$ at velocity $v_s \sim 10^9$ cm s^{-1}. The emission attenuation at such a large distance is considerable. Besides, the channel cross section with the excited molecule can be seen at a very small solid angle from the point of interest. For this reason, it is only the r_m/x^* portion of emitted photons that may be actually effective. Hence, we can evaluate the power of a 'point' photon source in expression (3.53) as $F \approx \omega \pi r_m^2 n_c v_s \left(r_m/x^*\right) = \omega \pi r_m^3 n_c/\tau^*$. Thus, the actual density of initial electrons at point r_0 in front of the head is, by order of magnitude, equal to

$$n \sim n_c \frac{\omega r_m^3 \exp\left(-\varkappa r_0\right) \exp\left(-\nu_a/\varkappa v_s\right)}{4 r_0^3 \varkappa v_s \tau^* \ln \varkappa_2/\varkappa_1} \tag{3.54}$$

For our example with $\nu_a \approx 10^{10}$ s^{-1}, $\omega = 10^{-2}$, and $\tau^* = 10^{-8}$ s, we get $n_0 \approx 5 \times 10^{-8} n_c \approx 1.6 \times 10^7$ cm^{-3}, a value two orders of magnitude larger than the minimum density estimate. Thus, we did not make a serious error, having taken $n_0 \sim 10^6$ cm^{-3} in Section 3.1. Estimates obtained for other conditions indicate that the streamer head radiation in air does provide an ionization wave with the required number of seed electrons.

3.5   The streamer inception

In our discussions of the streamer, we have naturally assumed it as already existing. Now we will try to understand under what conditions and how a streamer is generated. This problem was the focus of analysis by the authors of streamer theory. Trying to find the criterion for gap breakdown, they believed it to be equivalent to the gap overlap by a streamer. This, however, is not always the case. Still, the conditions for the formation of a streamer, as the basic structure in a spark discharge, are very important.

There are, at least, two conditions to be met for a streamer to appear. First, the gas ionization must give rise to a high conductivity plasma, which would pass a sufficiently large charge flow from the head and considerably reduce the external field. Second, the streamer must provide its propagation not only due to a strong field in front of its head but also to an adequate number of seed electrons to avoid a delay in the ionization restart. For the first condition to be met, the field in the vicinity of the smaller electrode (it is the anode, as before) must be fairly high, otherwise the electrons produced by avalanche ionization would be simply pulled out onto the anode, leaving behind a weakly charged positive ion trace. There would be no plasma channel with zero external field whatsoever. In order to find the external field and the gap length necessary for a plasma channel to arise, we will consider the ionization in moderate fields with a small number of initial electrons. Generally, a streamer may arise from a plasma source of any origin, if this source is placed in a strong field and its parameters meet certain requirements. A simple and common source to produce an adequate plasma is a single electron avalanche originating from one seed electron. A single avalanche is, therefore, an elementary structure inherent in the breakdown process. Below, we analyze its properties and establish the criterion for avalanche transformation to a streamer—the avalanche-streamer transition. Incidentally, it was via avalanche experiments and analysis of the effects observed that the authors of streamer theory eventually arrived at the criterion for 'breakdown', more exactly, for the avalanche-streamer transition. Our treatment of this issue is somewhat different from what has been suggested so far.

3.5.1   Requirements on streamer inception

Consider, for simplicity, an electrically neutral conducting sphere of radius R placed in a uniform field E_0. For a self-propagating plasma channel to arise, the external field inside the plasma source must be pushed out while that outside it enhanced. It is also necessary that the source size R be larger than the ionization path length α^{-1} in a field equal to or slightly higher than the external field. Otherwise the enhanced field outside the plasma will drop to the value of unperturbed field, before a sufficient number of new electron generations is produced in the developing ionization wave. Because of the sphere polarization by the external field, positive and negative surface charges appear on its respective hemispheres. For the reverse field induced by these charges to cancel the external field, the surface charge density must be $\tau q_s \sim \varepsilon_0 E_0$. To push out the latter field from the whole sphere space, the surface charge distribution must be nonuniform. A rigorous solution to this classical problem gives $\tau q_s = 3\varepsilon_0 E_0 \cos\psi$, where ψ is the angle between the vector radius of a given point on the sphere and the vector of $\boldsymbol{E_0}$.

The total charge of one sign on each respective hemisphere is equal to $eN = 3\varepsilon_0 \pi R^2 E_0$. So the plasma source must contain at least $N = 3\varepsilon_0 \pi R^2 E_0/e$ electrons capable of driving the external field out of it. Bearing in mind that the source radius is to be larger than α^{-1}, with $\alpha \approx \alpha(E_0)$, we arrive at the minimum amount of electrons necessary to generate a streamer:

$$N_{e\ \min} \approx \frac{3\pi\varepsilon_0 E_0}{e\alpha^2} \qquad (3.55)$$

As an illustration, consider atmospheric air again. A field lower than $30\ \mathrm{kV\ cm^{-1}}$ is known to be unable to initiate ionization because of electron attachment. Analysis of experimental data on breakdown in a plane air gap of 1 cm long leads one to the conclusion that the effective ionization coefficient, equal to the difference between the ionization and attachment coefficients (Section 2.3), is $12.4\ \mathrm{cm^{-1}}$ at breakdown field $E_0 = 31.4\ \mathrm{kV\ cm^{-1}}$. For calculations, one can take $\alpha \approx 10\ \mathrm{cm^{-1}}$ for atmospheric air at $E_0 \approx 30\ \mathrm{kV\ cm^{-1}}$. Expression (3.55) yields $N_{e\ \min} \approx 1.7 \times 10^9$.

The minimum plasma density in the initial source must be as high as

$$n_{e\ \min} \approx \frac{3N_{e\ \min}}{4\pi R^3} \approx \frac{9\varepsilon_0 E_0}{4eR} \qquad (3.56)$$

The restriction on the plasma density becomes more rigorous for smaller source radii. For minimum permissible size $R \sim \alpha^{-1}$, the density is to be $n_{e\ \min} \approx (9/4)\,\varepsilon_0 E_0 \alpha/e$, or $4 \times 10^{11}\ \mathrm{cm^{-3}}$ in atmospheric air. Calculation similar to that made in Section 3.4 shows that the number of photons emitted by such a plasma source is large enough to create the initial electron density n_0 necessary for the propagation of an ionization wave.

If a plasma source in a uniform field meets the above requirements, ionization waves will start from it in both directions along the field, each leaving behind a plasma channel. The reason is that the fields at both heads will become increasingly stronger, as the channel moves through the gap. A wave cannot propagate in the direction transversal to $\boldsymbol{E_0}$, since the radial component of the dipole field (a polarized sphere is a dipole) is very weak.

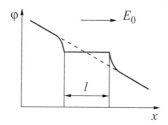

FIGURE 3.4
Enhanced field at the ends of a long conductor in an external field. Dashed line, unperturbed potential in absence of conductor.

The reason for a continuous field enhancement at the streamer head is illustrated in Figure 3.4. A thin conductor of length l and radius $r \ll l$, aligned with uniform field $\boldsymbol{E}_0$, only slightly perturbs this field at distances $\Delta x > r$ from its ends along the axis. A perfect conductor is under the same potential everywhere. By virtue of the symmetry condition, this potential coincides with the external potential at the conductor center. At the conductor ends, the potential differs from the unperturbed one by the value $\Delta U = E_0 l/2$ but returns to it at distance r from the ends; therefore, the field at the ends is, in order of magnitude, $E_{\mathrm{m}} \approx \Delta U/r \approx E_0 l/2r$. Numerical calculations of maximum fields at the ends of a long metallic rod aligned with a uniform external field E_0 are described by the interpolation formula

$$\frac{E_{\mathrm{m}}}{E_0} = 3 + 0.56 \, (l/r)^{0.92} \qquad 10 < l/r < 2000 \qquad (3.57)$$

which is in fairly good agreement with the evaluation above.

3.5.2   A single electron avalanche

Consider an avalanche in a uniform external field $\boldsymbol{E}_0$, initiated by a single electron produced at time $t = 0$. The x-coordinate will be counted off from the point of its production in the direction of $-\boldsymbol{E}_0$. We will ignore the possible formation of negative ions unimportant for further discussion. The number of electrons N_{e} and of positive ions N_+ increases in the avalanche as

$$\frac{\mathrm{d}N_{\mathrm{e}}}{\mathrm{d}x} = \frac{\mathrm{d}N_+}{\mathrm{d}x} = \alpha N_{\mathrm{e}} \qquad N_{\mathrm{e}} = N_+ + 1 = \exp\,(\alpha x) \qquad (3.58)$$

All electrons produced move in a group with drift velocity v_{e}, while positive ions practically do not change their position. Ignoring, for the time being, the electron group expansion, we can describe their trajectory with the equalities $x = v_e t$, $y = 0$ and $z = 0$, where y and z are the transverse coordinates. The densities of electrons and ions are expressed by the δ-functions:

$$n_e = \delta(x - v_e t)\,\delta(y)\,\delta(z)\exp(\alpha x)$$

$$n_+ = \int_0^t \alpha v_e n_e dt = \delta(y)\,\delta(z)\,\alpha\exp(\alpha x) \qquad x \le v_e t \qquad (3.59)$$

The 'center of mass' of ionic charge distributed along the axis [equations (3.59)] moves with velocity v_e at the distance α^{-1} behind the point of electron grouping (at $N_e \gg 1$). This means that the avalanche charges form a moving dipole. An equivalent dipole possesses the charges $N_e \approx N_+$ located at the distance α^{-1} from one another. This is already a prerequisite for a possible annihilation of external field between the dipole charges. The field at the dipole center, at the distance $\alpha^{-1}/2$ from the charge centers, vanishes, when the field $E' = eN_e 4\alpha^2/4\pi\varepsilon_0$, created by the electronic and ionic charges individually at this point, becomes equal to $E_0/2$. This occurs when the number of electrons in the avalanche becomes as large as $N_e = \pi\varepsilon_0 E_0/2e\alpha^2$, coinciding in order of magnitude with equation (3.55). This is because the parameters of both dipoles are more or less the same.

However, a linear charge arrangement cannot annihilate a field even on a line, let alone in a volume. For a field to be pushed out of a volume, the charge arrangement in it must also be spatial. So of primary importance is the problem of the avalanche head expansion and the range of this expansion. There are two expansion mechanisms: diffusion and electrostatic repulsion of electrons. Let us first consider diffusion. If, in addition to drift, electrons diffuse in space, their density $n_e(x, r, t)$, where $r = \sqrt{y^2 + x^2}$ is the radial coordinate, satisfies the continuity equation (2.24) without the recombination term. The electron cloud spreads around the center $x_0 = v_e t$ with $r = 0$. The solution to the equation for the diffusion with drift and sources, derived from formula (2.24), is written as

$$n_e = \frac{1}{(4\pi D_e t)^{3/2}}\exp\left[\frac{(x - v_e t)^2 + r^2}{4 D_e t} + \alpha v_e t\right] \qquad (3.60)$$

The electron density shows a Gaussian decrease away from the center. The radius of the sphere, on which n_e is e-times lower than the density at the center, grows with t or x_0 in the characteristic diffusion law:

$$r_D = (4 D_e t)^{1/2} = \left(4\frac{D_e}{\mu_e}\frac{x_0}{E_0}\right)^{1/2} = \left(\frac{8\bar{\varepsilon} x_0}{3 e E_0}\right)^{1/2} \qquad (3.61)$$

Here, D_e/μ_e has been expressed, using Einstein's ratio (2.7), through the mean energy of random electron motion, $\bar{\varepsilon}$. Ions accumulate at every point

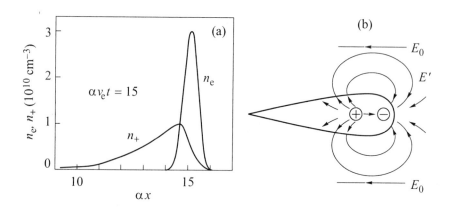

FIGURE 3.5

A single avalanche developing from a single electron in a uniform field E: calculated distributions of n_e and n_+ at the moment of $\alpha v_e t = 15$, with $4D_e\alpha/v_e = 4.8 \times 10^{-3}$ corresponding to $E = 30\ \text{kV cm}^{-1}$, $\alpha = 10\ \text{cm}^{-1}$, and $T_e = 3.6\ \text{eV}$ (a); avalanche outline and schematic lines of force of external field E_0 and of avalanche space charge field E' (b); circles, the 'centers of mass' of charges.

as the integral of equation (3.59), which can only be taken numerically (Figure 3.5). The 'center of mass' of space charge in the ion trace still lags behind the electron cloud center by a distance α^{-1}. The schematic representation of an avalanche in Figure 3.5 is consistent with photographs taken by a Wilson chamber. Figure 3.5(b) illustrates the structure of the field induced by an avalanche. This is a typical dipole field, whose vector is summed up with that of the external field. One can see the predisposition for annihilation of the external field between the electronic head and the center of mass of the ion trace.

The rate of the head diffusional expansion decreases with time or the avalanche path length ($dr_D/dt \sim t^{-1/2} \sim x^{-1}$) in spite of the growing number of charges in the avalanche. The electrostatic electron repulsion, on the contrary, becomes stronger so that the diffusion gives way to the repulsion at a certain moment of time. The velocity of the boundary of an electron 'sphere' of radius R is determined by the boundary electron drift in the field created by the dipole charges (at $R < \alpha^{-1}$, it is mainly the field of a negatively charged sphere, $E' \approx eN_e/4\pi\varepsilon_0 R^2$):

$$\frac{dR}{dt} = \mu_e E' = \frac{\mu_e e \exp(\alpha x)}{4\pi\varepsilon_0 R^2} \qquad x = \mu_e E_0 t \qquad \alpha \approx \alpha(E_0) \qquad (3.62)$$

By integration, we find the formula for the avalanche head expansion $R(x)$, then the field E', and the electron density n_e in the sphere:

$$R = \left(\frac{3e}{4\pi\varepsilon_0\alpha E_0}\right)^{1/3} \exp\left(\frac{\alpha x}{3}\right)$$

$$E' = \frac{E_0\alpha R}{3} \qquad n_e = \frac{3N_e}{4\pi R^3} = \frac{\varepsilon_0\alpha E_0}{e} \tag{3.63}$$

The head field E' grows with the radius, while the electron density remains constant due to simultaneous charge multiplication; the number of electrons grows in proportion to the head volume. By equating the diffusional expansion rate, which can be calculated from formula (3.61), and the increasing repulsional expansion rate, we can define the moment, when the diffusion is replaced by the repulsion. It corresponds to $\alpha x \approx 14$ and $N_e \approx \exp(\alpha x) \sim 10^6$ in air at $E_0 = 30$ kV cm^{-1} ($\alpha \approx 10$ cm^{-1} and $\bar{\varepsilon} = 3.6$ eV, the latter derived from the results of Section 2.1). The head field is then only 2–3% of the external field, so that the resulting field is still far from being zero. This is also clear from a comparison of $N_e \sim 10^6$ and $N_{e\,\min} \sim 10^9$ necessary for the field annihilation. Thus, the process of pushing the external field out of an avalanche takes place during the head expansion associated with repulsion rather than diffusion. This conclusion is generally supported by experiment. Photographs of avalanches show a rapid expansion of the head, which is likely to have been caused by electron repulsion.

3.5.3   Avalanche-streamer transition

It is clear from the foregoing that an avalanche transforms to a streamer, when the sum $2E'$ of the head field E' and a similar field of the ion trace reaches the external field value, compensating it in the region between the avalanche charge centers. According to formulas (3.63), this happens when the head radius becomes as large as $R \approx 3\alpha^{-1}/2$. The number of charges in the avalanche is then $N_{e\,\min} \approx 9\pi\varepsilon_0 E_0/2e\alpha^2$, nearly coinciding with the value obtained from expression (3.55). For air at $E_0 \approx 30$ kV cm^{-1} and $\alpha \approx 10$ cm^{-1}, it is $N_{e\,\min} = 2.5\times10^9$; the gain in an avalanche ready for the transition is $\alpha x \approx 22$, after it has covered the distance $x \approx 2.2$ cm. Thus, the following ratio may be taken as a criterion for an avalanche-streamer transition:

$$N_{e\,\min} = \frac{9\pi\varepsilon_0 E_0}{2e\alpha^2} \qquad \alpha(E_0)\,d = \ln(N_{e\,\min}) = \ln\left(\frac{9\pi\varepsilon_0 E_0}{2e\alpha^2}\right) \tag{3.64}$$

where d is the path length an avalanche is to cover in field E_0 in order to become a streamer. Two factors are important here: a sufficient abundance of electrons to be produced in the avalanche and a radius as large as α^{-1}.

Since the gain depends only slightly on the parameters under the sign of logarithm, the critical parameter may be $\alpha d \approx 22$. In earlier publications by Loeb and Meek, the diffusion radius of formula (3.61) was substituted into the condition $E' = E_0$, yielding the well-known Meek 'breakdown' criterion, $\alpha d \approx 20$, close to the one above. This is clear, because the radius enters expressions of the type (3.64) only under the logarithm. What is really important and why we have focused on the details of the avalanche-streamer transition is that an avalanche in earlier treatments ceased to exist at a much smaller diffusion radius, which was taken to be the streamer radius. In reality, the latter is much larger and has the α^{-1} scale. Since the streamer parameters strongly depend on its radius, the problem of an accurate evaluation of the transition radius is essential for spark discharge theory.

Following the transition, the ionization process undergoes some changes. For example, before the transition, all electrons were involved in the ionization with nearly the same probability characterized by $\alpha \equiv \alpha (E_0)$. But after the transition, the electrons behind the head front are affected by a weaker, even partly annihilated field and cannot be involved in the ionization. In other words, the individual avalanche, as it was described above, does not exist any more. The electron production in an avalanche has been shown to increase with distance x exponentially, $N_e \sim \exp (\alpha x)$, whereas the total number of electrons in a developing streamer is just proportional to the distance covered by the wave, $N_e \sim n_c x$, where n_c is the final ionization density in the wave [formula (3.11)]. The reason for this difference is that an electron cannot multiply infinitely in a wave, as it was in an avalanche, but it quickly leaves the strong field region to become inactive.

Since the electronic head radius is comparable with the distance between the electron and ion cloud centers, these clouds partly overlap to form a plasma. The field in this region is weakest. Earlier, when the head radius was much smaller than the distance α_0^{-1} between the electron and ion charge centers, the head expanded radially, primarily under the action of its own field. Now, we have $R \sim \alpha_0^{-1}$, and the ion field quenches the radial component of the electron field (Figure 3.5), thus slowing down the head radial expansion. The quantity $R_{\mathrm{max}} \sim \alpha_0^{-1}$ may be said to represent the maximum radius of an electron avalanche.

In contrast, the fields before the head front and behind the ion charge center are enhanced (Figure 3.5), because here the dipole field is added to the external field. Naturally, the electrons multiply much faster than in the initial avalanche. For example, in atmospheric air $\hat{\alpha} = \mathrm{d} \ln \alpha / \mathrm{d} \ln E \approx 4$ at $E_0 \approx 30 \; \mathrm{kV\,cm^{-1}}$, that is, a 1% field increase brings about a 4% increase in the ionization rate. This gives rise to two ionization waves, one of which travels in the same direction and the other back, along the ion trace. This is the way a streamer is formed. If an avalanche-streamer transition occurs in a uniform field far from the electrodes, the streamer develops in both directions: back to the cathode (cathode-directed streamer) and forward to the anode (anode-directed streamer).

Nearly all the cathode-directed streamer theory we discussed above can be extended to anode-directed streamers. There is no difference in their interpretation, if the streamer velocity v_s is much higher than the electron drift velocity v_e. Some difference arises only at a minimum streamer velocity comparable with v_e. Since they are either added or subtracted, depending on the direction of streamer propagation, the results turn out to be different. The contribution of photon ionization to the anode-directed streamer is not as essential as to the other one. Avalanche ionization can be initiated by distant diffused electrons or, in an intensive wave, by runaway electrons (Section 2.2).

In strongly nonuniform fields, streamers often arise in the strong field region, near the smaller electrode, and may be cathode- or anode-directed, depending on the sign. But in a weakly nonuniform field, a streamer has been observed to arise far from the electrodes and to develop in both directions. When the field is moderate or the interelectrode space d is small (αd is less than the critical transition value), the avalanche reaches the anode, the electrons enter it, and only the positive space charge of the ion trace remains in the gap.

3.5.4 Streamer head radius

We have pointed out that the problem of streamer head radius still remains to be solved. Today, there is no adequate theory, nor is there convincing experimental evidence to determine the head size reliably. We will offer some considerations that might indicate an approach to this problem. Clearly, theoretical or numerical methods can only be useful within a two-dimensional (axially symmetric) approach. One-dimensional models typical of existing theories cannot, in principal, yield a solution; the radii of the head and initial streamer are to be given. The two-dimensional numerical calculation of a rather arbitrarily stated problem made in [3.8] has not been analyzed adequately, but it does deserve attention. In that work, the initial plasma source was given arbitrarily at the anode, and the channel propagation in a strong uniform field was calculated. The channel was shown to preserve its initial radius at the short path length, for which the calculation was made. But it is hard to imagine a 1 m streamer of a millimeter radius 'remembering' about its origin. Its radius is, no doubt, established by some instantaneous processes occurring near the head, irrespective of the channel length.

Let us approach this task differently and ask what size the radius cannot have. Suppose the radius is very large, and the ionization wave front has a small curvature. Such a structure cannot be stable and viable. If a plasma perturbation of a smaller radius is caused by a fluctuation at the wave front, the field at this protrusion will be much stronger even at the same potential as on the rest of the front. The protrusion will move on, overtaking the small curvature front. A new, thinner and faster channel will eventually tear loose from the old one. So it is the instability condition that seems to determine the upper limit of possible head radii.

Assume now the radii of the protrusion and a new streamer, which has started from it, to be very small. The radial field at a small head will be so strong that the ionization wave will move not only forward but also laterally, expanding the head. Thus, the lower limit of possible radii is restricted at least by the condition of the head ionization expansion. There is another reason why a very small radius is unacceptable: avalanches produced in front of the head flow into it. No matter whether these are discrete or overlapping avalanches producing a continuously ionized medium, the electronic head (or the electron gas of the continuous medium) must expand laterally relative to the streamer axis. The streamer radius cannot be smaller than the transverse dimension of avalanches, which form the head. The calculations in Section 3.5.3 gave the maximum radius of an avalanche head, R_{max}, determined by electron repulsion, to be of the order of ionization path length α^{-1}; it varies inversely with the field, in which the plasma travels. If the field corresponds to the conditions of the avalanche-streamer transition ($R_{max} \sim \alpha^{-1} \approx 0.1$ cm for air), the radius of the streamer head accepting powerful but not expanding avalanches can hardly be smaller than $R_{max} \sim 0.1$ cm. This circumstance and some indirect experimental evidence have made us choose $r_m \approx 0.1$ cm for the analysis of general streamer properties. But we emphasize again that the considerations presented here require more theoretical work, numerical simulation and, if possible, an experimental support (see *Supplement*, p. 279).

3.6 Is a streamer breakdown possible?

This question may be stated more precisely: Can an ionization wave lead directly to a breakdown? Streamer breakdown has been a subject of common interest for a long time and still is, at least, for two reasons. It is a fundamental problem in spark discharge theory. Historically, it was the streamer mechanism that was suggested by the pioneers of spark theory to explain many properties of the spark discharge. Even its qualitative analysis may serve as a starting point for systematization and explanation of vast experimental material. This question is also of practical importance. The streamer that an ionization wave leaves behind, when propagating through a gas, represents a record fast discharge process. If the streamer overlap is equivalent to breakdown, the gas may lose its insulation strength very fast and, as fast, short circuiting may occur. This, for instance, could be important for many engineering problems: to speed up the operation of spark switching of electric energy flows, of arresters or spark switching equipment.

Even the analysis of a much simplified situation can give an idea of whether the electrode short circuiting by a streamer may be equivalent to a breakdown. Suppose an ionization wave closes a gap so quickly that the channel electrons behind it will have no time to be lost through attachment or recombination. The high wave velocity, assumed to be constant, indicates a higher ionization degree of the gas, which will also promote a breakdown. Let us ignore further ionization by a weaker longitudinal field behind the wave. Since the channel is cold, there is no thermal ionization. Suppose further the channel radius to be constant. Estimation of the ionization expansion has shown that this effect is quite possible, but it can hardly increase the radius of the ionization region by an order of magnitude. The radius for the case described in Section 3.2.4 increased only by a factor of 2.3; besides, this took time. On the assumptions made, the electrical resistance of a channel of length d after the overlap will be

$$R = R_1 d = \frac{d}{\pi r_c^2 e n_c \mu_e} = R_{1\ \min} d \frac{c}{v_s} = Z_B \frac{d}{2\pi r_c} \frac{c}{v_s} \qquad (3.65)$$

Here, we have used the formula $R_{1\ \min} = \left(\pi r_c^2 e n_{c\ \max} \mu_e\right)^{-1}$ for the minimum linear resistance (Section 3.2.3) and the proportionality of v_s to the electron density n_c derived from formula (3.14). The quantity $n_{c\ \max}$ corresponds to the maximum possible wave velocity $v_{s\ \max} = c$. Account has been taken of light velocity being $c = (\varepsilon_0 \mu_0)^{-1/2}$ in practical units. Besides, the parameter $Z_B = (\mu_0/\varepsilon_0)^{1/2} = 120\pi\ \Omega$ has been introduced as a kind of reference resistance. After the gap has been bridged, the current in it is

$$i = \frac{U_0}{R} = 2\pi \frac{v_s}{c} \frac{r_c E}{Z_B} \qquad (3.66)$$

where $E = U_0/d$ is an average field strength in the gap.

The gap overlap by a spark can be regarded as breakdown and short circuiting, if this greatly reduces the electrode voltage, as compared to U_0. This will happen, if the channel resistance R is much lower than the external circuit resistance, which is usually close to that of a high voltage source, R_S. As a rule, laboratory sources operating in the pulse mode have $R_S \approx 10^3$–$10^4\ \Omega$ or $R_S/Z_B \sim 10$, while high power equipment may have R_S/Z_B

smaller than unity. This means that even at a very high plasma density and streamer velocity ($v_s \sim c$), only the overlap of a very short gap of $d \leq 10r_c \sim 1$ cm may immediately result in breakdown without additional ionization. Observing a streamer in longer gaps, the experimenter will not be able to notice characteristic breakdown features immediately after the ionization wave contacts the opposite electrode, because the voltage between the electrodes will remain nearly the same and the channel current will be far from a usual short-circuit value of several or tens of kiloamperes. For example, the probability of a long air gap with a strongly nonuniform field being bridged by a streamer is high at $E = U_0/d \approx 10^4$ V cm^{-1}. At $r_c \approx 0.1$ cm, the maximum current after the overlap ($v_s = c$) will not exceed 20 A. Real current will be 10–100 times lower, because the velocity we actually have is $v_s \sim 10^8$–10^9 cm s^{-1}.

Therefore, only in very short gaps and at maximum plasma conductivity can an overlap by a streamer be regarded as breakdown. In longer gaps, the outcome cannot be predicted so easily. Breakdown will necessarily require a dramatic drop of the channel resistance. As a rule, the only way to do so is via a fast and powerful plasma heating. The temperature must rise to 7000–10000 K for the ionization to acquire a thermal character, as in the arc, where it provides a descending CVC ($\partial U/\partial i < 0$). Only with a descending CVC can the gap voltage drop, no matter what the generator resistance is (Section 1.4).

Whether a channel can be heated up and how this occurs depends on many factors: the gas properties, gap length, voltage pulse amplitude, and so on. Three situations may arise, depending on the medium characteristics.

First, a streamer breakdown can sometimes be observed in monatomic (inert) gases free from the harmful effect of attachment and energy loss due to molecular vibrations. High speed filming can register the flight of an unbranched streamer. After its contact with the cathode, the voltage in a gap of several tens of centimeters will drop immediately or a few microseconds after the streamer plasma has been heated uniformly. A streamer breakdown in argon with a strongly nonuniform field requires an average external field of $E = U_0/d \approx 0.65$ kV cm^{-1}; this is a very low insulation strength.

Second, it is more difficult to produce a streamer breakdown in positive molecular gases, say, in nitrogen. Although such a streamer has a conductivity close to that in inert gases, a serious obstacle is a great energy loss for the excitation of molecular vibrations. At low temperature, vibrations in nitrogen relax for 10^{-3}–10^{-2} s, which is eternity, as compared to the lifetime of streamer plasma, which rapidly decays due to dissociative recombination. For example, the characteristic recombination time is $\tau_r \sim (\beta n_c)^{-1} \sim$

1 μs (monatomic plasma recombines much slower) at the electron density $n_c \approx 10^{13}$ cm^{-3} and recombination coefficient $\beta \approx 10^{-7}$ cm^3 s^{-1}. There is still no reliable experimental evidence for the streamer-type breakdown in long gaps with strongly nonuniform fields in molecular gases. It may, probably, occur after a complete removal of negative impurities from the gas.

Third, negative molecular gases, air being the most common of them, provide the adverse conditions for a streamer breakdown. In addition to the same energy losses as in pure nitrogen, electrons in air are also lost through attachment (for 10^{-8}–10^{-7} s). A streamer breakdown seems to be feasible in air, but it requires special conditions: the ionization wave must close the gap for about 10^{-8} s or faster to allow the gas to be heated by conductivity current before the attachment could decrease the plasma conductivity. If the gas temperature can rise to 1000–2500 K during this time, the attachment will be suppressed, so that the conditions for a breakdown will become more favorable. For this, higher streamer velocities of at least 10^9 cm s^{-1} and short gaps of $d \leq 10$ cm are necessary. The voltage must be as high as several hundreds or thousands of kilovolts with a nanosecond rise time, so that the streamer would travel in a strong field and acquire maximum velocity at the shortest path length possible. However, in spite of the principal possibility, there seem to be no reliable experimental data to support the streamer breakdown in air or nitrogen.

Thus, a streamer breakdown should not be regarded as a common mechanism, but it is principally possible. It is a unique phenomenon requiring special conditions: a short gap and a very high voltage with a steep pulse rise. It is sometimes observed in argon but almost never in air.

3.7 An ionization wave in spark breakdown

3.7.1 Gas heating in a long spark

Experimental data available on spark breakdown in long gaps exclude the possibility for a long, streamer-type cold channel to develop in air. It is sufficient to apply 3–4 MV for a 100–200 m gap to be overlapped (the longest spark produced so far), that is, the average field in the spark channel is 200–300 V cm^{-1}. No ionization is possible in such a weak field at normal gas density; it requires about 30 kV cm^{-1} in cold air. It takes a spark a few milliseconds, at an average velocity of 2–3 cm μs^{-1}, to run a distance of 100 m, while in cold air the electrons nearly completely disappear within a few fractions of a microsecond. If the air in a long spark really remained cold, this would be simply a trace of recombining ions with a negligible conductivity. These contradictions are more appreciable in lightning, where the average field is several times lower, while the overlap time is an order of magnitude greater than in the longest laboratory spark. Therefore, there are grounds to believe that the plasma in a propagating spark is, in its basic characteristics, far from a cold streamer plasma but much closer to a steady electric arc. The field strength in a positive air column of atmospheric pressure lies in the range of $10-100$ V cm^{-1} and decreases with increasing current, in accordance with the falling CVC. The upper arc gradients are close to those measured in a long spark during its formation. The similarity between the parameter values becomes more remarkable in longer gaps.

There are no mechanisms capable to compensate for electron losses in a cold air spark at a low field. The gas must be heated up, so that the favorable effects of the heating become evident even at moderate temperatures. At $T \approx 1000$ K, the rate of negative ion destruction greatly increases, retarding the drop in the plasma conductivity. At $T \approx 1500$–2500 K, the VT relaxation of nitrogen molecules is accelerated, and the energy stored in vibrations transforms to the translational energy of the gas, intensely raising the temperature. Finally, above $6000-8000$ K, thermal ionization becomes a more pronounced and, eventually, dominant mechanism of electron production in the spark. Even prior to thermal ionization, the ionization due to electrons that have gained energy directly from the field is accelerated. The acceleration is associated with a lower gas density because of thermal expansion during the heating (we remind that the ionization rate is determined by the E/N ratio, where N is the number of molecules per unit volume). A long spark has enough time for thermal expansion to occur.

Thus, everything is in favor of a long spark channel being inevitably heated. Moreover, every new portion of the developing channel must be heated rapidly enough to overtake processes leading to the conductivity losses. This is the reason why the streamer mechanism can rarely lead directly to a breakdown. As was shown in Section 3.3, the energy input into a developing streamer is very small. At voltages of about 1000 kV commonly used in laboratory experiments, the energy is too small to heat a molecular gas, where its losses are so great. Estimations made with formula (3.40) show that the temperature does not rise above 100 K. Still, Nature has found a way out of this 'energy crisis', as will be seen below.

3.7.2   Leader as a bunch of streamers

We have started this chapter with the concepts of ionization wave and streamer. Beginning with seed electrons produced by photons in the gas or by processes involving photons, the primary plasma of a high electron density, $10^{12}-10^{14}$ cm^{-3}, is generated by ionization waves. This, of course, does not exclude further ionization in the spark or around it, but the fact is that even the latter process occurs via ionization wave propagation. The problem of additional ionization, contributing to the plasma density, is not of principal importance here. What is important is that the plasma with the 'first' and highest ionization degree is created by ionization waves, or streamers, which are the fundamental structural elements in a spark discharge.

The most important conclusion that follows from the analysis of ionization wave potentialities is that the production of free electrons is only part of the problem, and not the most difficult one. The electrons produced must be confined. It is the conditions for electron confinement and plasma conductivity maintenance which determine the breakdown voltage in a long gap. As for ionization waves, they could be excited even by a lower voltage.

It was pointed out earlier that electrons can be preserved by heating the air to, at least, several thousands of degrees, starting with the moment it transformed to a plasma. For this, a great energy must be released per unit channel length by a sufficiently large charge flow through the channel cross section. This is the charge that is continuously being pumped into the growing channel capacitance. (If the channel starts from the anode, the pumping of positive charges is performed by electrons drifting toward the anode and exposing the ions.) One can see this from formula (3.22). A unit channel length must possess a large capacitance C_1 capable of housing a much larger charge $C_1 U$ than that accommodated by a unit length of a single streamer with normal moderate heating. A direct increase of the cylindrical channel radius, which can somewhat raise the linear capacitance [equation (3.18)], gives no effect. On the contrary, the released energy density is reduced due to increasing volume, in which the energy dissipation proportional to r^2 considerably exceeds the effect of the weak, logarithmic capacitance rise.

Nature has found a way out by combining the currents of numerous streamers. Generally, the total capacitance of many closely spaced channels is smaller than the sum of capacitancies of individual channels (see Section 6.7.1). The unfavorable effect, however, is amply compensated by the combined effect of numerous streamer currents flowing through a thin channel. Qualitatively, this is the most effective scheme; it operates as follows. A great number of primary ionization waves start from the charged head of a hot channel, the head being a strong field source (see Chapter 6). This process is quite similar to the streamer inception at the anode due to the avalanche-streamer transition with the only difference that the role of a high potential anode is played by a hot channel head. The waves run some distance and stop in a low field region. The current of an individual streamer cannot heat up its channel; so, its conductivity decreases in time. But the total current of all streamers, combined in a thin unbranched stem from which 'streamer branches' outgrow, heats up this common region. Owing to the high conductivity of the heated region, the head electron charge quickly flows back from the head toward the anode, somewhat increasing the length of the hot charged channel. This process is repeated many times, creating the effect of quasicontinuous movement of a heated plasma channel with a bunch of cold streamers in front of the head.

Since the ionization waves are produced with a high frequency all the time, the streamers in front of the head have different lengths. Short streamers still preserve their conductivity, contributing their current to the common stem. Long streamers have already lost their electrons, so their contribution is small, but they do preserve their charge even after the head has run far ahead. As a result, the hot channel turns out to be surrounded by a relatively large region of space charge, or charge cover (see Chapter 6). This effect is equivalent to a higher resultant capacitance, as compared to the relatively small capacitance of a thin hot plasma column. This scheme is profitable for the heating, since the combined current flows through a small cross section of the channel, providing a high energy density in it. Meanwhile, charge is pumped into numerous long streamers and becomes distributed within a large streamer space in front of the head. As the head moves on, it forms a charge cover around the channel. The system capacitance in this case is characterized by a large effective cover radius close to the length of developing streamers, as evidenced by experiment. In spite of the logarithmic dependence $C_1 \sim (\ln l/r_{\text{eff}})^{-1}$, the capacitance proves to be large relative to that of a thin conducting channel. The energy density released into the common stem is inversely proportional to the small squared channel radius r_{c}, so that the heating is really strong: the weak currents of numerous streamers 'club together' to heat the thin leader channel.

This two-step process, in which a hot channel propagates with a velocity an order of magnitude lower than that of the streamers that heat it up, represents the *leader process*. In classical interpretations, a *leader* is this hot channel that propagates through a space filled by primary ionization waves, or *streamers*. The family of streamers around and in front of the head is termed as the *streamer zone*. The region of residual streamer charge surrounding the leader channel is called the *leader charge cover*.

We would like to emphasize a point of principal importance: the main function of streamers in the streamer zone is to increase the leader capacitance, or the energy input into the leader. The transverse dimension of the charge region may be several orders of magnitude larger than the leader radius. So we have an appreciable increase in the capacitance and energy input, heating the plasma by lower voltage than would be necessary for a separate unbranched streamer following a primary ionization wave.

The efficiency of the two-step leader scheme of a long spark has been confirmed by many experiments. The leader development has been observed not only in all negative gases studied, but also in pure positive molecular gases, for example, in nitrogen. Exceptions are possible with pure monatomic positive gases, in which electrons behind the ionization wave front disappear so slowly that even a relatively cold channel may retain much of its initial conductivity for tens of microseconds. If the channel is able to bridge the gap during this period, it has a chance to be heated up by the closed conductivity current to form an arc. The chances are that the heating will become greater, since there are no energy losses for vibration excitation.

4

Experimental Techniques

Most data available today on spark discharges have been obtained by simple experimental methods. This simplicity was, however, forced by the reality. When applied to parameter measurement in gaps of tens or hundreds of meters long at voltages of several megavolts, even a conceptually primitive technique acquires a great many sophisticated details. The literature on high voltage engineering is fairly extensive. There are handbooks on voltage sources and measuring equipment (see, for example [4.1]), which discuss these problems and give exhaustive instructions to the experimenter. We will not dwell on them here but rather focus on some general aspects of high voltage techniques and on measurement of basic discharge parameters.

In this chapter, our principal aim is to provide reliable tools for data check-up, to point out possible error sources and to analyze those aspects of experimental designs that may interfere with data treatment or result in misleading conclusions.

4.1 High voltage sources

Long sparks are generally produced by voltages of 3−5 MV from several, sometimes many, sources connected in series, each of which contributes tens or hundreds of kilovolts. The source used may be a transformer, if one needs alternating voltage, or a rectifier unit in case of constant voltage, or a precharged capacitor to generate short pulses. Under no-load conditions (in our case, in the absence of discharge), the output voltage of a circuit with n identical sources (stages) is $U = nU_0$, where U_0 is the voltage of an individual stage. Experimenters most often use voltage pulses produced by voltage pulse generators (VPG), which are more simple and less costly than stationary ones but have an important advantage—lower internal resistance.

The latter circumstance removes the external circuit effect on the events occurring in the discharge, permitting their observation as they occur.

4.1.1   The Marx generator

Research laboratories commonly use the Marx-type VPG [Figure 4.1(a)]. In the charge accumulation mode, stage capacitors C_1–C_n are connected to a common constant voltage source U_0 through a resistor series, R_1 and R_2. The capacitors become gradually charged at different rates. The latter fact does not matter, because even the most remote capacitors will eventually acquire U_0. After the charging, spark gaps, made controllable in modern VPG, come into action. The spark channels create a new circuit with a series connection of capacitors. The output voltage is multiplied by the number of stages n, while the equivalent capacitance is, respectively, reduced n-times, since the capacitor charges remain the same: $U_g = nU_0$; $C_g = C/n$, where C is the capacitance of a single VPG stage.

The generated pulse duration is determined by the stage capacitance and is maximum in the no-load mode, when the VPG is not connected yet to the gap or there is still no measurable current in the discharge. In this case [Figure 4.1(b)], the capacitors of each stage are discharged through their parallel resistors R_1 and R_2, except for the first and the last capacitors, because their discharge circuit has only one resistor. The voltage on each capacitor decreases as

$$C\frac{dU}{dt} + \frac{R_1 + R_2}{R_1 R_2}U = 0 \qquad U = U_0 \exp\left(-\frac{t}{T}\right) \qquad T = \frac{R_1 R_2}{R_1 + R_2}C \quad (4.1)$$

The pulse (waveform) duration is, by convention, taken to be the time necessary for the voltage to drop by half the amplitude value, $t_w \approx 0.7T$. Resistances R_1 and R_2 should not be taken too high. The current flowing

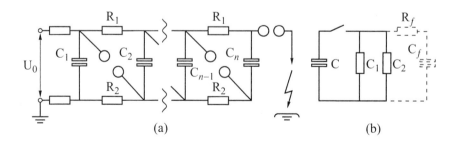

(a)   (b)

FIGURE 4.1
The basic circuit of a voltage pulse generator (a) and the design circuit for pulse parameter calculations for the no-load regime (b).

through them must sustain an arc in all spark gaps, otherwise the capacitors may become disconnected. The probability of disconnection greatly increases after the capacitors have partly discharged and the current has decreased. Commonly chosen resistances are $R_1 = R_2 < 20$–30 kΩ, so the pulse duration of a zero-load VPG is $t_w \approx 10$ ms at a typical stage capacitance $C \approx 1$ μF. This is sufficient to produce a spark of several hundred meters long. An important condition, however, is that the time for the voltage effect should not be greatly reduced by the flowing discharge current.

As a spark elongates, the source charge flows into it, and the external circuit has current i (Section 3.2.2). This current accelerates the VPG capacitor discharging, sometimes very appreciably. For example, if discharge current of 10 A flows through the VPG stages (this value is quite possible for a long leader), the capacitors with $C = 1$ μF, charged to 100 kV, can give off half of their charge to the gap for the time $\Delta t = CU_0 / (2i) = 5$ ms. The gap voltage then drops by half.

The pulse distortion is very undesirable, since the VPG may be destabilized: it may generate pulses of variable duration depending on the gap current variation. As a result, many discharge parameters, including the spark propagation rate and current, will also vary. The experimental conditions may then get out of control, so that the experimenter will eventually have the wrong idea of the natural dispersion of spark parameters.

The situation described is not at all far-fetched; it may even become unpredictable, because sporadic discharges may develop from some laboratory equipment and connective wires during megavolt-range experiments. They are capable of loading the VPG by their current but remain practically uncontrollable. This is a reason why the parameters of very long sparks measured at different laboratories and in different experiments often disagree, especially breakdown voltage data which show a large spread.

Consider the problem of pulse front which affects many spark characteristics, including breakdown voltage. The pulse rise time is usually controlled by changing the parameters of the RC-circuit connected to the VPG output in parallel with the gap [R_f and C_f in Figure 4.1(b)]. After the generator spark gaps start up, the voltage of capacitor C_f, known as front capacitor, rises exponentially; at $R_f C_f \ll t_w$ this happens in much the same way as in a RC-circuit connected to a constant voltage source:

$$U(t) = nU_0 \left[1 - \exp\left(-t/R_f C_f\right)\right] \qquad (4.2)$$

The time constant $R_f C_f$ characterizes the rise time t_f, which is close to $3R_f C_f$ at $t_f \gg t_w$. The problem of rise time control is equally complicated, when very small or very large t_f values are required. In case of very small t_f, one has to bear in mind that minimum C_f is limited by the total capacitance of the gap electrodes, connective wires and measuring equipment, primarily the voltage divider.

In megavolt-range experiments, C_f cannot be made lower than a few hundred picofarads. The value of R_f is also limited: it cannot be smaller than the total resistance of sparks in the gaps. Besides, to reduce the resistance to zero is not particularly advantageous, since the process of charging the capacitor C_f breaks into an oscillatory mode at $R_f < 2\sqrt{L/C_f}$, where L is the inductance of the current-carrying elements. The rate of the capacitor voltage rise is then defined not so much by resistance as by inductance L, which is the sum of spurious inductances of the VPG stage capacitors and the front capacitor plus the bus inductances and those inherent in the current reverse flow through the ground.

If a VPG designed for extremely high voltages is to be mounted in a usual laboratory building, its dimensions must be determined by the air insulation and can be rarely less than 25–30 m. For such a long current path, inductance L turns out to be as large as a few tens of microhenries, while the minimum time necessary for the voltage to reach an amplitude value equal to about half the oscillation period, $T = 2\pi\sqrt{LC_f}$, is close to 100 ns. Rise times of a genuine nanosecond range are practically unfeasible in a VPG of conventional design. Experimenters are striving to reduce the rise time by using specially designed capacitors with a low spurious inductance. Sometimes, they just use cable segments and mount a VPG in a chamber filled with a highly insulating gas. Here, the insulation distances determining the source dimensions can be made smaller by employing short spark gaps, which pick up more rapidly and produce sparks of lower resistance.

Pulses with smooth fronts of 100–1000 μs duration would seem to create no problems. They can be generated by simply enlarging the front capacitor C_f or resistor R_f. But the capacitor insulation must be good enough to withstand the total VPG voltage. For this reason, a special capacitor stack is rarely used as C_f but its function is normally performed by the whole assemblage of a high voltage divider, a large measuring sphere gap with a 1–3 m diameter connected to the VPG, and buses connecting the generator to the discharge gap.

Incidentally, the high voltage electrode capacitance is also part of C_f, so that the voltage waveform may vary with the electrode. The total capacitance becomes as low as 1–2 nF. This means that the generation of smooth pulses requires the use of a high resistance front resistor with $R_f \approx t_f/(3C_f) > 0.1$ MΩ. In fact, it separates the VPG capacitance from the front capacitance, $C_f \ll C_g$. When a discharge starts to develop in the gap, the relatively small C_f charge is quickly transported by discharge current i to the developing spark. The capacitor voltage appreciably decreases, because the current supply from the VPG is limited by high resistance R_f. The pulse shape in the gap may change so radically that the spark may slow down or even entirely disappear.

4.1.2 The Fitch circuit

In recent years, spark investigators have given much attention to the Fitch circuit capable of providing steadier pulses than the conventional Marx design. In this circuit, the capacitors of an even number of stages are *a priori* connected in series. Each capacitor gang (encircled by a broken line in Figure 4.2) includes a pair of capacitors C_1 and C_2, diodes D_1 and D_2, an inductance coil L, resistors R_1 and R_2, and a controllable spark gap. The capacitors are charged by constant voltage supply U_0 through diode D_1 and inductance; the two cell capacitors have opposite signs, so that the total cell voltage is zero. The other diode, D_2, is not involved in charging the cell but only passes charge current to the next cell. After the gaps break down in all the cells, the capacitor C_2 and inductance L form an oscillatory circuit, in which the current flows through an open diode, D_2. A half period $T = 2\pi\sqrt{LC_2}$ later, the capacitor C_2 becomes completely recharged, so its voltage is not canceled by that of C_1 but is added to it to give the cell voltage $2U_0$; the line voltage is $U = nU_0$, where n is the total number of capacitors. At this moment, the current in the circuit LC_2 ceases, because the diode D_2 turns out to be locked. This completes the formation of a voltage waveform with an amplitude nU_0 at the generator output. All the diodes are locked, and the output voltage drop under no-load conditions is determined by the capacitor discharging through the respective resistors. Then the source voltage is multiplied by the number of capacitors, exactly as in the Marx circuit.

The voltage rise from zero to the maximum takes $t_f \approx T/2 = \pi\sqrt{LC_2}$.

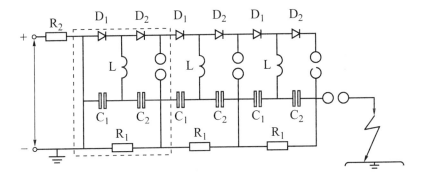

FIGURE 4.2
The Fitch circuit of a high voltage pulse generator.

By varying the coil inductances, one can control the rise time within wide limits, achieving, if necessary, millisecond values. No front capacitors are required for this. The pulse front is formed by the main VPG capacitors, and there are no series elements between the VPG and the gap that would lose voltage. This greatly reduces waveform distortions by discharge current, providing a high stability of the generated voltage parameters, even in high power spark experiments. The no-load pulse duration in this design is practically unlimited. Resistors R_1 and R_2, which determine the pulse duration, only affect the charging time of capacitors C by voltage U_0; therefore, the resistors can be taken to be very large. These advantages of the Fitch design compensate for its complexity and high cost associated with the use of high voltage diodes.

4.1.3   Alternating voltage sources

Alternating voltage sources of commercial frequencies are widely employed in technical laboratories and on spark test beds. They are assembled in series of 2–4 transformers of 350–1000 kV. Transformers operate either in the standard mode with their primary coils supplied by a commercial ac circuit, or in the pulse mode, when a precharged capacitor battery is discharged into the coil. In the latter case, one can generate very smooth pulse fronts of 2–4 ms, a result unattainable with a Marx circuit. Such slowly rising pulses are attractive owing to their similarity to surge pulses, or switching waves, which affect the insulation of high voltage equipment.

The deformation of transformer cascade output voltage caused by discharge current is appreciably greater than in any VPG circuit due to a high equivalent cascade inductance L, through which the energy is supplied to the discharge gap. In low power cascades normally employed in laboratories, L may be close to 10^3 H in order of magnitude. Like in the Marx circuit discussed above, the electric field energy is stored by the capacitors of the transformer cascade, current-carrying elements, and measuring equipment connected to the gap. The total cascade capacitance C rarely exceeds a few nanofarads, while the wave resistance $Z = \sqrt{L/C}$ of the oscillatory circuit LC lies within 500–1000 kΩ.

Practically any discharge current i exceeding 1 A is larger than the circuit current $i_1 \approx U/Z$, which varies from a few fractions of an ampere to several amperes. Situations have been observed when discharge current caused such a dramatic voltage drop that the leader process stopped and resumed only in several hundred microseconds, after current i_1 in the oscillatory circuit LC had recharged the capacitor C. Analysis of transformer cascade data requires much caution, because it is very easy to take purely 'circuit effects' for manifestations of the genuine discharge behavior.

4.2   Voltage measurement

4.2.1   A measuring sphere gap

The simplest and most reliable way of measuring voltage is by means of
a sphere gap. If the distance between two spheres is smaller than their
radius, the field distribution over the gap is nearly uniform. Breakdown
voltage in such gaps is practically independent of the waveform evolution
in the microsecond and longer time intervals. The statistical spread of
breakdown voltage data is less than one percent. This allows an accurate
gap calibration to be made. Tables of breakdown voltages as a function
of intersphere distance for gaps of $1-1.5$ m radius are published in special
reference books together with corrections for air density and humidity. The
voltage being measured is considered to be equal to a table value, when it
produces a 50% probability breakdown. Sphere gaps are still used to test
voltage dividers—the measurement simplicity and data reliability remain
an attractive feature. However, one should treat with caution earlier data,
when voltage was only controlled by a sphere gap. Most of those studies
were done with a rise time of about 1 μs, when the waveform was seldom
smooth but often showed damped oscillations associated with the VPG
spurious inductance. A sphere gap could pick up at the inductive overshoot
amplitude, while the process of interest was actually initiated at a lower
voltage.

4.2.2   Voltage dividers: a general description

Voltage dividers permit the use of ordinary low voltage equipment (oscil-
loscopes, peak voltmeters, analog-to-digital converters with computers) for
the registration of high voltage pulses. These instruments are indispensable in a modern laboratory. In a resistance divider, the voltage to be measured is distributed between series resistors R_1 and R_2 while in a capacity
divider between series capacitors C_1 and C_2 (Figure 4.3). In the first case
the voltage division ratio U/U_2 is $k_R = (R_1 + R_2)/R_2$, while in the second case $k_C = (C_1 + C_2)/C_1$. Both expressions are strictly valid only for
a purely speculative situation, when the divider arms can be depicted in a
simplified manner, like in Figure 4.3(a). In reality, a high voltage resistance
divider is fairly large, so one has to allow for the spurious capacitances of
the resistors and measuring instruments connected to the low voltage arm
R_2 [dashed line in Figure 4.3(a)]. The spurious parameters of a capacitance
divider are the capacitor leakage resistances and the input resistance of the
same instruments [dashed line in Figure 4.3(b)]. Therefore, both types of
divider can be described by a common circuit design [Figure 4.3(c)].

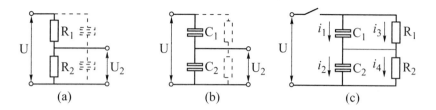

FIGURE 4.3
The basic circuits of resistance (a) and capacitance (b) voltage dividers, and their common design circuit (c).

Even a superficial examination of the circuit design shows that the voltage division ratio is generally time-dependent. This can be illustrated with reference to a rectangular voltage pulse of infinite duration, $U(t) = U$ at $t \geq 0$. At small times, when the divider arms obey the inequalities $t \ll R_1 C_1$ and $t \ll R_2 C_2$ and the current through the resistors has not much changed the C_1 and C_2 charges, the voltage distribution is characterized by the charge equality $U_1 C_1 = U_2 C_2$. The actual division ratio k_D is then close to an ideal ratio for the purely capacitive variant ($k_D \approx k_C$). In the steady-state mode at $t \gg R_1 C_1$ and $t \gg R_2 C_2$, the capacitors are completely recharged, there is no current through them, and the voltage of the divider arms provides equal currents through the series resistors: $U_1 = i R_1$ and $U_2 = i R_2$. The actual ratio now approximately corresponds to the ideal ratio for the resistance divider ($k_D \approx k_R$). Thus, the division ratio smoothly varies over the measurement period from the initial value $k_C = (C_1 + C_2)/C_1$ to the steady-state value $k_R = (R_1 + R_2)/R_2$. The similarity between the voltage $U(t)$ applied to the gap and the actual voltage being measured on the low voltage arm, $U_2(t) = U(t)/k_D$, is violated. The registered pulse will be distorted, that is, its waveform will be different from that of $U(t)$.

Consider in some detail the transient process in the general diagram of a divider in Figure 4.3(c). It can be described by the following equations:

$$i_3 R_1 + i_4 R_2 = U \qquad i_1 + i_3 = i_2 + i_4$$

$$\frac{1}{C_1} \int_0^t i_1 dt = R_1 i_3 \qquad \frac{1}{C_2} \int_0^t i_2 dt = R_2 i_4 \qquad (4.3)$$

Calculate this set of equations with respect to the output voltage to be measured, $U_2(t) = R_2 i_4$:

$$\frac{dU_2}{dt} + \frac{R_1 + R_2}{(C_1 + C_2) R_1 R_2} U_2 = \frac{U}{R_1 (C_1 + C_2)} \qquad (4.4)$$

For a rectangular voltage pulse of infinite duration, $U(t) = U$ at $t \geq 0$, formula (4.4) integrated with the initial condition $U_2(0) = C_1 U/(C_1 + C_2)$ gives

$$U_2(t) = y(t) U \qquad y(t) = \left(\frac{1}{k_C} - \frac{1}{k_R}\right) \exp\left(-\frac{t}{T_D}\right) + \frac{1}{k_R}$$

$$k_C = \frac{C_1 + C_2}{C_1} \qquad k_R = \frac{R_1 + R_2}{R_2} \tag{4.5}$$

$$T_D = \frac{R_1 R_2 (C_1 + C_2)}{R_1 + R_2}$$

where T_D is the time constant of the divider.

If the time variation of $U(t)$ has a complicated character, the solution to equation (4.4) for the registered signal $U_2(t)$ can be obtained using Duhamel's integral in one of its equivalent forms:

$$U_2(t) = y(0) U(t) + \int_0^t y'(\tau) U(t - \tau)\, d\tau \tag{4.6}$$

$$U_2(t) = y(t) U(0) + \int_0^t U'(t - \tau) y(\tau)\, d\tau \tag{4.7}$$

where $y(t)$ is the divider transient function, which coincides, by definition, with the solution to equation (4.4) at unit voltage [$U(t) = 1$ at $t \geq 0$], that is, the function prescribed by formulas (4.5).

4.2.3 Voltage dividers: special cases

Here we will discuss some special cases essential for designing spark experiments. Consider a resistance divider with a small spurious capacitance of the high voltage arm C_1 and appreciable capacitance C_2 loading the low voltage divider output. In this case, if the conditions $C_1/C_2 \ll R_2/R_1$ and $R_1 \gg R_2$ are met, the main formula of (4.5) takes a simplified form

$$U_2(t) \approx U \frac{R_2}{R_1} \left[1 - \exp\left(-\frac{t}{R_2 C_2}\right)\right] \tag{4.8}$$

Such a divider distorts the rectangular pulse front, changing the steep step to an exponent with the time constant $R_2 C_2$; the horizontal segment of the pulse is transferred without distortions. If a rectangular pulse has a limited duration, $t_w \gg R_2 C_2$, its back edge will be distorted as much as the front. The distorted pulse can be easily restored by representing the pulse of finite duration t_w as the sum of two infinite rectangular pulses, $+U$ and $-U$, time-shifted by $\Delta t = t_w$ [Figure 4.4(a)]. The deformed pulse front can distort the transmitted pulse amplitude if t_w is comparable with

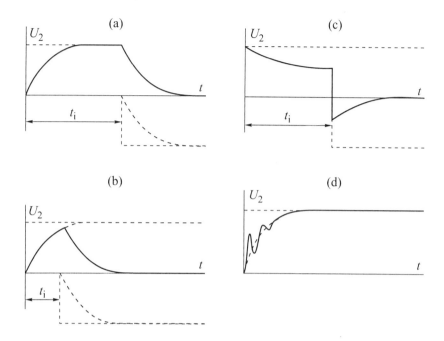

FIGURE 4.4
Distortions of a registered voltage pulse at the voltage divider output: the pulse front and back edge (a) and amplitude (b) distorted by a resistance divider; the pulse arm and back edge distorted by a capacitance divider (c); the pulse front distorted by the spurious divider inductance (d).

or smaller than the time constant R_2C_2 [Figure 4.4(b)]. The amplitude measurement error will increase with decreasing t_w relative to R_2C_2.

The leakage resistance R_1 of the high voltage arm in capacitance dividers is often extremely large, especially when the function of C_1 is performed by a measuring sphere gap or a standard capacitor with gas insulation. The inequalities $R_2/R_1 \ll C_1/C_2 \ll 1$ or $k_R \gg k_C \gg 1$ are usually fulfilled in this case. For an infinite rectangular pulse, expression (4.5) yields

$$U_2\left(t\right) \approx U\frac{C_1}{C_2}\exp\left(-\frac{t}{R_2C_2}\right) \tag{4.9}$$

It is now the horizontal pulse segment that becomes distorted: voltage U_2 is the lower, the greater is the discharging rate of the measuring arm capacitor C_2 through the leakage resistance R_2. The pulse front remains unaffected and its 'infinite' slope is transferred by the divider without distortions.

The back edge distortion of a rectangular pulse of finite duration t_w is manifested in that the voltage U_2 being registered passes through zero, giving rise to an exponentially decreasing 'tail' of the opposite sign [Figure 4.4(c)]. This can be easily demonstrated by representing a finite pulse as the sum of two time-shifted infinite rectangular pulses of the same amplitude but opposite signs. As was pointed out, the back edge remains unaffected. But the experimenter would make a serious error if he tried to measure the residual voltage after its partial drop, say, due to the gap breakdown, when he is to find the average spark conductivity. The measurement error may be as large as several hundred percent.

To conclude, let us turn to the idea of a combined, capacitance-resistance divider. It follows from formula (4.5) that at $k_C = k_R$ or at $C_1/(C_1 + C_2) = R_2/(R_1 + R_2)$, the factor with the exponential term vanishes, the behavior of U_2 immediately changes to a steady state, and the divider ratio becomes constant. Such a divider has a stepwise transient function, $y(t) = 1$ at $t \geq 0$, and can transmit any waveform without distortions. Capacity-resistance dividers with parameters chosen in accordance with $R_1 C_1 \approx R_2 C_2$ equivalent to $k_C = k_R$ are often used in laboratory practice. Their application, however, does not remove all difficulties associated with the registration of rapidly varying voltage pulses. Even an ideally adjusted divider may distort waveform because of the finite arm inductances and the spurious capacitances between the construction parts and the ground. These limitations of the 'ideal' circuit are to be borne in mind when measuring rapidly varying voltages (for instance, with a nanosecond rise time), especially in the megavolt range, when the divider has to be made as long as several dozens of meters to satisfy the electrical insulation requirements.

We will illustrate this with reference to the effect of spurious inductance. It should be introduced into the circuit design [Figure 4.3(c)] in series with a high voltage arm capacitor C_1. The oscillatory circuit LC_1 gives rise to damping oscillations of frequency $f = 1/(2\pi\sqrt{LC_1})$ in the region of both pulse edges. They overlap the 'steady' pulse, $U_2(t) = U(t) R_2/(R_1 + R_2)$, distorting it [Figure 4.4(d)]. The oscillations are damped by shunting resistances of the capacitors comprising C_1 (which is often made up of series capacitors to provide the desired electrical insulation) and by the wire resistance, which is quite large because of the skin-effect at high oscillation frequency. Sometimes, a special resistor with a resistance comparable with the oscillatory circuit impedance $\sqrt{L/C_1}$ is introduced into the divider circuit to achieve a more effective oscillation damping. This kind of damping affects unfavorably the quality of voltage transmission by the divider.

There are situations in which a divider has to be placed at a distance of 10−20 m from the discharge gap. Then one actually measures the voltage near the divider rather than in the gap, and the difference is as large as the self-induction emf in the conducting wires. In some cases this inductive component becomes a substantial source of additional instrumental errors and should be taken into account in the assessment of measurement validity.

The experimenter is not always free to choose a divider from the criterion of minimum distortion. He may have to allow distortions for the sake of accomplishing the principal experimental task. It may be important, for instance, that minimum energy be supplied to the spark from the divider capacitors. In this case, it is advisable to replace both the capacitance and combined dividers available by resistance dividers with very high resistances in order to slow down the discharging of spurious resistor capacitances. Clearly, the voltage registration precision will be lower, and one should take care not to cross the threshold, which separates an experiment with predictable errors from a principally false experiment.

4.2.4   Registration devices

A divider signal can be registered by a peak voltmeter, if one wants to measure the pulse amplitude only. The waveform is recorded by an oscilloscope or transmitted to a computer in the digital code with a subsequent registration by modern information recorders. Today, we can only speak historically of instrumental distortion. Modern electronics is capable to meet practically any requirements on the frequency response and input resistance of a device necessary for long spark studies. Still inaccessible, perhaps, are superhigh precision measurements with an error less than one percent, but these do not have much sense because of the imperfections of voltage dividers.

Some caution is necessary when making measurements with a peak voltmeter. Its principal design is similar to that of a peak detector and includes a diode and a storage capacitor. The diode is open, while the capacitor voltage is less than that to be measured. When these voltages become equal, the diode is locked, retaining the capacitor voltage close to the pulse amplitude. The capacitor has a sufficiently large capacitance to permit the voltage 'read-off' by a common digital or arrow voltmeter. The experimenter may be faced with two critical situations. First, some imperfections in the VPG or divider design may result in pulse front overshoot associated with high frequency oscillations that may be excited in circuits with spurious inductance. The peak voltmeter will record the voltage at the moment of overshoot, whereas the discharge will actually develop at a lower voltage. This situation is entirely identical to that for a sphere gap discussed in Section 4.1. Second, high frequency oscillations may also arise when the voltage drops abruptly during a breakdown not only due to the divider spurious inductance. In experiments with high voltages and large electrode capacitances typical of breakdown in large-radius sphere gaps, the gap becomes a very powerful source of electromagnetic radiation, which is received by a nearby divider functioning like an antenna and added to the voltage being measured. Voltage overshoots may be recorded by a peak voltmeter. To avoid this kind of error, selection circuits may be effectively used, because they respond to the changing sign of the derivative dU/dt and 'seal' the diode detector circuit, as soon as the pulse amplitude has been reached, until the moment of taking the readings.

Electromagnetic noise in a high voltage laboratory makes one take care of the equipment connecting the divider to the registering devices. Until recently, this problem, along with cable matching (to eliminate reflected waves), has determined the choice of experimental equipment to be used. Up-to-date electronics has resolved these problems by means of optical communication lines, analog-to-digital converters, and large storages.

The analog-to-digital signal conversion offers two possibilities. First, the digital information concerning the registered pulse can be stored in the on-line memory of the voltage divider. After the voltage registration devices have done their job, the computer can give a command to restore the voltaic connection with the divider and then to transmit the stored information. The time for doing this is long enough, because the pauses between pulses of a voltage source are much longer than the duration of a generated pulse. The communication line noise is no problem, since the line is disconnected from the devices and the voltage divider while high voltage is on. Besides, there is no problem of synchronizing the measuring equipment with the application of a voltage pulse to the gap. The on-line memory can be kept in the on-state to record and automatically erase the 'zero' information, until a useful signal exceeding the preset sensitivity threshold appears. The recorded information can be stored until required (see Section 4.8). The other possibility is to convert the digital information about a voltage pulse to a light signal and to transmit it through an optical fiber nearly perfectly protected from electromagnetic noise. Sometimes, an optical line directly transmits analog information, but the transmission is confined to a narrower dynamic range. (Difficulties arise when the experimenter must record voltages, whose maximum and minimum differ by several orders of magnitude.) At the receiver output, the light pulses are again converted to electrical signals and displayed on a screen or saved for later use.

4.3 Current measurement

The current measurement reduces to the measurement of voltage created by the current flow through a calibrated resistor known as a *shunt* or by its magnetic field (in case of variable current) affecting a nearby calibrated induction coil. Shunt measurements have no principal limitations—they are applicable to direct and alternating currents. Measurements from the coil magnetic induction emf are hard to make, if the current varies too slowly.

4.3.1 Shunt measurements

The current is found from the known shunt resistance and its measured voltage: $i = U/R$. The range of current values measurable in a spark discharge is very wide—from small fractions of an ampere in a corona to several hundred kiloamperes in a lightning return stroke or in a powerful spark. So the requirements on the shunt resistance are also different—from tens and hundreds of ohms at low currents to $10^{-3}-10^{-4}$ Ω at maximum currents. The smaller is the shunt resistance, the more remarkable is the effect of the emf of self-induction excited by time-varying current: $U = iR + Ldi/dt$, where L is the spurious shunt inductance. The inductive component of the voltage, Ldi/dt, defines the measurement error, especially large in the registration of the pulse front, where the current varies much faster than at the pulse center.

The shunt transient function, which characterizes the shunt response (that is, voltage) to the flowing current, should in principle be defined per unit current pulse with a rectangular front ($i = $ const $= 1$ at $t \geq 0$). But in a circuit consisting of series inductance and resistance, this approach has no physical sense, since the momentous introduction of the final current into the circuit would require an infinitely high voltage, which would have to overcome the shunt self-induction emf, also infinitely large at the moment of current turn-on. It is more reasonable to use the expression for a pulse with a steep but finite front slope and then go to the limit, if necessary. For instance, by representing the pulse front as the function $i(t) = I[1 - \exp(-\alpha t)]$, we find that the voltage being measured

$$U = IR[1 - \exp(-\alpha t)] + IL\alpha \exp(-\alpha t) \qquad (4.10)$$

will have a distorting overshoot at the front [Figure 4.5(a)]. Its amplitude will be the higher, the shorter is the front and the larger the spurious shunt inductance. The flat pulse segment corresponding to $t \gg 1/\alpha$ will be reproduced without distortion.

A shunt can be improved by reducing the inductive component of the output voltage. It follows from equation (4.10) that the effect of this component is substantial in a low resistance shunt, so complications arise in making shunts for measuring currents of tens and hundreds of kiloamperes, when R must be reduced to a few fractions of an ohm. There are two ways of limiting the inductive component. One is rather straightforward and is aimed at reducing the inductance of the material, from which the shunt is

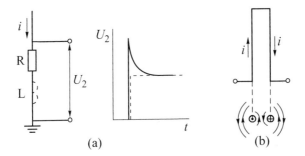

FIGURE 4.5
A shunt circuit for current measurement and an output signal distorted by
spurious inductance (a); the principle diagram of a loop shunt (b).

made. Various loop structures are employed for this purpose. The inductance of a loop is smaller than in a straight conductor of the same length, since it is determined by the magnetic flux going through the loop area [Figure 4.5(b)]. When the conductors are brought close to each other to make this area very small, the inductance decrease may be multifold. The limitation is associated with the conductor insulating coating, whose thickness is always finite. A shunt made from wire loops (there may be several loops connected in parallel) is known as a *loop shunt*. The application of loop shunts is restricted to the measurement of microsecond current pulses. For shorter pulses, the error due to the self-induction emf is unacceptably large.

The other, more effective way of raising the shunt precision is to reduce directly the self-induction emf in the circuit, whose output voltage U is being measured. This idea has been implemented in a *cylindrical (tubular) shunt*. Its resistance is created by a thin-walled cylindrical tube, whose ends are closed by massive metallic flanges. The current to be measured enters through the upper flange, flows through the tube and goes out of the bottom flange (Figure 4.6). In this design, the shunt inductance does not differ from that of a straight conductor of the same diameter. An important feature of a cylindrical shunt is the way in which its voltage is measured. Voltage is taken from the shunt wall in such a way that the conductors connecting the measuring device are located inside the tube. A special electrode, known as a potential electrode, is used for this purpose. It goes along the tube axis, comes in contact with the inner surface of the upper flange and goes out through a hole in the bottom flange without touching it. A coaxial connector connecting the communication cable and the measuring device, usually an oscilloscope, is introduced into the hole. The connector male

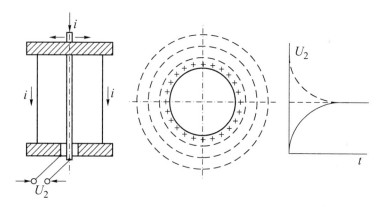

FIGURE 4.6
The circuit, magnetic field, and distorted electrical signal at the cylindrical
shunt output.

contact touches the potential electrode, while its screen touches the inner
surface of the bottom flange hole. The voltage, transmitted by the cable
and measured, is

$$U(t) = U_W(t) - M\frac{di}{dt} \tag{4.11}$$

where U_W is the voltage on the shunt wall and $-M di/dt$ is magnetic induction (more exactly, mutual induction) emf in the potential electrode. In
its turn, $U_W(t)$ is the total voltage drop on the metallic tube of the shunt,
when current $i(t)$ flows through it. If the tube wall is extremely thin and
current $i(t)$ is uniformly distributed throughout its thickness, we have

$$U_W(t) = Ri(t) + L\frac{di}{dt} \tag{4.12}$$

where R is the shunt wall resistance and L is its inductance determined
by the external (with respect to the wall) magnetic flux. Note that the
magnetic lines of force create, by virtue of symmetry, a set of concentric
circumferences with the cylinder wall cross section. They do not cross the
inner tube space (Figure 4.6), so that the magnetic flux through any inner
cross section will be zero, if we ignore the magnetic field due to a very
low current, as compared to $i(t)$, in the potential electrode closed via the
measuring equipment. Therefore, we can neglect the voltage drop caused
by this low current and take the inductive component $M di/dt$ to be equal
to $L di/dt$. The latter assumption is justified by the fact that the shunt wall
and the potential electrode are enveloped by the same magnetic lines, since
there is practically no magnetic field inside the cylinder; hence, $M = L$

(see Figure 4.6). Under these conditions, the voltage being measured

$$U(t) = Ri(t) + L\frac{di}{dt} - M\frac{di}{dt} = Ri \qquad (4.13)$$

is determined by the time-constant ohmic resistance of the shunt wall and is strictly proportional to $i(t)$. The inductive component of the wall voltage is not registered: it is totally compensated by the magnetic induction emf in the potential electrode, providing a high registration precision.

4.3.2 The skin effect in a tubular shunt

Current measurement by a tubular shunt produces a methodological error due to the skin effect. As a result, the time-varying current $i(t)$ is distributed nonuniformly throughout the wall thickness, loading the external conductor surface much more than the inner surface (the electromagnetic energy enters through the external surface). In the case of rectangular current pulses, $i(t) = I$ at $t \geq 0$, all the current flows through an infinitely thin external tube layer at the moment of turn-on and then gradually penetrates into the wall.

We would like to warn the reader against a purely formal account of the skin effect, when considering cylindrical shunts. A nonuniform current distribution through the conductor bulk increases its resistance. This would seem to increase voltage $U = Ri(t)$, so that one might expect a characteristic pulse distortion due to the front overshoot. In reality, quite the opposite is observed: the skin effect-related distortion is manifested as a smearing of $U(t)$, which increases with the conductivity and thickness of the cylinder wall (Figure 4.6). The reason for a possible misinterpretation lies in the unjustifiable substitution of general expressions (4.11) and (4.12) by the relation $U = Ri(t)$ valid only for the special case of an infinitely thin wall or for constant current. In the general case of a wall of finite thickness d, voltage $U_W(t)$ can be expressed, similarly to formula (4.12), as the sum of two components: the resistive component $U_R(t) \sim i(t)$ and the inductive component $U_I(t) \sim di/dt$. Let x be the radial coordinate in the circular cross section of the cylindrical shunt in question, counted off from the external cylinder surface inward. For any cylindrical surface characterized by the x-coordinate, the voltage drop $U_W(x,t)$ is the same along the cylinder length l, very much like conductors connected in parallel have the same voltage. However, the proportions of the components $U_R(x,t)$ and $U_I(x,t)$ for surfaces with different x-coordinates are different. At the moment of current alternation, when the skin effect reveals itself, the current density $j(x,t)$ and the voltage component $U_R(x,t) = lj(x,t)/\sigma$ (σ is the shunt wall conductivity) are maximal at the external surface, that is, at $x = 0$. On the inner tube surface, $x = d$, they are minimal and even equal to zero for a rectangular pulse. The inductive component $U_{\mathrm{I}}(x,t)$ is, on the contrary, maximal at the inner surface, because the magnetic circular lines enveloping the tube from the outside are added to those in the tube wall. Their number increases, as the x-coordinate approaches d.

Formally, it does not matter to which cross section we attribute the two components, $U_{\mathrm{R}}(x,t)$ and $U_{\mathrm{I}}(x,t)$, in the analysis of the measured voltage

$$U(t) = U_{\mathrm{R}}(x,t) + U_{\mathrm{I}}(x,t) - M\frac{\mathrm{d}i}{\mathrm{d}t} \tag{4.14}$$

because the result will eventually be the same. For the inner surface, however, expression (4.14) is simplified, because it is this surface and the potential electrode that are enveloped by all magnetic lines, inducing an absolutely identical induction emf, $U_{\mathrm{I}}(d,t) = M\mathrm{d}i/\mathrm{d}t$, in them. There is only one term left in expression (4.14) for $U(t)$

$$U(t) = U_{\mathrm{R}}(d,t) = \frac{lj(d,t)}{\sigma} \tag{4.15}$$

proportional not to current $i(t)$ but to its density in the cross section $x = d$, which increases with time as the current fills up the cross section.

The proportionality of $U(t)$ to the current density on the inner shunt wall, $j(d,t)$, not only accounts for the pulse front smoothing but also allows estimation of the error due to the skin effect. Indeed, the current penetrates into the metallic wall from the external surface together with the electromagnetic wave, whose propagation is described by the Maxwellian equations. In a well conducting medium, in which displacement current is negligible as compared to conduction current, the equations take the form

$$\mathrm{rot}\,\boldsymbol{E} = -\mu_0\frac{\partial \boldsymbol{H}}{\partial t} \qquad \mathrm{rot}\,\boldsymbol{H} = \sigma\boldsymbol{E} \tag{4.16}$$

By applying the *rot*-procedure to both equations and using the well-known formula of vector analysis for *rotrot*, we obtain the diffusion-type of equations

$$\frac{\partial \boldsymbol{E}}{\partial t} = \frac{\Delta \boldsymbol{E}}{\mu_0 \sigma} \tag{4.17}$$

for fields $\boldsymbol{E}$ and $\boldsymbol{H}$. The quantity $D = (\mu_0\sigma)^{-1}$ plays in them the role of a diffusion coefficient. Any region of the tube with the wall thickness much smaller than its radius can be regarded as plane, such that $\Delta = \partial^2/\partial x^2$.

Omitting a rather cumbersome solution to equation (4.17) for a wall of finite thickness, we will evaluate the time for the filling of a thin-walled conductor by current, using the formula for particle diffusion from a plane source in an infinite medium: $x = (2Dt)^{1/2}$. Obviously, the time scale of the field penetration into the wall and the filling of a thin-walled conductor by current is the quantity $T \sim d^2/(2D) = \mu_0\sigma d^2/2$. It is during this time that the inner wall current density, proportional to the shunt voltage measured [equation (4.15)], will grow from zero at the moment of the rectangular pulse turn-on to a value corresponding to constant current. Its order of magnitude will characterize the pulse front smoothing due to the skin effect.

In nonmagnetic heat-resistant alloys (manganine, constantan) with a specific conductivity $10^6 \, \Omega^{-1} \, m^{-1}$, the wall thickness $d \approx 0.1$–0.2 mm corresponds to $T \sim 10$ ns. This is approximately two orders of magnitude less than for loop shunts. Still, the capabilities of cylindrical shunts are sometimes insufficient. A streamer starting from an electrode of 1 cm radius may have current pulses of the nanosecond range. Excessive reliance on cylindrical shunts may be fraught with serious errors in both rise time and current amplitude measurements. Errors may be so large that the results will be unsuitable, for example, for a comparison with simulation data. The cylindrical shunt exists in several modifications, in which the harmful skin effect is compensated by 'injection' of a portion of the magnetic field into the shunt space through special slits in the walls. The number and width of the slits are chosen empirically. It is worth making an effort to improve the precision of cylindrical shunts, after all other error sources have been removed, for instance, signal distortions in the cable connecting the shunt with the oscilloscope. This error source should be taken into account in the analysis of streamer current data recorded by conventional means, without using a digital information transmission line. The distortion of a nanosecond pulse front during its transmission through a 10 m cable to a detector may be greater than that caused by the skin effect in a shunt.

4.3.3 Current measurement by a Rogovsky belt

The registration of current from its magnetic field is an attractive technique because this current does not flow through the measuring system. Its main element is a magnetic probe in the form of a coil with a magnetic flux created by the current to be measured. Theoretically, the current and the probe may be separated by any distance, and it is necessary to know only the mutual inductance M between the current circuit and the coil and to screen the coil from foreign magnetic fields. The current to be measured is determined by the time integral of the mutual induction emf E_m:

$$E_m = -M\frac{di}{dt} \qquad i(t) = -\frac{1}{M}\int_0^t E_m(t)\, dt \qquad (4.18)$$

For this, the coil is loaded on the integrator, whose output voltage $U_2(t)$ is proportional to the integral of $E_m(t)$ and, hence, of $i(t)$. The transmission gain of the measuring system $k = U_2/i$ can be calculated from known (or precalculated) M and the integrator electrical parameters, but, more often,

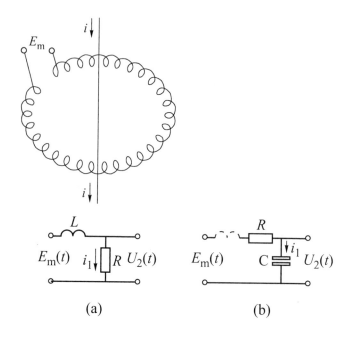

FIGURE 4.7
The Rogovsky belt and the design circuits of its integrating LR (a) and RC (b) chains.

the circuit is just calibrated by measuring the output voltage U_2 of the known current i with exactly the same current circuit geometry and probe position.

It is reasonable to use a toroidal inductance coil embracing a conductor or a line with the test current (Figure 4.7). Then the positions of the current and the coil relative to one another do not much affect the value of M. The toroidal design is known as a *Rogovsky belt*. This device usually uses, as an integrator, LR or RC circuits closed by a coil [Figure 4.7(a,b)]. In the first case, no special L coil is made: its role in the integration circuit is usually played by the probe coil, whose terminals are connected to an integrating resistor R. The resistor voltage is described as

$$L\frac{di_c}{dt} + Ri_c(t) = E_m(t) \tag{4.19}$$

where i_c is the coil current. If the resistance is so small that we have

$$L\frac{di_c}{dt} \gg Ri_c$$

during the whole period of measurement, then

$$i_c(t) \approx \frac{1}{L}\int_0^t E_m(t)\,dt = -\frac{M}{L}i(t) \qquad U_2(t) = Ri_c = -\frac{MR}{L}i(t) \quad (4.20)$$

that is, the output voltage $U_2(t)$ is proportional to $i(t)$ and the circuit transmission gain is $k = -MR/L$.

In RC integration, the Rogovsky coil is closed by a capacitor C and a resistor R connected in series, while the voltage is taken from the capacitor [Figure 4.7]. The voltage balance equation now takes the form:

$$L\frac{di_c}{dt} + Ri_c + \frac{1}{C}\int_0^t i_c dt = E_m(t) \qquad (4.21)$$

To integrate $E_m(t)$, one should choose such circuit parameters L, R, and C, at which the voltage of current i_c on the resistor R would be much larger than on the Rogovsky belt inductance and on the capacity. Then, we will have

$$i_c R \approx E_m(t) \qquad U_2(t) = \frac{1}{C}\int_0^t i_c dt = -\frac{M}{RC}i(t) \qquad (4.22)$$

Here, $U_2(t)$ is proportional to $i(t)$, but the transmission gain is $k = -M/RC$.

4.3.4 Rogovsky measurement errors

The principal source of errors in Rogovsky current measurement is the integrator design imperfections. The integration conditions for a LR circuit [equation (4.20)] or for a RC circuit [equation (4.22)] can be satisfied only approximately. To analyze the nature of pulse distortions during the measurement, we have to examine the exact solutions relating $U_2(t)$ to $i(t)$, using equations (4.19) and (4.21). Let us analyze equation (4.21) as a more general one and find the solution for a rectangular current pulse $i(t) = I_0 = $ const at $t \geq 0$. This will define the system transient function $[y(t) = U_2(t)$ at $I_0 = 1]$ and allow the calculation of $U_2(t)$ for arbitrary current $i(t)$ by the Duhamel integral. At $i(t) = I_0$, the substitution of the integral expression for U_2 from equation (4.22) into equation (4.21) yields

$$\frac{d^2 U_2}{dt^2} + \frac{R}{L}\frac{dU_2}{dt} + \frac{1}{LC}U_2 = 0 \qquad (4.23)$$

The general solution to equation (4.23) is the expression

$$U_2(t) = A\exp(p_1 t) + B\exp(p_2 t) \qquad (4.24)$$

where p_1 and p_2 are the roots of the equation $p^2 + Rp/L + 1/LC = 0$,

$$p_{1,2} = -\frac{R}{2L} \pm \left(\frac{R^2}{4L^2} - \frac{1}{LC}\right)^{1/2} \tag{4.25}$$

To define the initial conditions, we again assume that the rectangular current pulse to be measured is differentiable on its front at the moment $t = 0$. The unlimited rate of pulse rise allows all terms in equation (4.21) to be neglected, except those containing the magnetic induction emf. Then, for $t = 0$, we write $L\,di_c/dt = E_m(0) = -M\,di/dt$. The integration yields

$$i_c(0) = -\left(\frac{M}{L}\right)I_0$$

Equation (4.22) gives $i_c(0) = C\,(dU_2/dt)_{t=0}$. From equation (4.24), we have the following equality for the integration constants A and B:

$$i_c(0) = C\,(p_1 A + p_2 B) = -\left(\frac{M}{L}\right)I_0 \tag{4.26}$$

The second equation for A and B follows from the capacitor charge and voltage being zero at $t = 0$:

$$U_2(0) = A + B = 0 \tag{4.27}$$

From equations (4.26) and (4.27), we find the expression

$$A = -B = -\frac{M I_0}{LC\,(p_1 - p_2)} \tag{4.28}$$

Together with equations (4.24) and (4.25), it defines the registered voltage $U_2(t)$.

When a Rogovsky coil is loaded by a RC integrating circuit, the LRC contour formed must be strongly damped for the integration to be made; hence, $R \gg 2\sqrt{LC}$. Then, from equation (4.25), we obtain

$$p_1 \approx -\frac{R}{L} \qquad p_2 \approx -\frac{1}{RC}$$

and $|p_1| \gg |p_2|$. We can now approximately rewrite equation (4.24) as

$$U_2(t) \approx -\frac{M I_0}{RC}\left(e^{-t/RC} - e^{-tR/L}\right) \tag{4.29}$$

and, assuming $I_0 = 1$, obtain for the transient function

$$y(t) \approx -k\left(e^{-t/RC} - e^{-tR/L}\right) \qquad k = -\frac{M}{RC} \tag{4.30}$$

The smaller time constant $T_1 = L/R$ determines the pulse front distortion. The front smearing becomes greater with increasing integral inductance of the Rogovsky coil and decreasing resistance R. The larger time constant $T_2 = RC$ is responsible for the distortions of the flat portion of the pulse. To avoid distortions of extensive pulses, T_2 must be as large as possible. Thus, increasing the resistance R in the integrating circuit proves to be a universal means of limiting distortions of the pulse front and back edge. If one ignores the losses of the output voltage inversely proportional to RC, the current can be successfully registered in a wide time range. Modern amplifiers are capable of coping with this problem.

Similarly, the solution to equation (4.19) with E_m defined by equation (4.18) and $i(t) = I_0$ at $t \geq 0$ for a Rogovsky belt with a LR integrating circuit helps to express the coil current i_c as

$$i_c(t) = -\left(\frac{M}{L}\right) I_0 e^{-tR/L}$$

The output voltage and the transient function are represented as

$$U_2(t) = Ri_c(t) = -\left(\frac{MR}{L}\right) I_0 e^{-tR/L}$$

$$y(t) = ke^{-tR/L} \tag{4.31}$$

In the latter formula we have $k = -MR/L$.

The flat portion of the pulse is distorted greater at a smaller time constant L/R. The value of L/R cannot be increased much by increasing the coil inductance. Although the resistance reduction may seem to be a universal means, it cannot give much effect either. The reason for this is not the decreased output voltage U_2, proportional to R—this difficulty could be removed by enhancing the integrated signal. The matter is that the Rogovsky coil winding itself possesses a resistance, which defines the lower limit of R.

A LR-circuit has no advantages over a RC-circuit in measurement of slowly varying currents, but it is preferable in recording short rise times. According to equation (4.31), the pulse front is supposed to be transmitted with no distortions at all, but in reality, these are caused by spurious oscillations in the circuit formed by the coil inductance and its spurious capacitance. This capacitance is not too small, so the coil has to be protected with a metallic screen against stray pick-up. (The screen should not be closed to chop off eddy currents, which decrease the mutual inductance emf.) The grounded screen is placed close to the coil winding, increasing its spurious capacitance.

The period of spurious coil oscillations must be much less than the rise time to be measured. Spurious oscillations will then rapidly damp due to the coil resistance, which grows because of the skin effect at high oscillation frequency. They will be detectable only at the very beginning of pulse registration but will actually be unable to distort the pulse front profile.

Modern Rogovsky belt designs permit satisfactory measurements of current pulses with a nanosecond rise time.

We have mentioned that mutual inductance determining the level of the voltage to be measured is not calculated but is found from the calibrated measuring circuit. Calibration is not difficult to make in laboratory conditions, but it presents a problem when one attempts to make remote measurements of lightning current by registering its electromagnetic field. This is not easy, even if the distance between the point at which a lightning discharge has struck the ground and the inductance probe is known exactly. The reason is the indefinite discharge trajectory between the cloud and the ground, as well as the *a priori* unknown, time-variable current distribution along the spark.

4.4 Current data analysis

It is not easy to explain the current measurements in a spark discharge, whose head moves fast in space and the current distribution in the channel varies with time. One must clearly understand what magnitude is being measured and how it is related to the discharge current, a parameter of primary interest to the experimenter. Easy, unambiguous answers to these questions do not come ready. We will analyze a few typical situations to illustrate the uncertainties that arise. As a rule, they are of simple physical nature, but some details may escape even an experienced researcher.

4.4.1 Current escape from the measuring circuit

To simplify the experimental design, current detectors are usually placed where the potential relative to the ground is minimal or, better, zero. For this, a low voltage electrode, often a plane, is connected to the ground through a shunt, or a Rogovsky belt is mounted on a grounding conductor. Until the discharge channel has overlapped the gap and the charges drifting in the electric field have reached the grounded electrode, the current coming to the electrode surface from the gap represents displacement current. The current to be measured is

$$i = \varepsilon_0 \int_S \frac{\partial E}{\partial t} \, \mathrm{d}S \tag{4.32}$$

and the integration of the derivative of the near-ground field E is made over the whole plane surface, to which the current detector is connected. As a rule, field strength variation is due to space charge accumulation and redistribution, as the discharge develops. Consider, as an illustration, a time-varying point charge q, located at distance h above a grounded metallic plane. The field of charge q at point r on the plane is

$$E\left(r\right) = \frac{2qh}{4\pi\varepsilon_0 \left(r^2 + h^2\right)^{3/2}} \tag{4.33}$$

(factor 2 in the numerator allows for the image charge in the plane). Substitution of equation (4.33) into equation (4.32) yields the expression

$$i = \frac{dq}{dt}\frac{h}{2\pi} \int_0^{r_1} \frac{2\pi r dr}{\left(r^2 + h^2\right)^{3/2}} = \frac{dq}{dt}\left[1 - \frac{h}{\left(h^2 + r_1^2\right)^{1/2}}\right] \tag{4.34}$$

where r_1 is the plane radius. The term $\delta = h/\left(h^2 + r_1^2\right)^{1/2}$ in equation (4.34) describes a relative error resulting from the limited size of the measurement plane. Not all field lines of force are closed on the plane— some are closed on the grounded floor and on the grounded equipment of the high voltage laboratory. To keep the error δ on a tolerable level, the measurement plane radius must be sufficiently large:

$$r_1 = \frac{h}{\delta}\left(1 - \delta\right) \approx \frac{h}{\delta} \tag{4.35}$$

In other words, a reliable measurement (within a 10% error) of discharge current in the external circuit requires that the plane radius be an order of magnitude larger than the gap length. It is not easy to satisfy this constraint, and one often has to use a plane with $r_1 \approx h$. From equation (4.34), the current registration error in this case exceeds 70%, at least, at the initial discharge stages, while the channel head is still far from the plane. Such measurements should be regarded as purely phenomenological.

In a laboratory building of limited area, the shunt (or another probe) has to be carried over to the high voltage electrode to provide more accurate registration. The total discharge current will now flow through it, entirely eliminating this kind of error. On the other hand, it is much more difficult to measure current in a circuit element under high potential. Today, we can discuss two nearly identical solutions of this problem. One approach is to convert the probe signal to a light pulse, analog or digital, and to transmit it through an optic fiber line to the ground, where it is converted back to an electrical signal to be registered. Clearly, the lightguide insulation must endure maximum voltage applied in the experiment. The other approach consists in analog-to-digital conversion of the probe signal and its storage in the on-line converter memory (also under high potential). After the discharge is completed and the voltage drops to zero, the digital codes are transmitted to registering devices in any convenient way.

We would like to remind the reader of the usual practice of measuring currents before the modern electronic devices were introduced. We mean the 'inverted' electrode arrangement. A large-radius plane electrode was suspended at the necessary height and connected to a high voltage supply of the opposite sign, while a small-radius electrode triggering a discharge was mounted on a high thin rod. The shunt connected to the electrode picking up nearly all current in the early discharge stages turned out to be grounded, so that the signal could be transmitted to a meter through a common cable.

However, the conditions for discharge development in the conventional and 'inverted' designs are not identical. The grounded rod cannot be lifted too high, because its length is limited by the building height. So the electric field distribution over the discharge gap may be markedly affected by the grounded floor (some of the lines of force come out of the floor but not off the rod). Other conditions being equal, the 'inverted' field always proves to be more uniform than the field in a common electrode arrangement. This may affect the measurements.

4.4.2   Conduction current discrimination

Measurement of total external current $i(t)$ is insufficient for finding spark conductivity, released energy or gap space charge. Evaluation of these parameters requires the knowledge of the spark conduction current, which may considerably differ from total external current.

Since the current detector is usually connected to the external circuit, the measurements can, at best, show the conduction current i_{0c} at the contact of the spark with the electrode, from which it started. The current in the other channel cross sections may be quite different. We encountered a similar situation, when analyzing channel parameters behind the ionization wave front in Section 3.2.2.

A detector for conduction current measurement must be placed at the electrode the spark contacts. Generally, the total current through the detector, $i(t)$, may differ from the conduction current $i_{0c}(t)$ to be found:

$$i(t) = i_{0c}(t) + \frac{dq_i}{dt} + C_e \frac{dU_0}{dt} \qquad (4.36)$$

The last two terms in equation (4.36) can be formally regarded as the 'capacitive' component of total current defined by the variation rate of electrode charge q_i induced and accumulated by the electrode capacitance C_e at the gap voltage U_0. The experimental design for conduction current measurement must be such that the 'capacitive' current could avoid the detector. This can be achieved by using *sectioned electrodes* [4.2, 4.3].

Let us find, on the electrode surface, a measuring section of such a small area S_1 that the charge on it can be taken to be negligible relative to the

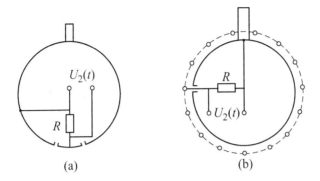

FIGURE 4.8
A sectioned electrode for conduction current measurements (a); a detector
modification including the electrode with a measuring grid (b).

total electrode charge. By connecting this section to the rest of the electrode with a low resistance shunt [Figure 4.8(a)], we make the section insulation good enough to endure the maximum shunt voltage. If a discharge
starts from the measuring section, which does occur with a certain probability, all conduction current will flow through the shunt, whereas only
a small portion of capacitive current transporting the charge to the electrode, S_1/S_e (S_e is the electrode area), will reach the shunt.

This way of eliminating capacitive current has proved fairly effective; it is
sufficient to have $S_1/S_e \leq 0.05-0.1$ for most measurements. Reduction in
the section area increases the measurement precision but at the expense of
a greater nervous tension of the experimenter, who has to wait wearisomely
for a discharge to start from a prescribed place. The hope that a discharge
will start from the electrode tip with maximum field rarely comes true
for pulse voltage because of sporadic initial electrons, whose multiplication
may excite a discharge.

When time is precious and the low probability of discharges is unacceptable, an electrode with a measuring grid may come in handy [4.3]. The
method of eliminating capacitive current is similar to that just described,
but a small area section is replaced by a small area grid made from thin
wire [Figure 4.8(b)]. The grid is separated from the electrode by a thin,
high insulation strength layer, for example, an epoxy compound or sput-
tered polyethylene. A shunt is connected to the grid in exactly the same
way as to the measuring section.

If the grid is made perfect and the insulation layer is a few fractions of
a millimeter, the field distortions will not practically affect even the spark
ignition conditions, let alone its development in the gap. It does not matter

now whether a discharge will be excited at any other point on the electrode.

4.4.3   Base and remote currents

Current measurement in the channel base and remote cross sections involves certain difficulties (Chapter 3). A shunt cannot, in principle, be introduced into a discharge channel. The position of an induction probe is also problematic, since high voltages require large insulation distances, but then the magnetic field at the probe will vary with the current distribution along the channel. The measurement cannot be tied up to any particular cross section either. Some promise is offered by induction probes with wireless or optic fiber lines for information transmission to measuring devices. A probe can be placed closer to the channel, but the distance between them must be measured very accurately, because it is related to mutual induction. A long spark path is subject to sporadic changes, so this distance may show a considerable discharge-to-discharge variation. It is quite natural that the researcher should try to extract as much information as he can from conventional methods, but he should be very cautious in treating his data.

Let us turn to an example, which, in spite of its sketchiness, reflects a realistic situation and shows that the information obtained may prove to be illusionary. Suppose a shunt mounted on a sectioned electrode is used to register the conduction current at the base of a spark being formed. Suppose further that, having reached length $l = 10$ m, the spark has temporarily stopped growing, and a pause of a few hundred microseconds makes the current flow cease entirely. The pause is then followed by the spark revival, so that the propagation rate 'immediately' reaches a value v. The conduction current in the ionization wave front, proportional to v (Section 3.2.2), also 'immediately' rises to $i_t = Av$, where A is a constant. A flash-like form of spark is sometimes observed in air gaps. Evaluate the shunt current response to this event, for it is the only kind of current that is actually accessible to measurement. When the self-induction emf is much less than the voltage drop due to the channel resistance, equation (3.36), similar to the diffusion or heat conduction equation, holds true:

$$\frac{\partial^2 i}{\partial x^2} - R_1 C_1 \frac{\partial i}{\partial t} = 0 \qquad (4.37)$$

where R_1 and C_1 are linear resistance and capacity. If the x-coordinate is counted off from the head, at the moment the current begins to grow after the pause ($t = 0$), we have

$$i\,(0, t) = Av = \text{const} \qquad i\,(x, 0) = 0 \qquad U\,(x, 0) = U_0$$

because the potential equalized along the channel during the pause.

Equation (4.37) can be solved with a Laplace transform, substituting $\partial i/\partial t$ by pi, where p is a subsidiary complex variable. Then, we obtain

$$\frac{\mathrm{d}^2 i}{\mathrm{d}x^2} - R_1 C_1 pi = 0$$

$$i\,(x,p) = B_1 \exp\left(-x\sqrt{R_1 C_1 p}\right) - B_2 \exp\left(x\sqrt{R_1 C_1 p}\right) \tag{4.38}$$

Consider the case of an infinite channel corresponding to short times, during which the perturbation due to the current restart at the head has just reached the base. Because of the limited current at $x \to \infty$, we have $B_2 = 0$ and

$$i\,(x,p) = B_1 \exp\left(-x\sqrt{R_1 C_1 p}\right) \tag{4.39}$$

Inverse Laplace transformation yields

$$i\,(x,t) \approx Av\left[1 - \Phi\left(\frac{x}{2}\sqrt{\frac{R_1 C_1}{t}}\right)\right] \tag{4.40}$$

where

$$\Phi\,(z) = \mathrm{erf}\,(z) = \frac{2}{\sqrt{\pi}} \int_0^z \exp\left(-y^2\right) \mathrm{d}y$$

is the tabulated Gaussian error integral. At $z \gg 1$, with an error lower than $1/\left(2z^2\right)$, we have $1 - \Phi\,(z) \approx \exp\left(-z^2\right)/\left(z\sqrt{\pi}\right)$, or at $t \ll x^2 R_1 C_1$ we obtain

$$i\,(x,t) \approx Av\frac{2\exp\left(-x^2 R_1 C_1/4t\right)\sqrt{t}}{\sqrt{\pi x^2 R_1 C_1}} \tag{4.41}$$

Let us take $R_1 = 5 \times 10^5\ \Omega\,\mathrm{m}^{-1}$ and $C_1 = 10\ \mathrm{pF\,m}^{-1}$, which are realistic values for a long air spark, and estimate the base current at $x = l = 10$ m. Over the time interval of 10 μs after the discharge revival, the shunt current will still be 10^{-6} of the head current, or practically undetectable. Only in 50 μs will the current rise to about $0.03i_t$ and become measurable. Such a smooth pulse rise in the shunt will hardly help the experimenter to reconstruct the real picture of stepwise discharge acceleration.

4.5   Gap charge measurement

Space charge measurements are necessary for determining the electrical field in a gap, as discharge processes develop. How important and complicated this task is can be seen even from the analytical evaluations presented in Chapter 3. All of them were eventually aimed at finding the lon-

gitudinal field at the ionization wave front and behind it or the radial field
at the channel lateral surface. Total gap charge can be found by integrating the conduction current measured on the base electrode:

$$q_0 = \int_0^t i_{0c}(t)\,dt \qquad (4.42)$$

Measurements make sense only before the ionization wave has overlapped
the gap, when all charge current flows through this electrode. Conduction
current should be registered by sectioned electrodes (Section 4.4.2).

Space charge evaluation from total electrode current often leads to very
large errors. To estimate the error, we assume that charges $+q$ and $-q$
($-q$ is electron charge) are due to ionization at distance x from an anode
of radius r_a. Electron drift with velocity v_e towards the anode induces
external current, which is defined by Shockley's theorem (Section 2.7) as

$$i(t) = qv_e(x)\frac{r_a}{r^2}$$

The current registration over the time of electron drift to the anode, followed by integration, gives the charge

$$q_{reg} = qr_a \int_0^t \frac{dr}{r^2} = q\left(1 - \frac{r_a}{r_a + x}\right) \qquad (4.43)$$

although charge $+q > q_{reg}$ does remain in the gap, after all electrons have
arrived at the anode. The relative $(q - q_{reg})/q = r_a/(r_a + x)$ grows with
smaller distance between the site of charge creation and the electrode. At
$x = r_a$, the measured charge is one half the actual gap charge.

Naturally, equation (4.42) contains only quantitative information about
the difference between positive and negative charges at a given moment
of time, but it does not reveal charge distribution over the gap. One can
imagine a situation, when the gap charges have separated to form two
oppositely charged regions, thereby distorting the initial field distribution
(Figure 4.9). The researcher, however, is unable to recognize the existence
of these regions from the conduction current registration, because none of
the charges have escaped the gap. Indirect information on this situation
can be obtained from field measurements at the electrode surfaces.

The formation of any charged regions results in the gap field redistribution, and this can be verified in experiment. Theoretically, it is possible to
find the charge distribution pattern in the gap by covering the electrodes
with an infinitely large number of field detectors (see Section 4.6) and by
solving the inverse electrostatic problem with full information about the
field variation on the gap boundary surfaces. Something like this has been
done experimentally but with a finite number of detectors [4.4]. Only in this
case may one hope to obtain a qualitatively plausible charge distribution

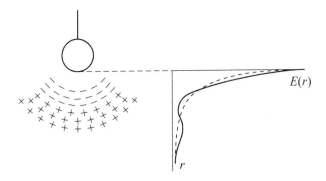

FIGURE 4.9
The gap field distorted by a bipolar charge layer.

pattern. Suppose that the readings are taken simultaneously from n field detectors, each of which registers $E_k(t)$. Since the detectors are located on the electrodes, the vectors $\boldsymbol{E}_k$ normal to the metallic surface are known exactly. Let us divide all space charge into n regions in such a way that the charge in each region could be characterized by a yet unknown numerical parameter. This may be the mean space density ρ_m, if we assume the charge be uniformly distributed over the m-th region, or the mean surface density σ_m, if we assume the charge to be localized at the external boundary of this region. Finally, we can perceive the charge distribution along the m-th segment of the axis with linear density τ_m or build a simple model of point charge q_m, concentrated at the 'center of mass' in each region. The field E_k at the k-th point of measurement is linearly related to unknown charge parameters. For example, in the point charge model, the field is

$$E_k = \sum_{m=1}^{n} \frac{q_m \cos \psi_{mk}}{4\pi\varepsilon_0 r_{mk}^2} \equiv \sum_{m=1}^{n} \beta_{mk} q_m \qquad (4.44)$$

where r_{mk} is the distance between the k-th detector and the charge localization point in the m-th region, while ψ_{mk} is the angle between the line connecting them and the normal to the electrode surface at the k-th point. The direct electrostatic problem is to calculate E_k from the known q_m, whereas the inverse problem is to find q_m from the known E_k by solving a set of n algebraic equations similar to equation (4.44).

The researcher is free to choose an appropriate charge distribution model and the way of subdividing the gap into calculation regions. His choice will define the expression for the coefficient β_{mk} with the charge or with its density. In making this choice, he is, however, often guided by circumstantial considerations, not always clear or justifiable. The chances for his inspired guesses grow, as the number of field measurement points is reduced. Nevertheless, the researcher is sometimes lucky to obtain reasonable results, especially for weakly branching sparks, whose paths are exactly known from optical registrations. The radius of the charge cover enveloping the channel is usually much smaller than the channel length; so, modeling the channel as a charged filament seems quite justifiable.

A simple and convenient way of analyzing the situation in question is to examine a straight channel approaching a plane electrode with field detectors mounted on it at the right angle to its surface. If τ_m is taken to be a yet unknown linear charge density on the m-th channel section, the geometrical coefficients are equal to

$$\beta_{mk} = \frac{2}{4\pi\varepsilon_0} \int_{x_{2m}}^{x_{1m}} \frac{x\,\mathrm{d}x}{\left(x^2 + r_k^2\right)^{3/2}}$$

$$= \frac{1}{2\pi\varepsilon_0} \left[\left(x_{1m}^2 + r_k^2\right)^{-1/2} - \left(x_{2m}^2 + r_k^2\right)^{-1/2} \right] \qquad (4.45)$$

The coordinates of the m-th section boundaries, x_{1m} and x_{2m}, are counted off from the plane (factor 2 in equation (4.45) accounts for the effect of image charges in the plane).

Equation (4.45) allows the formulation of precision requirements on field measurement for a reliable evaluation of channel charges. Clearly, the field right under the channel, at $r_k = r_1 = 0$, will respond most dramatically to charge variation. For this point, we have

$$\beta_{1m} = \frac{1}{2\pi\varepsilon_0} \left(\frac{1}{x_{1m}} - \frac{1}{x_{2m}} \right) \qquad (4.46)$$

Consider a field created by two neighboring channel sections of equal length Δx and charge densities τ_1 and τ_2. Let the section boundary closest to the plane be located at height $h \gg \Delta x$. Then, equation (4.46) gives the electrode field under the channel created by $\tau_1 \Delta x$ and $\tau_2 \Delta x$ charges:

$$E = \frac{\tau_1}{2\pi\varepsilon_0} \left(\frac{1}{h} - \frac{1}{h + \Delta x} \right) + \frac{\tau_2}{2\pi\varepsilon_0} \left(\frac{1}{h + \Delta x} - \frac{1}{h + 2\Delta x} \right) \qquad (4.47)$$

By averaging the charge distribution over the sections and using the average linear density $\tau = (\tau_1 + \tau_2)/2$, we will find the field created by charge $\tau 2\Delta x$ at the same point:

$$E' = \frac{\tau}{2\pi\varepsilon_0} \left(\frac{1}{h} - \frac{1}{h + 2\Delta x} \right) \qquad (4.48)$$

In order to reveal the charge nonuniformity from the measurements, their precision must be high enough to register reliably the difference $\Delta E = E - E' \approx \Delta x^2 \left(\tau_1 - \tau_2\right) / \left(2\pi\varepsilon_0 h^3\right)$. For this, the relative error in the field measurement must be *a priori* less than

$$\frac{\Delta E}{E'} \approx \frac{\Delta x \, (\tau_1 - \tau_2)}{h \, (\tau_1 + \tau_2)} \tag{4.49}$$

For example, if the average linear charge densities in adjacent sections differ by a factor of two, the precision requirements will be quite strict even in this case: $\Delta E/E' < \Delta x/3h$. At $\Delta x = 0.1h$, the measurement error must be about 1%, which is not easy to achieve in experiment.

There are few experimental studies where the charge distribution model used was brought in agreement with the attainable measurement accuracy. When analyzing experimental findings, one should keep in mind that they can give more or less reliable information only for those charged regions that are comparable in their characteristic dimensions with the distance to the measurement point. For this reason, detectors should be placed on both electrodes to create more or less identical conditions for the measurements in the spark head and base. Naturally, the calculation of β_{mk} coefficients should take into account not only the effect of space charges themselves but also of induced charges on the electrodes. Charge distribution evaluations acquire a greater reliability, if field registrations are accompanied by total gap charge measurements from conduction current. Then the sum total of all charges becomes known, and this knowledge can be used to check the calculations.

4.6 Electric field measurement

There are two problems differing in complexity: the measurement of field strength on the electrodes and in the gap, including the spark. The first problem is relatively simple, but the other one is still a challenge to researchers.

4.6.1 Field measurement on the electrodes

Field measurement on the electrode is almost always reduced to surface charge registration. The charge density σ is related to the electrical field strength at a given point on the surface as $\varepsilon_0 E = \sigma$ (we mean the normal field component, because the tangential component of a perfect conductor is zero). If the electrode is subdivided into small sections, such that the surface density within each section can be taken to be equal to an average value σ, then the average field at the section surface is

$$E = \frac{\sigma}{\varepsilon_0} = \frac{q_1}{\varepsilon_0 S_1} \tag{4.50}$$

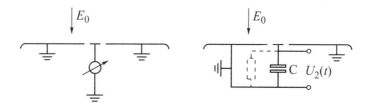

FIGURE 4.10
Registration of the electrical field near the electrode from its surface charge.

where q_1 is the surface charge of a section of area S_1. When no charge comes to a section from the gap and there was no charge or field on it prior to the registration, q_1 is defined as

$$q_1 = \int_0^t i_{\text{reg}}(t)\,dt \qquad (4.51)$$

where i_{reg} is the current through the section circuit (Figure 4.10). With $q_1(t)$ calculated from equation (4.51), the field $E_1(t)$ can be easily determined. Since we want to know the current integral rather than current itself, the detector voltage can be applied to an integrating element. But it is more reasonable to measure the charge through the section directly, without introducing a current detector into the circuit. For this, it suffices to connect the section to the external circuit by an integrating capacitor C (Figure 4.10). The capacitor voltage is proportional to the desired field:

$$U_1(t) = \frac{q_1}{C} = \frac{\varepsilon_0 S_1 E(t)}{C} \qquad (4.52)$$

The fact that the section potential will differ from the remaining electrode potential by the value U_1 will not affect the discharge development. The capacitor voltage of about 1 V is sufficient for reliable measurements, whereas the gap voltage in long spark experiments is by a factor of 10^5-10^6 larger.

Generally, the capacitor C introduced into the measurement circuit has a maximum possible capacitance in order to reduce the error from charge leakage through the resistance R_2 shunting the capacitor (Figure 4.10). The leakage resistance R_2 is the sum of spurious insulation resistances and capacitance of the section, but the main contribution is made by the input resistance of registering devices. The error reveals itself very much like that discussed in Section 4.2 for the capacitive voltage divider with finite lower arm resistance R_2. The transient function of the measurement circuit, $y(t) = \exp(-t/T)$, where $T = R_2 C$, indicates a distortion of the flat portion of the pulse, which may lead to a considerable error in the back pulse profile when the field abruptly drops, for example, in gap breakdown.

Let us assess the actual capabilities of a registration circuit in a typical experiment. Suppose we have chosen a measurement section of area $S_1 = 10$ cm^2 on the electrode. In field $E_i = 30$ kV cm^{-1} (a threshold value for ionization in atmospheric air), its charge $q_1 = \varepsilon_0 S_1 E_i \approx 2.6 \times 10^{-8}$ C is sufficient to raise voltage U_1 to 1 V at capacitance $C = q_1/U_1 \approx 26$ nF. The application of modern plastics for insulation of the section and high resistance measuring devices gives leakage $R_2 \approx 10$ MΩ and time constant $T \approx 26$ ms. The latter is an appreciable value for laboratory sparks, whose lifetimes are measured in microseconds. So the experimenter is rarely faced with difficulties, except, perhaps, in some exotic situations. One example is a discharge in a very hot gas, when leakage currents through ceramic insulation materials rise abruptly at temperatures above 1000 K.

4.6.2 Measurement of varying fields at electrodes

Measurement of slowly varying currents at the electrode surfaces may be necessary in studies of very long sparks, especially in electric field registrations at the ground during lightning discharges. The time of the lightning leader formation often exceeds 10 ms, while the near-ground field rises very slowly and is only $100-200$ V cm^{-1} most of the time. For this reason, one has to take care of additional ways of measuring the field constant component.

Very common are devices in which a constant electric field at the surface of the measuring section is modulated by moving mechanically a grounded electrostatic screen. In a simple design of fluxmeters, as these devices are termed, constant field E is modulated by rotating a grounded metallic slotted disc, acting as a screen, above the plane measuring section [Figure 4.11(a)]. The lines of force E are incident on the screen surface only

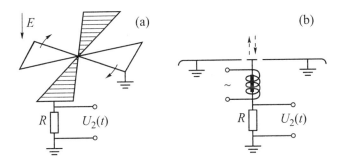

FIGURE 4.11
The basic circuits of a rotor fluxmeter (a) and a vibrational fluxmeter (b).

through the slots. For this reason, the section surface charge varies in time during the screen rotation. Surface charge variation induces current through the section. The current value is proportional to the rotation rate of the screening disc, to the section surface area and field E. The field should not change markedly over one screen revolution. Having stabilized the rotation rate, one can find the field E from the section current measurements.

It is difficult to calculate the actual fluxmeter sensitivity. The fluxmeter is usually calibrated in a gap with a known uniform electric field prior to the measurements. Modern fluxmeter modifications have no rotating elements, which increases their performance reliability. The field is modulated by a vibrating grounded string placed above the measuring section in the vertical plane (vibration frequency is about 1 kHz).

Sometimes, the measuring section itself is vibrated by being pulled away periodically with a magnet, after which it returns to its initial position in a hole of the grounded metallic base of the device [Figure 4.11(b)]. Such procedures alter the flux of vector E through the measuring section. This induces current, which changes the section surface charge proportional to this flux. The measurement of current then presents no problem.

Let us examine the measurable range of electric fields. The possibility of recording weak fields is principally unlimited—everything depends on the experimenter's skill and equipment precision. As for the upper limit, all attempts to measure fields above a threshold value of E_i have failed. At this threshold, a self-sustained discharge starts from the measuring section surface. In this case, the discharge current will flow through the capacitor C, and equation (4.52) relating U_1 to E will become meaningless.

A similar situation arises with the fluxmeter, where the capacitive current proportional to E is added to the discharge current, disturbing the device calibration. The threshold field in atmospheric air is $E_i \approx 30$ kV cm^{-1}; the actually obtained values are lower, because there is always a probability of local field enhancement at the measuring section edges, where they contact the insulation. A very weak discharge current may be sufficient to produce a large measurement error. For instance, at $E = 30$ kV cm^{-1} and $S_1 = 10$ cm^2, when the section surface charge is $q_1 \approx 2.6 \times 10^{-8}$ C, the transport of the same charge will require current $i \approx 0.3$ mA over the registration time $t \approx 100$ μs. This current produced a 100% error in the example above; it is by three or four orders of magnitude smaller than the streamer current at the anode of several centimeters in radius. In other words, even very weak stray discharges from a measuring section are to be suppressed. The situation can be somewhat improved by using sectioned electrodes with a grid (Section 4.4.2). Their surface is protected by a dielectric that suppresses discharges starting from the measuring section edges.

4.6.3   Field measurement in the gap bulk

In order to measure the gap field, one can take advantage of the procedure described and measure the surface charge on an auxiliary electrode inserted as a probe at a desired point in the gap. Polarization charge appears on the metallic body, which can be analytically related to the external field for probes of simple geometry. The respective formulas were written out in Section 3.5.1 for a metallic sphere of radius R in a uniform field:

$$q = 3\pi\varepsilon_0 R^2 E \qquad (4.53)$$

In order to measure the polarization charge, it is sufficient to separate the hemispheres of a probe electrode and to introduce an integrating capacitor C_1 between them. Its voltage $U_1 = q/C_1$ is proportional to the external field E being measured (Figure 4.12). After the connection of capacitor C_1, the hemispheres will turn out to be under different potentials, while the voltage between them, $U_1 \approx 1$ V (a value necessary for a reliable measurement), will be small relative to the RE value determining the voltage 'pushed out' of the sphere space in real conditions.

The technical problem is only to transmit the information about U_1 to a measuring device. We addressed a similar problem in Section 4.3 in connection with the measurement of current through a high voltage electrode. The same engineering solutions are applicable in this case too, but the use of a lightguide is to be preferred, because an electric pulse can be converted to a light signal by a simple electronic device. To minimize the

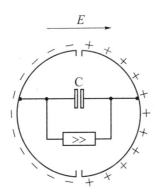

FIGURE 4.12
A sectioned detector for electrical field measurement in the gap.

gap field perturbation, one has to use as small a probe electrode as possible.

Field perturbation during the registration sets the upper limit for the measurable magnitudes. Polarization charge enhances the field at the probe surface, so it may become several times greater than the external field. For example, the maximum strength for a sphere is $E_{\max} = 3E$ and should not exceed $E_i \approx 30$ kV cm^{-1} in normal density air. Therefore, a spherical probe can measure fields up to $E_i/3 = 10$ kV cm^{-1}. The polarization effect is weaker for a cylinder, and fields as strong as $E_i/2 = 15$ kV cm^{-1} are accessible to measurement under the same conditions.

Replacement of metallic probes by dielectric detectors is a way to raise the level of measurable field strengths. Dielectric detectors are based on the Pockels or Kerr electrooptic effects of birefringence variation due to electric field. Their efficiency, however, is not very high because of the large permittivity ($\varepsilon > 10$) of crystals with acceptable electrooptic characteristics. The near-surface field of a crystal with $\varepsilon \gg 1$ is nearly the same as that of a metallic probe of identical geometry.

Besides, the polarization effect on the crystal edges is not easy to avoid. A dielectric detector is also preferable, because it can be made very small to produce minimum perturbation in the gap. In fact, this sort of detector contains nothing more than a crystal plate with two polished surfaces and lightguides attached to them. A polarized laser beam is supplied to the plate through one lightguide and removed through the other [4.5, 4.6]. The beam polarization altered by the crystal field is registered, when the beam passes through a polarization filter at the other end of the lightguide, outside the gap. Crystal detectors have a very fast response—they can operate in the picosecond range of field variation.

The detector to be used for measurements is precalibrated in a known uniform field. It is important to make the calibration accurately and to set exactly the upper measurable strength limit. Excitation of an electric discharge also distorts the measurements by a crystal detector but they are less dramatic. The crystal conductivity is extremely small, so charges created by ionization are deposited on the crystal surface, until the field in the ionization region has decreased to the threshold level E_i (only in this region but not on the whole crystal surface!). The field becomes stabilized at this level. If the precalibration was made carelessly, this purely 'instrumental' stabilization of the registered signal can be easily taken for something inherent in the discharge process under study.

The conclusion from this consideration is not particularly comforting. Electric field is inaccessible to measurement, where it is especially valuable and intriguing, primarily in the ionization region. We are pinning some hope on spectroscopic methods, in particular, on the Stark effect. However, this promise should not be exaggerated, because measurements under the conditions of a rapidly propagating ionization wave are very difficult to make because of its weak luminosity and unpredictable trajectory.

4.7 Optical registration of discharge

Lightning and laboratory sparks are so beautiful that it is not easy to resist the temptation of taking pictures of them. Most studies of discharges were started by photographing them, so it is not surprising that it was optical registration that provided science with most of the available information on long sparks. This information was largely phenomenological—the phenomenon was described but not explained. Yet, even today one should not ignore still photography as a method of studying spark discharges, because, with much skill, it can provide valuable material for theoretical speculations.

4.7.1 Still photography

Still photographs are, at least, necessary for measuring lengths and diameters of spark channels, which is not as easy as it might seem. Consider the spark length. Immediately the question arises as to what sort of length is meant. Suppose a spark has overlapped a gap of length d. This certainly does not mean that the real spark length l is also equal to d. A spark channel has various bends, sometimes quite large, and a photograph made from one point, that is, along one channel projection, cannot give the real length. Two projections are needed, not necessarily at 90° to each other, although the latter would be easier for processing. A special computer program can restore the real spark trajectory from these projections. With one projection, the measurement error may be too large. The error usually grows with gap size, because a long spark has more bends than a short one. Spark length measurements are rarely the goal of a discharge experiment. These data are used for the evaluation of other parameters, in particular, the average discharge rate from the known average time of spark propagation or the average field from the measured gap voltage. An error in length measurement will naturally lead to as large, or a larger relative error in the final result.

A word of caution is necessary as to still photography of very long sparks or lightning discharges. One has to photograph at a large distance, often with conventional rather than long focal optics. In a photograph, one can identify bends comparable in scale with the channel radius. Such 'microbends' do not much affect the channel electric or magnetic fields; they only slightly increase its capacitance and, hence, the energy input into the channel (Section 3.3). But in evaluating the temperature (and, therefore, the volume of the heated gas), one has to take into account even very small bends. Clearly, one should use photographs very cautiously, when gathering information on the energy balance. This is also true of plasma conductivity evaluations.

Spark radius measurements present still greater difficulties. First, it is necessary to provide a good spatial resolution. If one uses high quality optics capable to provide the film resolution of $N = 1000$ lines per centimeter, then in the measurement of a streamer radius $r_c \approx 0.1$ cm with a 10% error ($\delta \approx 0.1$), the spark image should not be reduced by more than $k_m = r_c \delta N$ relative to the real streamer size (in our example $k_m = 10$). When the camera is placed at distance D from the discharge, the objective focal length must be $f = D/k_m$. This presents a serious challenge to the researcher, because photographs in high voltage studies are to be made at a distance of several tens of meters, or hundreds of meters in the case of lightning. Second, it is very important to choose the right exposure and adequately develop the film, otherwise a halo, especially large in excessively long exposures, will reduce to zero the advantages of perfect optics. Problems may also arise in interpreting a still photograph, as the image sometimes allows several interpretations. Imagine a situation when an ionization wave of radius r_c' travels through a gap, leaving behind a thin hot channel of a much smaller radius r_c. Light radiation from the wave front illuminates $2r_c'$ film widths, so that the image obtained gives a false idea of the channel radius. This example clearly illustrates the reality. The leader head in air is 20–30 times wider than the initial channel width and is so bright that the leader itself is indiscernible against the bright background. Narrow-band light filters may sometimes be helpful. For example, thermal radiation from a hot channel is more visible in the red and infrared regions, whereas the source of light in a cold head is the excited atoms and molecules with maximum emission in the near-ultraviolet region.

It is worth considering the voltage cut-off method, which was very popular 40–50 years ago and was described in detail in the literature. A discharge was photographed, when the gap voltage was cut off nearly to zero by a controllable arrester, so that the spark was interrupted at fixed moments of time [Figure 4.13(a)]. A series of consecutive shots taken at various cut-off times provided something like a high speed movie film, which gave an idea of the discharge propagation rate and structural dynamics. Some investigators managed to take such excellent photographs that even today they can be regarded as standards of perfection [4.7]. Photographic skills could not, however, prevent distortions caused by *reverse discharge*.

A reverse discharge arises after the voltage cut-off in the space charge

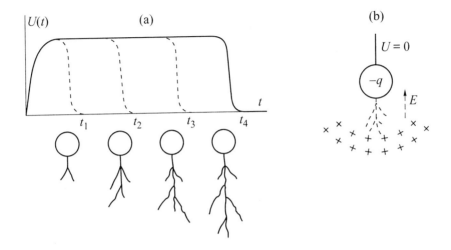

FIGURE 4.13
Registration of the optical discharge pattern by the gap voltage cut-off (a); a reverse discharge distorting the image (b).

field [Figure 4.13(b)]. The sphere of radius r_t in the diagram is the leader head, or tip, and the filament behind the head is the leader itself, assumed to be perfectly conductive. The positive charge in front of the head has been transported to the gap by numerous streamers that started from the head during the leader development. The electric field at the head with potential U is defined by the total head charge:

$$E\left(r\right) = \frac{Q}{4\pi\varepsilon_0 r_t^2} \qquad Q = 4\pi\varepsilon_0 r_t U + q_i$$

where the first summand in Q is the head charge positive at $U > 0$; the induced charge q_i has a sign opposite to that of the gap space charge and is negative in our case. After the voltage cut-off ($U = 0$), the field at the head reverses its sign and may trigger a discharge of the opposite sign. Naturally, the images of the forward (positive) and reverse (negative) discharges will superimpose in a still photograph, as if a careless photographer had used the same film twice. Very short exposures could spare the researcher all these difficulties, but it is hard to implement this idea. The point is that the time resolution in spark experiments lies within the micro- and nanosecond ranges. Very short exposures are feasible only with an electronic shutter.

4.7.2   Streak photography

The principle of this method is easy to understand, if one imagines a series of consecutive photographs of a glowing object made at minimum exposure and arranged chronologically one after another without spacings. A moving glowing point is represented on the film as an inclined line and an elongating spark as an illuminated triangle (Figure 4.14). Streak pictures are obtained by continuously shifting the discharge image along the film. A simple way is to continuously pull the film through in the focal plane before the objective projecting this image, but the speed will be too low. It is more effective to rotate a flat disk covered by a sheet film or to rotate the objective relative to a motionless film disk. It is still more effective to rotate both the film and the objective in opposite directions to attain maximum relative velocity. The inclination α of an illuminated line or triangle will indicate the velocity v of a point light source or an elongating glowing spark. If a real spark of length l has a film size $x = l/k_m$, where k_m is the image reduction factor, then $\mathrm{d}y/\mathrm{d}x = \tan\alpha = v_\mathrm{d} k_m/v$, where v_d is the display rate of the image and $y = v_\mathrm{d} t$ (Figure 4.14). A reliable measurement of the inclination α is made, if it is close to $45°$ ($\tan\alpha \sim 1$). This means that the display rate must be approximately $v_\mathrm{d} \sim v/k_m$. For instance, if a streamer has length $l = 30$ cm, propagation rate $v = 10^9$ cm s^{-1}, and the film image length $x = 3$ cm, the display rate must be about 10^8 cm s^{-1}.

Since mechanical devices are unable to attain such a high display rate, experimenters normally use electron-optical image converters. An image is transferred from the converter photocathode onto its fluorescent screen by a three-dimensional electron flux: electrons flying out of the photocathode

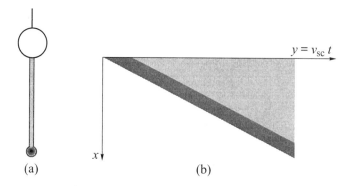

(a)   (b)

FIGURE 4.14
The principle scheme of continuous streak photography. For comparison, still photography (a) and streak photography (b).

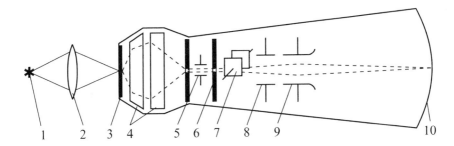

FIGURE 4.15
An electron-optical image converter: 1, object to be filmed; 2, objective;
3, photocathode; 4, beam acceleration and focusing; 5, electronic shutter plates;
6, shutter diaphragm; 7, 8, horizontal and vertical deflecting plates;
9, compensating plates; 10, screen.

are focused and accelerated by the electric field created by a set of electrodes
in the converter tube (Figure 4.15). Deflecting plates shift the image along
the screen at fixed velocity. This is done by applying to them a linearly
rising (or decreasing) voltage pulse [Figure 4.16(a)], as in an oscilloscope.
Another pair of plate shutters is capable of deflecting the electron flux so
much that it does not enter the diaphragm slit, entirely missing the screen.
The image on the screen disappears (is shut). This allows regulation of the
exposure by means of a shutter rectangular pulse. A similar voltage pulse
but of the opposite sign is applied to correcting plates, which regulate the
field in the tube in such a way that the image remains immobile on the
screen, while the electron flux is shifted within the diaphragm by a shutter
pulse. This prevents the image 'smearing' during the shutter performance.

The converter screen diameter seldom exceeds 10 cm, but the image
display rate can be made as high as $10^{10}-10^{12}$ cm s^{-1} (this is phase velocity,
it can exceed the speed of light!). The shortest exposure may be a few
fractions of a nanosecond, permitting 'instantaneous' photography of record
fast discharge events. No continuous image display is made in this case, but
a stepwise voltage pulse is sometimes applied to the deflecting plates instead
of a linearly rising pulse. Each step 'instantaneously' shifts the image to a
new position on the screen. The electronic shutter remains closed at the
moment of shift. Pulses opening the image are strictly synchronized with
the plane sections of display pulses [Figure 4.16(b)]. As a result, several still
shots are projected onto the screen, producing a superhigh speed filming
effect.

The electron flux from the photocathode to the screen can be enhanced in
much the same way as in a photomultiplier. A series of amplification stages
provide a $10^{3}-10^{4}$-fold increase in the image brightness—an important fac-

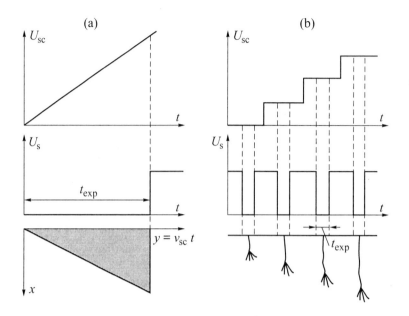

FIGURE 4.16
Operation of an electron-optical image converter in a continuous (a) and
shot-by-shot (b) regime.

tor in the registration of fast and weak discharge structures. The opposite
is of equal importance, namely, the protection of the converter screen from
excessive excitation at the final or extremely bright spark stages capable
of complete image illumination. An automatic regulation of brightness can
eliminate the risk of image illumination, but it is easier to do this by making the exposure shorter.

Shot-by-shot registration with a stepwise image display is used quite
rarely: there is not enough space for large shots on a small converter screen.
One also fails to obtain reliable information by repeated filming of typical
discharge events, because, in actual fact, spark trajectories are not recur-
rent. Normally, a continuous image display in streak pictures is preferred,
although they are not easy to decode, especially in the case of branched
sparks. Let us examine a typical error in spark velocity estimations from
streak pictures, no matter whether they were taken electronically or mechanically.

The decoding principle was illustrated above for a simple situation, when
a discharge image is strictly normal to the time axis (display direction).
One can see in Figure 4.14 that the average channel velocity over a time

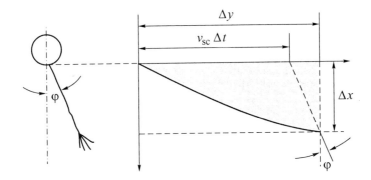

FIGURE 4.17
Evaluation of spark velocity from continuous streak pictures.

interval Δt close to the moment t can be found from the formula

$$v = \frac{\Delta l}{\Delta t} = \frac{\Delta x k_m v_d}{\Delta y} = \frac{k_m v_d}{\tan \alpha} \qquad \tan \alpha = \frac{\Delta y}{\Delta x} \qquad (4.54)$$

Here, Δl is the elongation of a real channel over the time Δt, $\Delta x = \Delta l / k_m$ is the elongation of its image on the film, Δy is the displacement of the point corresponding to the channel head along the time axis $y = v_d t$ in a streak picture, and α is the inclination of the head trace in the picture with respect to the normal at a given moment of time t.

In actual reality, however, the channel may be tilted by an unknown angle φ to the vertical axis, such that even its static image will be represented by a tilted line. If this circumstance is ignored and the image is interpreted only with equation (4.54), the velocity estimate will contain an error, which will vary with the channel tilting direction. We will regard the angle φ as positive, when the channel is tilted toward the time axis, as is shown in Figure 4.17. The head image displacement along the time axis now is

$$\Delta y \approx v_d \Delta t + \Delta x \sin \varphi \qquad (4.55)$$

where the second summand describes the image displacement along the y-axis without the display procedure. Strictly, now $\Delta l = \Delta x k_m \cos \varphi$, but if the angle φ is small, the difference between $\cos \varphi$ and 1 will be the second-order small value in φ, whereas $\sin \varphi \sim \varphi$ the first-order value, so that the channel velocity will now be calculated from the formula:

$$v \approx \frac{\Delta l}{\Delta t} \approx \frac{\Delta x k_m v_d}{\Delta y - \Delta x \sin \varphi} = \frac{k_m v_d}{\tan \alpha - \sin \varphi} \qquad (4.56)$$

The relative error in velocity measurement ignoring the channel tilt is

$$\frac{\Delta v}{v} \approx \frac{\sin \varphi}{\tan \alpha} \tag{4.57}$$

When the channel is tilted toward the display direction and $\sin \varphi > 0$, the velocity calculated from equation (4.54) ignoring the tilt will be underestimated. Nevertheless, in a carefully designed experiment, in which the display rate is chosen such that $\alpha \sim 45°$, $\tan \alpha \sim 1$ and $\sin \varphi$ is small compared to $\tan \alpha$, the velocity estimate will be correct in order of magnitude. A fatal situation may arise, when the channel is tilted in the direction opposite to the time direction φ, $\sin \varphi < 0$: both components of displacement along the y-axis in equation (4.55) are subtracted, compensating one another. Then, the resulting displacement Δy will be very small, the display line will be nearly vertical ($\tan \alpha \to 0$), and the channel velocity, calculated with the tilting ignored [equation (4.54)], will be overestimated many-fold, as compared to the experimental reality. Even the order of magnitude may be wrong.

To avoid such troubles in streak picture processing, it is important to have a still discharge photograph taken by a camera, whose image plane coincides with the streak picture plane. A photograph of this kind will allow a reference line to be drawn to correct the distance Δy. This is not always easy to do, so one can map a still image right on the converter screen. This can be done automatically by applying, to the deflecting plates, an oblique voltage pulse changing into a flat-top pulse. The beam will scan the screen at constant velocity during the linear voltage rise time, and when the voltage stops growing, a still discharge picture will appear at the end of the screen—a reference line. We emphasize again that even a very careful processing technique can provide information only on one velocity vector projection, usually without an account of the spark bends, let alone microbends.

Streak photography performed through a narrow slit allows one to isolate one discharge fragment and to follow changes in its geometry. A classical example is the observation of leader expansion [4.8]. For this purpose, a leader was photographed through a narrow transverse slit, and the image display was performed along the leader axis. What one can see on the converter screen or on a film is a characteristic wedge, whose width corresponds to the leader diameter and the angle at its base defines the expansion velocity. Naturally, the requirements on optical resolution remain the same as for the measurement of leader dimensions from a still photograph. Streak photography through a horizontal slit is also useful for the registration of brightness dynamics in a particular channel fragment.

Finally, narrow transverse slits are sometimes used for velocity measurements of weak ionization waves, unattainable with electron-optical converters requiring more intense light, let alone with mechanical cameras. A series of equally spaced slits are made, and a highly sensitive photomultiplier is mounted behind each slit, whose signals are transmitted to an oscilloscope. When an ionization wave passes by a slit, its photomultiplier registers a radiation spike. The time delay of signals from different photomultipliers will indicate the passing wave velocity.

Streak photography through longitudinal slits is rarely used, but it may prove very helpful in some situations. Suppose it is necessary to control the velocity of a streamer that started from the leader head and propagates through the streamer zone. This task might seem unfeasible because of the great number of streamers traveling in different directions. Streak photography shows the streamer zone as a wide, continuous bright trace in front of the leader. The details are unidentifiable there. A narrow longitudinal slit cuts off all streamers, except those going along it. Their number is not large, so the image obtained can be decoded.

4.8 Synchronization

Synchronization was one of the most difficult problems in the earlier experiments. As automatic devices came into use, the difficulties were gradually reduced, but even today there are measurements, in which the experimenter's skill cannot be surpassed by any instrument. It is hardly possible to supply the investigator with a complete list of procedures that would provide the switching of a registration device at the right moment during an experiment. This strongly depends not only on a particular experimental design but also on the details of its implementation. So we will offer some general considerations, emphasizing possible error sources.

Today, all problems arising in the registration of electrical parameters can be resolved by means of on-line memory devices. No synchronization for such equipment is required. It can operate in a continuous registration mode, automatically erasing the 'zero' information, as the memory is filled up. A threshold device, which is probably the only synchronization element left, must shape the pulse, when the detector voltage exceeds a level taken to be the zero level in a particular experiment. This pulse will turn off the automatic erasing and store the information on-line until the next readout. No special fast response of the threshold device or switching circuit is necessary. The response time is defined by the on-line memory capacity rather than by the duration of the pulse to be registered: the measuring device must be adjusted to the information storage mode before the on-line memory is filled up. This facilitates the solution of many engineer-

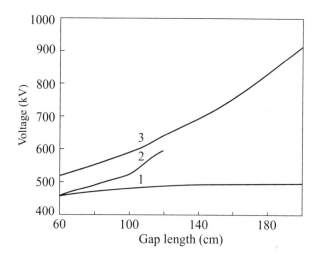

FIGURE 4.18
Ignition voltage of a positive corona flash in a sphere-plane gap ($r_a = 12.5$ cm):
1, calculation neglecting the suspension thread; 2, 3, measurements from [4.7]
and [4.9], respectively.

ing problems, because modern systems with large storage capacities do not
require fast response devices for recording even the shortest pulses.

The danger of making a measurement error lies in the wrong choice of
the zero (reference) level rather than in the response time of the threshold device. If the zero level chosen is too high, some of the information
will be lost. The other extreme is also open to hazards, which are not always detectable. One illustration is the registration of a large amplitude
pulse with a very short rise time. Suppose such a signal decreases rapidly
following its peak and then is damped slowly (Figure 4.18). The dynamic
range of measuring equipment is always limited. When the reference chosen is too low, the pulse top will be cut off due to the saturation of the
analog-to-digital converter. In this case, not only is the pulse amplitude
underestimated but its duration is considerably overestimated, because it
is arbitrarily defined by the half-amplitude value (Figure 4.18).

Synchronization devices are really indispensable in optical registrations.
They play a dual role. On the one hand, the image display voltage must
be applied exactly on time to the deflecting plates of an electron-optical
converter usually working in a single operation mode, and the electronic
shutter must be opened and shut exactly at the prescribed moment. On
the other hand, it is necessary to synchronize the converter filming with
the operation of other devices that register discharge electrical parameters.

The requirements to synchronization become higher with decreasing spark velocity and the lifetime of the process of interest.

The initial discharge can be observed by starting up a precharged high voltage pulse generator (VPG) on command from the image display unit of an electron-optical converter. The latter is driven by hand. The voltage pulse formed at the moment of start-up is used as a command to the control unit of the VPG spark gaps (Section 4.1). After their breakdown, high voltage is applied to the discharge gap of interest. Success is guaranteed, if the time for streak photography *a priori* exceeds the pulse rise time plus the statistical delay time of the discharge. Then the image of the initial streamer flash will, by all means, be caught in a shot and streak displayed. When a high registration rate is required, such that the display time is short, the initial flash may come too late and remain out of shot. It is hard to suggest anything else but a careful manual control of the delay time and photographing onset relative to the VPG start-up. A time delay device is necessarily included in the experiment control system, but the investigator has to rely on his own experience and intuition in controlling the delay time.

An electron-optical converter may be started up by a signal from a current detector or a sensitive photomultiplier. But the very beginning of the discharge process will be lost to registration, and one has to decide if this loss is acceptable. There is another procedure. In addition to the VPG pulse, a voltage pulse with a high rise time and large amplitude is formed by a special circuit. Both voltages are summed up in the discharge gap, and only then does the electric field appear strong enough to excite a discharge. The moment the additional pulse is applied can be taken as a reference time, while its rise time can be used to form a synchronizing signal. Before starting an experiment, one will have to make up one's mind as to how much the discharge characteristics will be affected by the fast partial voltage rise.

In studying intermediate discharge stages, the optical devices can sometimes be started up by a photomultiplier signal focused through a narrow slit at the necessary point in the gap. It is also useful to keep under control the discharge electrical parameters, especially if they have been subjected to qualitative changes.

A typical example of the situation described above is a so-called *final jump* of a long spark. It occurs at the moment a streamer comes in contact with the electrode of opposite sign, when the leader velocity rapidly increases by one or two orders of magnitude. After the contact, the current through the electrode rises abruptly, and this rise can be used as a synchronizing pulse. This approach has permitted detailed studies of final leader jumps, using streak photography, but such a situation should rather be considered as an exception to the rule. Fast events occurring in the gap interior produce very weak changes in the current through the detector mounted on one of the electrodes (Section 4.4.3), so the changes cannot be used for synchronization.

Synchronization of streak pictures with oscillograms of discharge electrical parameters is necessary for finding some relationships. Even if streak pictures are synchronized very precisely, it is difficult to align the time axes of an oscillogram and an optical image. The origin and endpoint in an oscillogram are marked quite accurately: the beam leaves a trace on the screen even after a zero signal. However, there is no image on the converter screen at all in the absence of discharge. To obtain reference points for the alignment of the time axes, special light labels are projected onto the converter screen together with the discharge image. The labels may represent radiation from a light diode, to which voltage pulses of fixed duration are applied.

To bring a streak picture and an oscillogram into coincidence, streak photography is performed in the shut-open-shut mode. In the normal mode, constant voltage is applied to the converter shutter plates locking-in the image. Then the anode voltage is compensated by applying a rectangular voltage pulse of opposite sign, resulting in the image projection onto the tube screen. The exposure is exactly equal to the applied pulse duration. When the pulse time is over, the image is locked in again by the remaining constant voltage. The pulse from the shutters is applied to one of the oscilloscope beams and recorded together with the discharge electrical parameters. This provides exact synchronization of oscillograms and streak pictures.

The result of the above procedure is especially good, when the voltage applied to the oscilloscope and converter tubes comes from a common sweep generator. If possible, the image on the converter screen should be opened a little earlier than streak filming begins. Then a still picture of the initial streamer section appears in the left-hand part of the shot, marking the starting point of streak filming.

4.9 Gap field data: theory and reality

Electrodes of simple geometry should be preferred for gap field measurements, because they make the data reproducible and easier for theoretical treatment and numerical simulation. But even with simple electrodes, the actual electric field distribution in a gap differs markedly from a model distribution. An experimental field is always affected by the electrode suspensions and nearby devices, which are either under high potential or are grounded. Field distortions may result in large measurement errors, not always readily detectable. We will illustrate this with the effect of a thread suspending a sphere of radius r_a above a grounded plane at distance d from the sphere. Maximum sphere field is related to voltage U by the expression

$$E_{max} = U r_a \left(1 + \frac{r_a}{2d}\right) \left[\frac{1}{r_a^2} + \frac{1}{(2d + r_a)^2}\right] \qquad (4.58)$$

Expression (4.58) takes into account the sphere image charge in the plane and gives only a 1% error at $d > 5r_a$. To maintain E_{max} constant at increasing d, an 11% voltage rise at $5 < d/r_a < \infty$ (Figure 4.18, curve 1) is sufficient.

Imagine now an experiment for measuring the streamer inception voltage U_{inc}. The field at which a streamer starts from the anode of radius r_a must depend only slightly on the gap length d. This follows from the results of the avalanche-streamer transition analysis made in Section 3.5.3, at least, for the case with $d \gg r_a$ and $r_a > 20\alpha^{-1}$ (α is the Townsend ionization coefficient). Therefore, one may expect U_{inc} to be nearly constant in a long gap, whereas the experimental relationship $U_{inc}(d)$ in [4.7] proved to be much stronger than that predicted by theory (Figure 4.18, curve 2). The theoretical upper limit of U_{inc} was attained in the experiment at $d/r_a = 10$. As far as the data of the work [4.9] are concerned, they show an increasingly larger voltage U_{inc}, the rise being doubled within the range $5 < d/r_a < 20$ (Figure 4.18, curve 3). This demonstrates the disagreement between experiment and theory to be even of a qualitative nature.

In another experiment [4.7], a sphere was suspended above a plane lying on the ground, so that the field distortion was largely due to the suspension charge. The distortion increased with the thread thickness. When the thread radius was equal to the electrode radius, the whole design looked like a rod with a hemispherical head, in which the field is given by expressions (3.20) and (3.21). Its maximum strength is half the value for an isolated sphere, the field slowly falling along the gap. The maximum strength in a sphere-plane gap is lower than its theoretical limit, even if the thread is very thin. For the thread radius $r_s \approx 0.01r_a$, the maximum field on a sphere in the experiment is $E_{max} < 0.95U/r_a$, although it is $E_{max} > U/r_a$ from formula (4.58). This is not surprising: the thread capacity and, hence, the charge only logarithmically decrease with lower r_s.

One can never entirely avoid the suspension effect on the field measurements. Nor can one reduce this error to a few percent by making the thread thinner. At high voltage, a thin thread begins to display a corona even before a discharge is ignited on the electrode, which is equivalent to making the suspension radius larger. As a result, the field in the gap rises and becomes equalized, while near the electrode it is reduced. The corona effect varies sporadically from one experiment to another and is poorly predictable, while the relationship $E_{max}(U)$ becomes nonlinear.

The effect of adjacent grounded objects on the field strength may also be appreciable. In the experiments [4.9] with a 'reverse' design consisting of a grounded sphere and a plane under high potential (Section 4.4.1), the principal source of field distortion was the building floor located too close to the sphere. Note that the authors of the work [4.9] were aware of this effect and made the necessary reservations, which, unfortunately, does not often happen. Most researchers neglect the reproducibility criterion for their gap field data and omit important details in the descriptions, so that it is impossible to assess the discrepancy between the theoretical and actual gap field distributions in their experiments.

5

Long Streamers

The rest of the book will be concerned with long sparks. Vast experimental data have accumulated on this type of discharge during the second half of the 20th century. Most of the data, however, describe it phenomenologically, and very few characterize the basic relations among the parameters of discharge processes. So it is quite difficult to build up a complete physical picture in terms of experimental findings only—such a picture would have so many blank spots that it would hardly be applicable for a qualitative check-up of hypotheses, let alone of quantitative estimations necessary for applications. Wherever possible, we will supplement experimental data with computer simulation results. Computer simulation, as a tool of numerical experiments, has become a common technique in discharge studies. A numerical experiment is capable of providing comprehensive and reliable information, if one strictly controls the initial model assumptions and operates within their range of applicability. As for errors, they are equally possible in any, including physical, experiments, and we have already discussed this problem at length.

The experimental data to be referred to in this book will necessarily be used to check theoretical speculations and will be compared with simulation results. The reader, however, should show a sound skepticism: he should not overestimate the importance of either coincidences or discrepancies between the calculations and the measurements. Very often, important experimental details that must be taken into account are not mentioned by the authors in their publications, while the functional relations among the parameters of interest may be so strong that even a slight variation in the initial conditions may radically change the final results. Difficulties grow like a snowball, when the data by several authors are to be united. The experimental conditions of all studies have to be analyzed carefully to make sure that they are comparable; otherwise, any theoretical treatment may prove meaningless.

5.1   Phenomenology and dimensions

A single long streamer can be observed very rarely, practically never in air. At high voltage, a streamer normally becomes branched. The discharge flash shown in Figure 1.3 was photographed in air under normal atmospheric conditions in a sphere-plane gap. The voltage applied to a spherical anode of 25 cm radius was 800 kV. A branched streamer flash is often termed a *corona flash*. Such a flash indicates the onset of a spark dis-

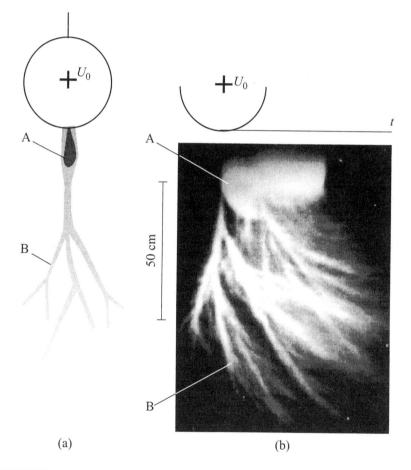

(a)   (b)

FIGURE 5.1
The stem and branches of a positive corona flash (a) and its image on a converter screen showing the streamer channel expansion (b). A, the stem; B, the corona branches.

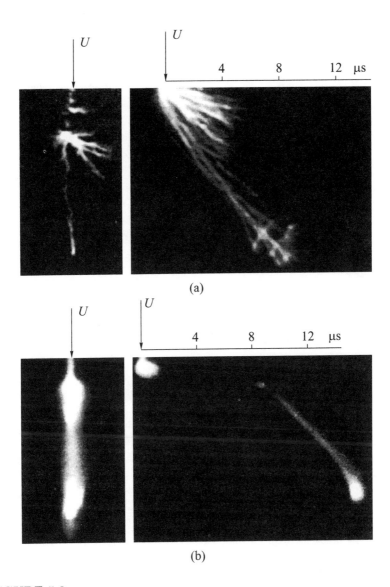

FIGURE 5.2
Still and streak photographs of unbranched cathode-directed streamers in pure argon (a) and helium (b) in a needle-plane gap of 25 cm long.

charge in air and many other gases in a long gap. Depending on the voltage pulse amplitude, branched streamers (corona branches) may either reach the cathode surface or end up somewhere in the gap interior. As a leader develops, numerous streamers start from its head (rather than from the anode), which is, in a sense, the 'extension' of the high voltage electrode. Streamers fill up a fairly large space in front of the leader head known as the leader streamer zone. A streamer zone may be as long as several meters even in laboratory conditions. To identify individual streamers within the zone, photographs are taken through a narrow vertical slit; otherwise, the streamer zone would look like a long diffuse glow in a photograph.

The unbranched streamer section is relatively short in air and is close to the anode radius under normal atmospheric conditions. Nearly half of this section, beginning from the electrode, represents the corona stem. The stem looks brighter than the rest of the channel (Figure 1.3). High quality photographs can show the stem fine structure [5.1]. Its interior is a very bright core expanding from the starting point in the direction of the streamer propagation. The stem envelope is not as bright and has a nearly unvarying radius (Figure 5.1).

Single long streamers were observed in pure electropositive gases such as argon and helium [5.2], in which a streamer discharge could develop at lower voltages than in air. The streamers did not branch, if the gap voltage was close to their ignition voltage (Figure 5.2). A higher voltage usually resulted in streamer branching.

5.1.1 Streamer radius

Measurement of a streamer radius is the most difficult task of all 'geometrical' measurements. In Chapter 3, we already discussed the ambiguity of the radius problem formulation and suggested that the initial radius must be of the same order of magnitude as α^{-1} (α is the Townsend ionization coefficient), at least, at the moment of avalanche-streamer transition. The radius may then increase many times over its initial value due to the ionization expansion of the streamer channel.

Experimenters have so far measured the 'optical' streamer radius only, using its image on a photofilm or on the display of an electron-optical image converter. Such measurements can give a general idea of the size of electron-filled space but provide little information on the electron density distribution over the channel cross section. Of special interest could be the near-anode region of long cathode-directed streamers, where the channel expansion should be especially evident, but in conventional photographs this region is screened by the stem that forms after the channel has developed. To separate the stem from the streamer channel, the electron-optical image converter must have a high spatial and time resolution and operate in a shot-by-shot mode. This complicates the experiment, because the spatial resolution of a converter with light amplification hardly exceeds 5–10 lines per millimeter, while photographs of high voltage streamers have to be taken at a large distance.

Nevertheless, the existence of both 'large-scale' streamers, with a radius many times larger than the theoretical diffusion radius, and ionization expansion of the streamer channel have been reliably confirmed by experiment. For example, direct measurements were made in neon at atmospheric pressure in a uniform electric field of 6.6–16 kV cm^{-1} [5.3]. A streamer arose in the gap interior and propagated toward both plane electrodes, being of the anode- and cathode-directed type. The measurement accuracy of geometrical parameters was better than 0.2 mm; the time matching error of the shots made with an image converter was less than 1 ns. The authors [5.3] give the following data on the maximum radius of a streamer of length l at $E = 6.6$ kV cm^{-1}:

l (cm)	0.2	0.6	1.0	1.4	1.8
r_c (cm)	0.045	0.07	0.085	0.11	0.12

The value of r_c, measured at about the moment of the streamer inception, is indeed close to $\alpha^{-1} \approx 0.0375$ and is five times larger than the diffusion avalanche radius r_c in classical theory. As the streamer becomes longer, its polarization charge grows, increasing the radial field, so the streamer radius again increases by nearly a factor of 3.

Thicker channels produced by a primary ionization wave were observed in pure helium [5.2]. The experiments were done in a gap of 25 cm long. The streamer started from a needle-shaped anode and propagated toward a plane cathode at the gap voltage of 11 kV. Its still and streak pictures are shown in Figure 5.2(b). Both pictures were taken by an image converter with two light amplification cascades; a multialkali cathode had nearly identical sensitivity in the visible and near-ultraviolet regions. The pictures cannot give a clear view of the channel expansion, but its average radius is over 1 cm (helium is a peculiar gas in this and other respects).

It is more difficult to make direct measurements of long streamer radii in air, where a voltage of 10^5–10^6 V is to be applied. Such voltage forces one to place the measuring equipment at a distance of several meters, or even several tens of meters, away from the discharge gap. Still, the channel expansion can be shown quite clearly. The image of a streamer in Figure 5.1(b) was obtained in atmospheric air. The streamer has started from a spherical anode of 25 cm radius towards a plane placed at a distance of 1.5 m. A continuous streak photography was performed with a rate chosen such that the streamer stem could be separated from the channel; the latter can be viewed in a still photograph. The fact of expansion of the initial, unbranched streamer is evident, its radius increasing with distance from the anode.

It is unclear why the radius of a long streamer has failed to become a subject of detailed studies, in spite of being a crucial streamer parameter. For example, Les Renardieres Group of international researchers completely ignored this issue in their long-term project aimed at the study of long sparks [5.4, 5.5]. In his well-known review [5.6], I. Gallimberti only makes reference to [5.7], where the streamer radius was found to be $(1-3) \times 10^{-3}$ cm; in his opinion, it was characteristic of a short streamer. Some indirect data on streamer radii can be found in [5.8].

5.1.2   Streamer length

In principle, streamer length has no limits. A streamer may grow as long as the gap and voltage source permit. In laboratory conditions, cathode-directed streamers observed in air were as long as 10 m at a gap voltage of 5 MV. Such a streamer was, of course, extensively branched [Figure 5.3(a)]. Sometimes, several flashes started from the electrode simultaneously. Longer streamers could be observed in lightning, but no lightning

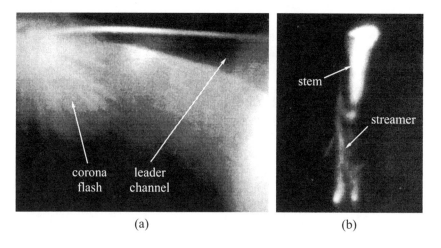

(a)   (b)

FIGURE 5.3
An initial positive corona flash in a sphere-plane gap ($r_a = 50$ cm). Continuous streak photographs taken in a 8 m gap at a voltage pulse of 1.7 μs front and 4.2 MV amplitude; the corona is followed by a leader (a). Still photographs taken with 50 ns exposure in a 50 cm gap; the stem is being formed in the streamer trace (b).

streamer zones have so far been photographed because of the poor filming conditions. At a distance of 500–1000 m, at which lightning is normally photographed, the short-wavelength streamer radiation (the blue and near-ultraviolet regions) is entirely dissipated in the air, especially if it rains.

Let us consider briefly the stem length Δl_s of a streamer flash. Other conditions being equal, the stem length is proportional to the triggering electrode radius. The experimental data [5.1] obtained in air for spherical anodes of 3.1–25 cm radii are described by the empirical formula:

$$\Delta l_s = 0.58 r_a \qquad (5.1)$$

All measurements were made in gaps of length $d \approx (3-4)r_a$ in strongly nonuniform electric fields. A voltage pulse of $1/50~\mu s$ was sufficient to produce a 50% probability breakdown. As the field was made more uniform by reducing the gap length at constant anode radius, the gap section covered by the stem became longer, while the stem length decreased. There is a reason to believe that the stem develops in a gap section, where the external field is still unperturbed by space charge and is high enough for the ionization to proceed. This is supported by measurements [5.9] made in a weakly nonuniform field between a spherical anode of 50 cm radius and a plane. The voltage pulse amplitude of $5/2000~\mu s$ was chosen so as to provide a 50% probability breakdown. The measurements gave the following stem lengths:

d (cm)	100	50	27	20
Δl_s (cm)	30	25	22	~20

Streak photography performed simultaneously with the measurements showed that a streamer should not necessarily have branches for the stem to develop. The latter may follow the path of a single unbranched streamer, as is seen in Figure 5.3(b).

5.2 Streamer velocity

Streamer velocity is probably the best described parameter. The first plausible data on long streamer velocities were obtained from still photographs taken with voltage cut-off or from the time delay of signals registered by photomultipliers receiving radiation from different points along the discharge gap (Section 4.7.1) [5.8]. These data were later refined and made much more detailed with electron-optical image converters [5.1, 5.4, 5.5, 5.11]. These measurements did not show a sufficient accuracy either. The point is that velocity data for long streamers can be obtained in a simple way by using an image converter operating in a continuous mode rather than in a shot-by-shot regime. Disadvantages of this method were already discussed in Section 4.7.2, the principal one being the absence of a reference line for time counts. In most experiments, however, the problem of a reference line was not even raised. Moreover, all velocity measurements only gave average values for long streamer sections, along one channel projection without account of its numerous bends.

There is no reason to present streamer velocity measurements here in a chronological sequence, especially because this was done in the well-known book edited by Meek and Craggs [5.8]. Earlier studies placed emphasis on the highest velocity (up to 10^9 cm s^{-1}) but gave little attention to the relationships between the streamer velocity and the field distribution in the gap, the gas composition or its state.

Let us start with minimum streamer velocity. The existence of a minimum was pointed out in [5.12]. Measurements made in atmospheric air showed that a streamer could not cover any noticeable distance at all, if its velocity was below 10^7 cm s^{-1}. On the other hand, if the velocity even slightly exceeded the critical value, a streamer could have a practically unlimited length. A slow development with a nearly constant, minimum velocity is characteristic of streamers that are formed within the streamer zone of a leader. Their velocity was measured from streak photographs taken through a narrow vertical slit, which exhibited only those streamers that moved parallel to it. This was a way to resolve the problem of reference line. The critical streamer velocity decreases, if electronegative components are removed from the gas. In technical nitrogen (about 1% O_2), for example, long-living steady streamers could be observed propagating at 5×10^6 cm s^{-1} [5.13] and in pure argon at 2×10^6 cm s^{-1} [5.2]. A cathode-directed streamer can propagate at the same velocity in pure helium [Figure 5.2(b)].

Maximum velocity measurements made so far do not have a sufficient accuracy. One of the first studies [5.10] using an image converter yielded the value of 5×10^9 cm s^{-1} for a cathode-directed streamer in normal density air. The measurement accuracy is, however, doubtful because of the absence of a reference line; it was not even mentioned by the author.

More reliable data can be found on maximum velocities of ionization waves in dielectric discharge tubes of a fixed radius [5.14]. There are no reasons to regard such waves as being essentially different from streamer waves, though they develop in somewhat refined conditions excluding branching and path bends. Experiments on maximum velocities used glass or quartz tubes of tens of centimeters long and a few fractions of a centimeter in radius (for example, 47 cm and 0.2–0.75 cm, respectively [5.14]) filled with various gases (air, nitrogen, inert gases, SF_6, CCl_4, CO_2, etc.) at pressures from below 1 Torr to atmospheric pressure. The applied voltage pulses had a rise time of 2–10 ns and an amplitude up to 250 kV. Normally, an average velocity for the total time of flight of an ionization wave through the tube was measured. It was found that the gas pressure, at which maximum velocity was registered, rose with the pulse amplitude. It was 15–25 Torr at 250 kV in air and nitrogen. No matter what gas was used, the velocity of an anode-directed streamer rose to 2×10^{10} cm s^{-1}, or two thirds of light speed. With further pressure increase, the velocity markedly decreased, but at atmospheric pressure it was still of the order of 10^9 cm s^{-1}. The velocity of cathode-directed ionization waves was 1.5–2 times as low. Therefore, the velocity of ionization waves in strong fields approaches that of light, and this can be considered as an experimentally established fact.

Continuous streak photography clearly shows a gradual decrease in the velocity of a streamer, as it propagates through a long gap with a strongly nonuniform field. Such evidence was, for instance, obtained in [5.10]. Observations [5.5] of a cathode-directed streamer in a sphere-plane gap of 4 m long (anode radius 12.5 cm) gave the following values for the average velocity along the path length $l_2 - l_1$:

$l_2 - l_1$ (cm)	6 − 0	13 − 6	23 − 13	30 − 23
v_s (cm s^{-1})	4.4×10^8	3.5×10^8	1.0×10^8	6.0×10^7

The positive voltage pulse was $200/10,000$ μs; the voltage remained practically constant at about 600 kV over the time of the streamer flash development. The anode field at the moment of the start was 40 kV cm^{-1} (less than U/r_a because of the effect of the sphere suspension thread). When the streamer length was 30 cm, its head was at the point of the gap, where the external, unperturbed field had decreased to 4 kV cm^{-1}.

It would be wrong to suggest that the velocity follows the external field variation in the gap and is unambiguously defined by its strength at the point of the streamer head location. The ionization wave velocity is largely determined by the head charge (Section 3.1). When the head is far from the anode, the direct effect of its charge is negligible. So it is more reasonable to find the relation of the velocity to the head potential rather than to the external field in the head vicinity. In the case of a well conducting channel with low voltage losses, one can also try to find directly the relation to the potential of the electrode, from which the streamer started. Such a situation is typical of positive gases, in which electrons slowly escape the streamer plasma due to the electron-ion recombination only. Measurements in pure argon [5.15] did show that a streamer propagating through a gap of a few tens of centimeters at a constant gap voltage has a practically constant velocity [Figure 5.2(a)], although the field is strongly nonuniform in case of a needle-shaped electrode. The velocity varies with the anode potential nearly linearly, $v_s \approx (U_0 \text{ [kV]} - 9.3) \times 10^5$ cm s^{-1}. An attempt to reproduce this experiment in air would lead to a great velocity variation along the gap.

The streamer velocity in negative gases changes mainly for two reasons. First, as the channel becomes longer, its voltage drop increases, the head potential falls, decreasing the wave front field responsible for the ionization. Second, even if the head potential remains constant, its charge diminishes as the streamer becomes more and more branched, since the branches screen one another (see Section 6.7.1). Streamers propagating through negative gases are multiply branched, and it is necessary to continuously raise the gap voltage to make a streamer move at a constant velocity in a strongly nonuniform field. Measurements made in air give an approximately linear relationship between the rate of voltage rise A and the streamer velocity v_s. The average velocity is $v_s \approx 5 \times 10^7$ cm s^{-1} at $A = 500$ kV μs^{-1} and $v_s \approx 3 \times 10^8$ cm s^{-1} at $A = 3500$ kV μs^{-1}. The scatter in experimental data here is great, but this is not surprising. The measurements [5.12] were made through a narrow vertical slit, so it is hard to attribute the measured value to a particular generation of branches.

There is a minimum critical rate of voltage rise, at which a steady development of a cathode-directed streamer is still possible: $A_{\min} \approx (4-5) \times 10^{10}$ V s^{-1} for atmospheric air. The existence of $A_{\min}$ is closely related to the minimum streamer velocity $v_{s\,\min}$. If we assume that the average field in a streamer channel also has a minimum value, $E_{c\,\min}$, then $U_{a\,\min} \approx E_{c\,\min}l + U_{t\,\min}$, where $U_{a\,\min}$ and $U_{t\,\min}$ are the minimum possible potentials of the anode and the head for a streamer of length l. Hence, $A_{\min} \approx dU_{a\,\min}/dt = v_{s\,\min}E_{c\,\min}$. It will be shown below that the value of $E_{c\,\min} \approx 4-5$ kV cm^{-1} derived for atmospheric air agrees with experiment.

Velocity data for long anode-directed streamers are more scarce. There are reasons to believe that, other conditions being equal, they can develop higher velocities than cathode-directed streamers. Some information can be derived from a comparison of experimental data of [5.4, 5.5] and [5.16] obtained in similar conditions by similar techniques. The authors of [5.16] present continuous streak photographs of a long anode-directed streamer in a 5 m air gap between a rod and a plane. The streamer started from the rod cathode with a hemispherical head of 30 cm in radius at voltage of 1250 kV, which remained constant during the measurement. The unperturbed field at the cathode was 32 kV cm^{-1} at the moment of the start; it was high enough to accelerate the streamer to a velocity of 1×10^9 cm s^{-1}. As soon as the streamer became 50–60 cm long 100 ns after the start, the velocity decreased 3–4 times as compared to its maximum value.

It is interesting that earlier registrations [5.17] were also made by an electron-optical converter with about the same accuracy. A streamer starting from a spherical anode of 12.5 cm in radius in a field of 39 kV cm^{-1} at 390 kV was only accelerated to $(2-2.5) \times 10^8$ cm s^{-1}. This again supports the direct relation of the velocity to the head potential (rather than to the field strength) comparable with the electrode potential, if the streamer is not very long. In the experiment described in [5.17], the gap length was only 32 cm, while its voltage (390 kV) was three times lower than in [5.16].

There is a definite relationship between the streamer velocity and the rate of voltage rise for both types of streamer. Measurements were made in the streamer zone of a negative (anode-directed) leader when its length was relatively small, a few tens of centimeters [5.18]. Streamers started from the head of a leader which can, in a sense, be regarded as the extension of a rod cathode; the voltage loss in a short, well conducting channel can be neglected. The following average velocity values were obtained:

A (kV μs^{-1})	1500	700	400	80	30
v_s (cm s^{-1})	1×10^9	7×10^8	5×10^8	3×10^8	$\sim 1 \times 10^8$

Note that the velocities of cathode-directed streamers in the streamer zone of a leader measured in [5.18] by the same method were about twice as high as in [5.17]. It is hard to say whether this difference is due to some measurement errors or to the yet unknown behavior of streamers within the zone.

5.3   Streamer channel field

No direct measurements of electric field in a streamer channel are available. Probe methods successfully used to study other types of discharge, for example glow discharge, are inapplicable because of the small radius of the channel, its unpredictable trajectory, and the short time of development. Nearly all data available on streamer fields are based on average field evaluations from the overlapped gap voltage.

The principal factor determining the average field in a channel is the gas composition. As electronegative components are removed from the gas, the voltage fall in the streamer becomes slower. This is not surprising since there is no electron attachment in the absence of negative molecules, and the channel conductivity does not decrease so much for the time of the streamer flight through the gap. The lowest values of channel longitudinal field, E_c, have naturally been observed in inert gases. For instance, a streamer overlapped a 25–40 cm gap at an average field of 400–450 V cm^{-1} in pure argon and helium at atmospheric pressure [5.2]. When a small percentage of oxygen was added, the channel field increased multifold (Figure 5.4). The same was true of other gases: any acceleration of electron escape from the streamer plasma resulted in higher E_c values; on the contrary,

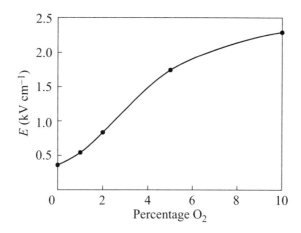

FIGURE 5.4
Average electrical field in a cathode-directed streamer versus O_2 content in Ar_2.

the channel field became lower when the electron density decreased slower.

Average longitudinal fields high enough for streamers to overlap long (to 10 m) air gaps were registered in many studies [5.18–5.22]. The average field in a cathode-directed streamer, irrespective of its length, can be taken to be constant and is 4.5–5 kV cm^{-1} under normal atmospheric conditions (the authors of [5.15] give 4.65 kV cm^{-1}). Any deviations from these values are possible, if the conditions for electron loss are varied. For instance, when air is saturated with water vapor, E_c grows from 4.7 kV cm^{-1} at humidity of 3 g m^{-3} to 5.6 kV cm^{-1} at 18 g m^{-3}. On the other hand, when air is replaced by technical nitrogen (less than 1.5% of H_2O and O_2), long cathode-directed streamers can be observed at 1.5 kV cm^{-1}.

The values of E_c can be changed not only by varying the gas composition but also by changing its temperature and density [5.20, 5.21]. The latter effect is supported by measurements made in a 0.5 m rod-plane air gap, the rod having a hemispherical head with a 0.5 cm radius. The air density was varied either by decreasing its pressure at constant temperature of 290 K or by heating the gas to $T = 900$ K at atmospheric pressure (Figure 5.5). In the first case the rate constants of electron production and loss remained nearly constant, while the reaction rates changed only due to the lower density of neutrals, so the values of E_c were found to be substantially higher (up to 40%) than in the case of air heating. Temperature affects the rate constants of reactions responsible for negative ion decay and the proportions of simple and complex positive ions in the streamer plasma.

Other conditions being equal, anode-directed streamers require stronger

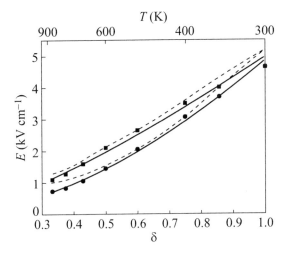

FIGURE 5.5
Average electrical field in a cathode-directed streamer versus the relative air density variation due to heating at $p = 1$ atm and to decreasing pressure at 300 K. Solid lines, experimental data; dashed lines, simulation results.

fields than cathode-directed ones. This fact is absolutely reliable, because it has been established by direct measurement of voltage necessary for a cathode-directed streamer to overlap the gap. Unfortunately, an average anode-directed streamer field cannot be found by dividing this voltage by the gap length. The structure of a streamer flash starting from a negative electrode is nonuniform, and anode-directed streamers do not overlap the whole gap if it is large: some of the streamers are of the cathode type and are capable of starting from the gap interior [5.16, 5.18]. We will discuss the structure of a negative corona flash below. Here, it should only be noted that its complexity has led to a considerable ambiguity of field data. Some of them were derived from questionable, sometimes purely speculative assumptions. As a result, there is a great scatter in the data, and some of them are not strictly experimental. With the voltage balance equation for the streamer zone of a negative leader in 6 m gaps, the authors of [5.18] estimated average E_c as 13 kV cm^{-1} ranging between 10 and 16 kV cm^{-1}. In [5.16], E_c in the streamer zone was found to be 11 kV cm^{-1}, while in [5.17] it was found from streak picture analysis as being equal to 8 kV cm^{-1} for a 32 cm sphere-plane gap ($r_a = 12.5$ cm) and was considered by the author as the minimum possible value of E_c.

The available information is inadequate to understand the reason for the data scatter. It is quite possible that E_c for anode-directed streamers is not a constant value but varies with their length. Besides, one should not ignore the substantial difference in the streamer states at the moment of corona ignition and in the streamer zone; this may result in different values of field strength. Still, the estimations given in [5.17] seem most reasonable; at least they do not contradict reliable optical records [5.17] and definitely indicate a uniform structure of the anode streamer investigated. It is noteworthy that the authors of [5.16] described in detail the field distribution in the gap, which makes their results suitable for numerical evaluations. The average length of anode streamers in that experiment was $l = 65$ cm at $U = 1250$ kV. With the negative channel charge, the external field could be driven partially or totally out of the channel, but average field enhancement in it should be ruled out. Hence, the upper limit is $E_c = [U - U_0(l)]/l$, where $U_0(l)$ is the potential of the external unperturbed field in the head vicinity equal to about $0.35U$ under the conditions described, that is, $E_c < 12.5 \, \mathrm{kV \, cm^{-1}}$.

5.4 Current and charge data analysis

The available measurements of streamer current are quite reliable, and the problem is to interpret them adequately rather than to make registrations themselves. A 5% relative error in nanosecond pulse registrations is an ordinary thing. Such a high accuracy is quite necessary: it makes the records usable for evaluation of some streamer parameters unaccessible to direct measurement. To reduce errors in the registration of total current, as an algebraic sum of conduction and capacitive currents, a detector is placed on the side of the electrode that triggers the streamer (Section 4.4.1). In some experiments [5.23], a detector was connected to a small area measuring section, from which a streamer was to start; in that case, only conduction current could be registered (Section 4.4.2).

Three groups of experiments can be identified with respect to their goals. Some measure currents through the electrode surface, from which a single streamer has started. The results provide information, often implicitly, on the ionization processes at the streamer base and on the linear charge density along the channel. Another group of experiments is aimed at measuring the current of an initial, multiply branching streamer flash (corona flash). These can provide information on the energy input into the discharge, on the amount of space charge in the gap and on the resulting field redistribution. In the latter case, conduction current data are analyzed together with records of capacitive current through those anode and cathode surface areas, which have zero conduction current. Finally, the third group of experiments is to measure current in a streamer channel, which has over-lapped the gap. These data are used to find the average channel conduc-tivity in order to decide whether the gap overlap would lead to breakdown.

5.4.1 Initial current rise due to the base ionization

The time for the observation of a single streamer is very short, because it necessarily starts branching. Fortunately, maximum current is often attained before the appearance of the first branch. Therefore, the current rise time can be said to be a characteristic of a single streamer. A high rate of current growth was observed even in the first precise registration [5.24]. Later [5.24], the pulse rise time was found to be 8 ns for both anode- and cathode-directed streamers starting from high voltage electrodes of 0.8 cm in radius at voltage pulses of 0.7/100 or 1/100 μs and 145 kV amplitude. Nanosecond current pulses have been registered on large electrodes [5.25, 5.11, 5.5], and the rise time was observed to increase with the radius. For example, the streamer current on an anode with 12.5 cm radius reached the amplitude value for 30−40 ns [5.25, 5.11].

Typical current oscillograms for a cathode-directed streamer in air are presented in Figure 5.6. The pulse often has two humps, because the next streamer propagates from the anode with a short time delay relative to the first one. When several streamers start simultaneously from different points on the anode, no humps are, naturally, observed. It is only the pulse amplitude that grows, the growth being multifold in case of a well developed flash. The amplitude rises, when the gap voltage becomes high

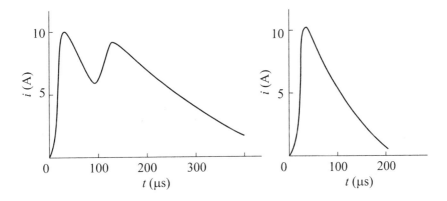

FIGURE 5.6
Characteristic oscillograms of current through the spherical anode surface ($r_a = 12.5$ cm) for the case of a cathode-directed streamer (see text).

enough to trigger a flash. This is due not only to the current rise in a single streamer, but also to the larger number of simultaneous streamers; so, the total rise may be quite substantial. At a fixed voltage, the current amplitude is larger and the rise time is shorter the smaller the triggering electrode radius (the field here is stronger). For instance, at $r_a = 0.8$ cm and $U_a \approx 145$ kV, the amplitude exceeded 10 A [5.24]; the same value was obtained in [5.5, 5.25] at $r_a = 12.5$ cm and $U_a \approx 550-600$ kV.

If a current detector is placed on an electrode section, which is small relative to the total electrode area, it will only register the conduction current, and one can use the current oscillograms to get some information on the ionization processes in the streamer. Indeed, the registered current $i_c = \pi r_c^2 e n_e v_e$ markedly increases, as long as the number of electrons per unit channel length, $N_e = \pi r_c^2 n_e$, grows. Thus, the pulse rise time of conduction current through the anode surface approximately coincides with the ionization time at the anode.

Another important conclusion can be drawn from current oscillogram analyses. At the moment of current peak, the longitudinal channel field at the anode must be close to the ionization threshold, $E_i \approx 30$ kV cm^{-1}, in normal air. It cannot be much higher, because in that case the ionization and the current would continue to grow. It cannot be much lower than E_i either, since there is no mechanism for the production of a larger number of electrons at $E \ll E_i$: negative ions have not yet accumulated at the anode, while the ionization due to electron-excited molecules requires a field almost as strong as E_i. Hence, with the account of the moderate dependence of $v_e(E)$, it is reasonable to assume the drift velocity at the current peak to be $v_e \approx v_e(E_i) \approx 1.4 \times 10^7$ cm s^{-1}.

The linear electron density near the anode can now be calculated from the measured conduction current amplitude: $N_e = \pi r_c^2 n_{ea} \approx i_{c\,max} / [e v_e(E_i)]$. For long streamers, it is $N_e \approx 5 \times 10^{12}$ cm^{-1} at the characteristic amplitude $i_{c\,max} \approx 10$ A. By dividing N_e by the assumed area of the channel cross section, we can find the respective plasma density and judge about the plausibility of our assumption concerning r_c. Some researchers have reported very small streamer radii, down to 3×10^{-3} cm. This value, as the lower limit, is obtained when the avalanche-streamer transition is only analyzed in terms of the diffusion mechanism of the avalanche head expansion (Section 3.5.2). With this radius, the plasma density $n_e \approx 10^{17}$ cm^{-3} turns out to be incredibly large. Even the assumption of $r_c = 10^{-2}$ cm leads to a very large value: $n_e \approx 10^{16}$ cm^{-3}. On the contrary, at $r_c \approx 0.1$ cm, the value of $n_e \approx 10^{14}$ cm^{-3} seems more probable and agrees with other estimates.

Let us look again at the current oscillogram in Figure 5.6. Its analysis leads one to some general conclusions about the streamer process at the initial, unsteady stage, while the streamer length l still remains smaller than the triggering anode radius r_a. This experiment offers a good opportunity to study the initial stage. Since the radius here is relatively large, $r_a = 12.5$ cm, the streamer development at the anode takes long enough for its parameters to be registered.

The initial streamer process cannot be fitted into the concept of quasistationary growth, when the streamer velocity and plasma density are largely determined by the head field, with the field behind the head being so weak that it cannot maintain the ionization (Chapter 3). As a result, most electrons are produced in a narrow layer $\Delta x < r_c \ll r_a$ for the time $\Delta t < r_c/v_s$; in case of strong waves with $v_s \gg v_e$, $\Delta t < 1$ ns. The oscillogram in Figure 5.6 contradicts this pattern. The current rise time of 35 ns indicates that the ionization process at the anode takes a long time. Over this period, the streamer channel has elongated by 10 cm (the wave front has covered a distance of about $10^2 r_c$ from the anode), but the channel field at the anode still remains high enough for the ionization to go on. The fact that the field has not been driven out by the plasma polarization indicates an extremely low electron density in the plasma.

Indeed, the field at the ionization wave front at $l < r_a$ is strongly affected by the anode charge q_a and only weakly depends on the head charge Q (the latter is still too weak to compete for effectiveness with the anode charge). The maximum field at the wave front is actually of the order of unperturbed anode field $E_a \approx q_a/(4\pi\varepsilon_0 r_a^2)$. In the experiment concerned ($E_a \approx 40$ kV cm^{-1}), it is much lower than fields $E_m \sim 10^5$ kV cm^{-1} typical of strong streamer waves discussed in Chapter 3. If those formulas were extrapolated to such weak front fields $E_m \approx E_a = 40$ kV cm^{-1} (the Townsend coefficient $\alpha \approx 20$ cm^{-1}, ionization frequency $\nu_i \approx 3 \times 10^8$ s^{-1}, and drift velocity $v_e \approx 1.5 \times 10^7$ cm s^{-1}), the plasma density in such a wave would be only equal to $n_c \approx 2 \times 10^{11}$ cm^{-3}. Even though all electrons were removed from a thin channel of radius $r_c \ll l$, positive charge would have such a low density, en_c, that it might be insufficient for quenching the anode field at the channel base.

This situation can be demonstrated by calculating the field at the base center of a long uniformly charged cylinder ($\rho = en_c$):

$$E' = \int_0^l \int_0^{r_c} \frac{-2\pi en_c x r \, dx dr}{4\pi\varepsilon_0 \left(r^2 + x^2\right)^{3/2}} \approx -\frac{en_c r_c}{2\varepsilon_0} \tag{5.2}$$

The reverse field near the anode is nearly doubled by the channel image charge in the sphere, such that the greatest possible field reduction at $r_c \approx 0.1$ cm would be $\Delta E \approx 2E' \approx en_c r_c/\varepsilon_0 \approx 36$ kV cm^{-1}. The actual effect, however, will be weaker, because it will take much time, comparable with the total time of the streamer current rise, $\Delta l/v_e \sim 20$ ns, to draw all electrons out of the channel section of length $\Delta l \approx (2\text{--}3) r_c$ making the

major contribution to the field E'.[1] Another factor maintaining a high field
at the anode is the current rise associated with the streamer acceleration.
It drags most of the positive charge toward the head, so that the potential
difference at the anode increases, retarding the field decrease.

Thus, until the streamer has acquired a sufficient length $l > r_a$, its current grows from a small initial value, $i_0 \approx \pi r_c n_c v_e (E_a) \approx 15$ mA, to an
amplitude value, $i_{max} \approx 10$ A, due to the ionization occurring under the
action of a field close to the external anode field, $E_a \approx 40$ kV cm^{-1}. Indeed, for the observable current rise time of 35 ns at $\nu_i \approx 3 \times 10^8$ s^{-1}, the
plasma density increases by a factor of $i_{max}/i_0 \sim 10^3$. Our current data
interpretation provides a general idea about the initial streamer development. A more detailed treatment of ionization wave acceleration and of
the time variation of streamer current and field will require a comprehensive analysis of processes within and behind the streamer head. We cannot
rely on the analytical theory only but will also use numerical simulations.

5.4.2 Single streamer charge and radius

Current oscillogram analysis allows one to evaluate the charge incorporated
into a streamer channel and its linear density. This can be done by calculating the conduction current integral over the time between the streamer
start and its first branching. The integration is performed either graphi-
cally from a current oscillogram or directly in the experiment. In the latter
case, a current detector is replaced by an integrating unit, say, a storage
capacitor (Section 4.5). The accuracy of charge measurement is normally
as good as that of current, or a little better.

The estimation of linear charge density τ for an unbranched streamer section is not, in itself, very informative. The charge distribution at the anode
is strongly nonuniform. Because of the anode surface charge, the charge
density at the channel base is zero and gradually increases by one or two orders of magnitude at a distance of several electrode radii from the electrode.
However, charge measurements along the unbranched streamer section are
of use, because they can help clarify one of the most debatable issues—that
of the streamer radius under the conditions of megavolt experiments.

The analysis involves a comparison of the experimental value of total
charge transported to the gap from a voltage source (what is meant by this
will be specified below) and its calculations from various assumptions of

---

[1]Note that the upper limit of the head and total channel charge contributions to
the wave front field is given by the same integral (5.2). For the values discussed, this
contribution is smaller than 18 kV cm^{-1}, which supports the statement above that the
field at the head is largely created by the anode charge at an early stage of streamer
development.

the channel radius, since the channel capacitance varies with r_c. The approximate theoretical capacitance of a perfectly conducting isolated channel is inversely proportional to $\ln(l/r_c)$. In reality, this dependence on r_c is also weak, but the discrepancies in the radius estimates are expressed in orders of magnitude. Variation in r_c by an order changes the calculated capacitance by tens of percent; this considerably exceeds charge measurement errors controllable in modern experiments.

By solving the electrostatic problem of charge transport to a perfectly conducting channel from an anode, one can determine this charge exactly by taking into account the anode charge we ignored in Section 3.2.1. One can further assert that, at a given anode potential U_a, a channel of finite conductivity will hold a smaller amount of charge, since its potential will decrease toward the head. Conversely, the same charge can be incorporated into a real channel with an average potential lower than U_a only if the channel capacitance is larger than in the ideal case. But channel capacitance grows with radius. Therefore, if the measured charge is equal to the calculated charge, one can be sure that the radius assumed in the calculations will not be larger than the real one. In other words, we arrive at the low limit of channel radius.

It is clear that channel charge must be calculated as accurately as possible by taking account of all charges, including that induced by the channel on the anode. The account of induced charge can even make one change the experimental design and the calculations. The charge incorporated directly into a channel can be found by time integration of the anode conduction current at the channel base. But if one also calculates the anode charge induced by the channel, total current oscillograms can be taken experimentally. Its integral will yield the total charge variation in a system consisting of an electrode and a channel (in fact, this is the difference between the absolute values of channel charge and that induced on the electrode, since they have opposite signs). The measurement of current through a source is easier to make, so its accuracy is higher.

Unlike the ideal model of an isolated conducting filament (Section 3.2.1) valid for the filament length l much larger than the anode radius r_a, real experiments deal with $l \sim r_a$ for initial, unbranched streamer sections. Figure 5.6, for example, is for a streamer of $l = 14$ cm at the current peak and $r_a = 12.5$ cm. In this case the positive linear charge density monotonically grows from the base to the head. Assuming the growth to be linear, $\tau = ax$, we can find the unknown factor a (a similar simplification is often used for an approximate solution of a complex problem). Let us find the expression for the potential at a point on the channel surface, say, at its center. The potential at this point, as at any other point of a perfect conductor, is equal to U. Let us identify three potential components: φ_1

from the intrinsic channel charge, φ_2 from its image charge on a spherical anode (that is, induced charge potential), and φ_3 from the sphere charge $Q_a = 4\pi\varepsilon_0 r_a U$.

The first component taking into account $l \gg r_c$ is

$$\varphi_1 = \frac{a}{4\pi\varepsilon_0} \int_0^l \frac{x\,dx}{\left[(x - 1/2)^2 + r_c^2\right]^{1/2}} \approx \frac{al}{4\pi\varepsilon_0} \ln\left(\frac{l}{r_c}\right) \qquad (5.3)$$

From the Shockley-Rameau theorem (Section 2.6), the unit charge δq at a distance r from the center of a sphere of radius r_a induces a charge $\delta q_i = -\delta q r_a/r$, which is 'localized' at the point $r_i = r_a^2/r$ inside the sphere on the radius connecting its center with δq (this follows from the equipotentiality of the sphere surface). The anode charge induced by the whole channel is

$$q_i = -\int_0^l \frac{r_a \tau(x)\,dx}{r_a + x} = -a r_a \left[l - r_a \ln\left(\frac{r_a + l}{r_a}\right)\right] \qquad \tau\,dx = \delta q \quad (5.4)$$

It is distributed nonuniformly along the section limited by $r_{i1} = r_a^2/(r_a + l)$ and $r_{i2} = r_a$ on the streamer axis. Let us neglect this nonuniformity and introduce the average linear induced charge density $\tau_i = q_i/(r_{i2} - r_{i1})$. Calculations similar to those with equation (5.3) yield the approximate potential from the induced (image) charge:

$$\varphi_2 = -\frac{a(r_a + l)}{4\pi\varepsilon_0 l}\left[l - r_a \ln\left(\frac{r_a + l}{r_a}\right)\right]\ln\left(\frac{3r_a + l}{r_a + l}\right) \qquad (5.5)$$

Finally, the potential from charge Q_a at the channel center is

$$\varphi_3 = \frac{U r_a}{(r_a + l/2)} \qquad (5.6)$$

By summing equations (5.3), (5.5), and (5.6) and by equating the result to the anode potential U, we find

$$a = \frac{4\pi\varepsilon_0 U(4r_a + l)}{l(2r_a + l)\left\{\ln\frac{l}{r_c} - \left(1 + \frac{r_a}{l}\right)\left[1 - \frac{r_a}{l}\ln\left(1 + \frac{l}{r_a}\right)\right]\ln\frac{3r_a + l}{r_a + l}\right\}} \qquad (5.7)$$

The total channel charge is $q = al^2/2$. The charge transported to the anode through the voltage source is $q_0 = q + q_i$, where q_i is defined by formula (5.4). The values of q_0 calculated for various r_c must be compared with the experimental value. At the current peak in Figure 5.6, $q_{exp} = 1.6 \times 10^{-7}$ C ($r_a = 12.5$ cm) at $l = 14$ cm. This value corresponds to anode voltages $U = 550\text{–}600$ kV, varying with the thread thickness, and to unperturbed anode field $E_a \approx 40$ kV cm^{-1}. The calculated and measured charge values coincide at $r_c \approx 0.15\text{–}0.25$ cm, which is the low limit of the channel radius. The actual field at the anode is $E \sim 10$ kV cm^{-1}, so that the voltage drop in the channel is appreciable. Hence, the channel radius should be taken to be even larger. At the commonly used radius $r_c = 10^{-2}$ cm, the calculated charge is as small as 9×10^{-8} C, while at $r_c = 3 \times 10^{-3}$ cm used in [5.26, 5.6], $q_0 = 8 \times 10^{-8}$ C, or half the measured value. This discrepancy cannot be accounted for by an experimental error. As for the above theoretical value, it was well supported by numerical simulations. Thus, the current and charge measurements suggest that the average streamer radius is, at least, a few decimal fractions of a centimeter for air at $10^5 - 10^6$ V.

5.4.3 Flash charge and current

The current behind the peak in Figure 5.6 characterizes a branched streamer (streamer flash). It decreases much slower than it rose at the pulse front. A pulse with a half amplitude lasts $0.1 - 1\mu s$ in air, and its duration decreases with lower voltage and smaller gap. The time of current registration at the base approximately coincides with that of streamer development. This is clear, because the process of spark formation requires energy for the gas ionization and excitation, for changing its internal energy and that stored by the redistributed gap field. All the energy is borrowed from a high voltage source. However, the time values may not necessarily be identical. A streamer is capable of propagating for some time owing to its own energy accumulated by its space charge. In the experiment [5.26], the gap voltage was rapidly cut down to zero, but the streamer did not stop immediately. A possible streamer development at zero external field was first mentioned in [5.26]. This process necessarily involves space charge redistribution and is unable to last long. The current through the source may be zero or even alter its sign—some energy returns from the gap to the source.

The current of a branched streamer cannot yield information on the currents of individual branches. One cannot even estimate the average current of a branch, because the number of branches cannot be exactly counted from a photograph. Current registration is more often used to find the space charge contributed to the gap by a streamer flash. This parameter is important for the breakdown process as a whole: the flash charge is usually comparable with the electrode charge, which means that the field distribution changes substantially during a flash, affecting the later discharge stages.

One always speaks of total charge, since its distribution among the branches, the more so along each branch, is as impossible to measure as the current distribution. Experimenters prefer to integrate the conduction current of a streamer flash, using a storage capacitor instead of a shunt. Naturally, the capacitor must be connected to a measuring electrode section

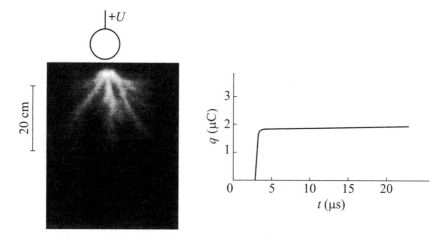

FIGURE 5.7
Still photograph of a corona flash and its charge oscillogram.

of small area. Practically, only conduction current will then flow through the capacitor, and its charge will be equal to the total space charge in the gap. A characteristic charge oscillogram is shown in Figure 5.7 together with a photograph of the respective streamer flash. Over 95% of the charge was formed for the time shorter than 1 μs. After that, the gap current dropped below the threshold sensitivity of the measuring system, in this case 5×10^{-3} A. Note for comparison that the current pulse amplitude is larger than this threshold value by 2 or 3 orders of magnitude.

The measured charge values show a wide spread, its range increasing with shorter pulse front duration. The effect of the pulse front is especially dramatic, when the streamers develop at a maximum steepness. It is under these conditions that the charge distributions of streamer flashes in Figure 5.8(a) were obtained from the data of [5.27]. The experiment had a 'reverse' design, in which a 3×3 m^2 plane cathode was suspended at a height of 5 m and was under high potential and an anode (a 3 m vertical rod with a hemispherical head of 1.5 cm radius) was placed on the ground. The measurements were stable only for very smooth voltage pulses with the rise time $t_f > 100$ μs. At $t_f = 11$ μs, for instance, the spread is $0.2-1.3$ μC with the average value of $q \approx 0.5$ μC. One can get an idea of how large is the flash charge by comparing it with the charge of the hemispherical head: the latter is by an order of magnitude smaller. As the anode radius is increased, the portion of the charge from the streamer flash somewhat decreases, still remaining substantial. At the same rise time of 11 μs, the average charge q from a spherical anode with a 3.25 cm radius (about 1.5 μC)

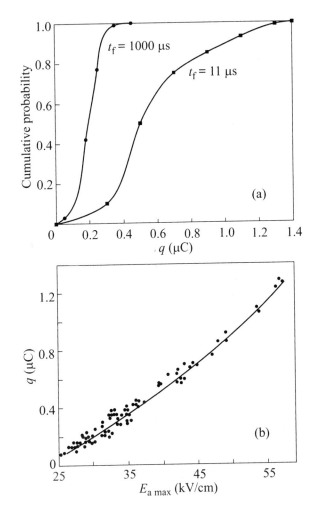

FIGURE 5.8
Positive corona charge in a 2 m rod-plane gap ($r_a = 1.5$ cm) at voltage pulses of varying rise time: statistical distributions (a) and variation with maximum anode field at the moment of corona ignition (b).

was three times the total charge of a full sphere. According to [5.4, 5.5], the average values of the charges in question became identical (~ 35 μC) at the spherical anode radius of 30 cm; but when r_a increased to 50 cm, the flash charge was only 70% of the anode surface charge (~ 65 μC).

Simultaneous measurements of the anode charge and maximum field have showed that the flash charge is directly and closely related to the triggering field [5.23, 5.4]. The relationships $q\,(E_{a\,max})$ are plotted in Figure 5.8(b).

It is now clear why the charge data spread was so large in experiments with a rapidly rising voltage. Because of the statistical deviation, a discharge arises at different voltages and moments of time, and the streamer flash contributes to the gap a charge growing with the triggering field. The field dependence of the streamer current amplitude was discussed in Section 5.4.1; here a similar dependence is manifested for the current pulse as a whole.

The relationship between the near-electrode field and the contributed charge can be followed for various gap lengths and electrode geometries. For example, the authors of [5.4] suggested empirical formulas for a flash from a 121 cm conical anode with 30 cm base radius in a gap of length $d = 5$ m: $q\,[\mu\mathrm{C}] = 3.23 \times 10^{-7} E_{\mathrm{max}}^{3.25}\,[\mathrm{kV\,cm^{-1}}]$; for $d = 10$ m, the formula was: $q = 3.45 \times 10^{-7} E_{\mathrm{max}}^{3.28}$. The factor and power index of E_{max} were practically independent of d. These formulas were experimentally proved to be valid in the range $E_{\mathrm{a}} = 70-110\,\mathrm{kV\,cm^{-1}}$. The relation $q\,(E_{\mathrm{max}})$ was also observed for anode-directed streamers but it was not as evident. Other things being equal, the statistical deviation of the time delay and, hence, of the field strength initiating anode-directed discharges is much smaller than for cathode-directed streamers, so it is hard to find experimentally a significant field range. The data concerning anode streamers at $E_{\mathrm{a\,max}} = 31-34\,\mathrm{kV\,cm^{-1}}$ were approximated by the authors of [5.16] with a straight line: $q\,[\mu\mathrm{C}] = 2.6\,(E_{\mathrm{a\,max}}\,[\mathrm{kV\,cm^{-1}}] - 29.5)$. In these experiments, a flash started from a rod cathode with a hemispherical head of 30 cm radius toward a plane anode at $d = 5$ m. The linear dependence $q\,(E_{\mathrm{max}})$ is very approximate due to a large scatter of charge values at fixed E_{max}; the observed scatter at $32\,\mathrm{kV\,cm^{-1}}$ is nearly as large as one order of magnitude. The charge measurements in streamer flashes (corona flashes) have been made on electrodes of various geometry, primarily for applied problems of electrical strength of long gaps.

5.4.4   Flash charge data analysis

Flash charge measurements are made either to find the charge location or to define the field distortion caused by the gap field charge. These are actually two aspects of the same problem—the principal electrostatic problem of the field-charge relationship mentioned in Section 4.5. It is not difficult to find the answer to the pragmatic question of field distortion at a particular point or in a limited area. Field can be registered simultaneously with charge registration, and even the triggering electrode surface is accessible to measurement (Section 4.6). Moreover, it is this surface that is of special interest, since all repeated discharge structures develop in a distorted field, and their behavior depends on the nature of the distortions.

Figure 5.9(a) shows some typical field strength oscillograms taken on

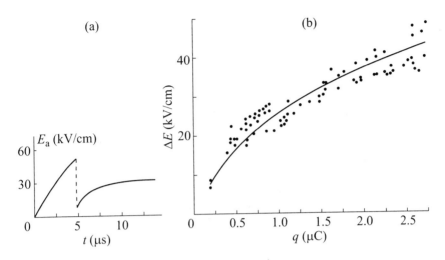

FIGURE 5.9
Decrease in the anode field due to ignition of a positive corona flash in a 2 m
rod-plane gap ($r_a = 1.5$ cm): a characteristic field oscillogram (a) and ΔE as a
function of incorporated charge (b).

the anode during streamer flash development. Prior to the flash, the field
$E_a(t)$ changes similarly to the gap voltage $U(t)$, because the field is only
created by the electrode charges. Positive charge is incorporated into the
gap during the flash, somewhat enhancing the gap field in front of the
streamers but decreasing it near the anode. The anode field strength drops
dramatically for the fractions of a microsecond while the current exposing
space charge passes through the surface; it becomes as low as a few dozens
of kilovolts per centimeter but never changes its sign. This is clear, because
in a negative field, electrons could no longer arrive at the anode, leaving
a positive ion charge behind them. Field ΔE decreases with increasing
charge q in a gap with electrodes of a given geometry [Figure 5.9(b)]. It is
important for further discussion that a strong flash incorporates so much
charge into the gap that the field drops below the ionization threshold, E_i,
across the whole anode, producing a pause in the discharge development.
The pause lasts until the ions creating the charge become incorporated into
the gap or until the charge effect is compensated by the voltage rise. If the
gap voltage remains constant or changes slowly, the duration of the pause,
Δt, may be as large as the time of ion drift at a distance comparable with
the anode radius ($\Delta t \sim 10^{-3}$ s at $r_a \sim 10$ cm and average field 5 kV cm^{-1}).

An anode-directed flash decreases the field at the cathode as much as the
opposite flash does at the anode. The residual field turns out to be lower
than the ionization threshold, interrupting the development of a negative

spark. This process has, however, a specificity at negative voltage. If the rate of voltage rise is not very high, say 200 μs, and the cathode radius is less than 10–12 cm, cathode streamers look as if they were degenerate: they represent a sequence of frequently appearing structures less than a centimeter in length. If the charge is registered with a microsecond resolution, the oscillogram shows a continuous charge increase proportional to the voltage, while the cathode field becomes practically stabilized. In a similar way, it is stabilized on a thin wire displaying a so-called avalanche corona.

Let us see what information can be derived from flash charge registrations about its distribution. The value of the charge measured is informative in itself. It indicates the minimum distance of this charge from the electrode. If the electrode field variation is measured simultaneously, one can also find the exact location of the 'center of mass' of the charge incorporated into the gap. Indeed, let us calculate the field variation on a spherical electrode, having replaced the distributed charge q by a point charge of the same value located at distance r from the center. For simplicity, we will assume this charge to be positive and the electrode to be an anode of radius r_a.

Charge q induces, on the sphere, charge $q_i = -qr_a/r$ (Section 2.6), which is located on the x-axis connecting q with the sphere center at distance $r_i = r_a^2/r$ from the center. The latter assertion is a sort of mathematical abstraction. It is with this localization of the induced point charge that the sphere remains equipotential. The field on the x-axis is a sum of field E_a created by the capacitive charge of the sphere $q_a = 4\pi\varepsilon_0 r_a^2 E_a$ and two fields of two point charges—positive charge q and negative induced charge q_i. The resulting field on the x-axis on the outer sphere surface, that is, at the point nearest to q, is defined as

$$E' = E_a - \frac{q}{4\pi\varepsilon_0 r_a^2 \psi(z)} \qquad \psi = \frac{(z-1)^2}{z+1} \qquad z = \frac{r}{r_a} \qquad (5.8)$$

It was mentioned above that during the incorporation of a positive charge into a gap, the field at the anode cannot become negative, otherwise the electrons would be unable to leave the gap and the process would cease. This may happen only at $\psi(z) > q/q_a$. For instance, at $q = 10q_a$ typical of experiments with medium-sized anodes, the space charge 'center of mass' must be located at a distance larger than 12 anode radii from the nearest point on the anode surface ($z > 13$). Only a relatively small charge $q < q_a \Delta r/2r_a$ may be located at the small distance Δr. Under the experimental conditions of Figure 5.9(b), the charge $q = 1.5$ μC decreased the near-anode field by $\Delta E = E_a - E' = 32$ kV cm^{-1} ($r_a = 3.25$ cm). According to formula (5.8), the equivalent point charge was located on the radius $r = 6.7r_a = 22$ cm. This value is close to the experimental distance between the external boundary of streamer branches and the anode.

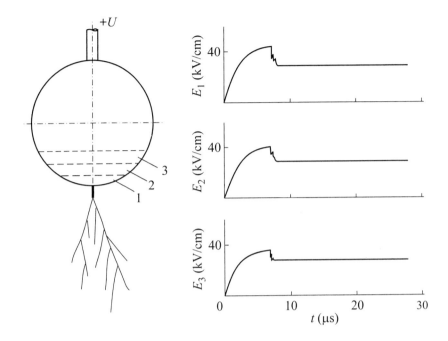

FIGURE 5.10
Typical electric field oscillograms for various points on a spherical anode at the moment of positive corona ignition.

More detailed information on spatial charge distribution can be obtained by measuring the field at several points on the anode in different cross sections normal to the gap axis. Simultaneously registered oscillograms are illustrated in Figure 5.10. The observation points lie at different distances from the streamer branches, with the field falling differently: the value of ΔE is larger the nearer the detector to the anode apex, from which a streamer flash started.

Using a numerical solution, one can relate the measured values of ΔE at n observation points to n via yet unknown parameters characterizing the charge distribution in the gap (say, average charge densities in n-disks, into which the flash volume is divided in an approximate model). One can then derive these densities by solving a set of algebraic equations. This was the way the experimental data were treated in [5.23, 5.27]. The result obtained does not, in principle, differ from the simplified estimates above: most of the space charge is located on the visible flash periphery. This does not, of course, mean that the charge of each individual streamer is concentrated at its end. Branching just increases the total number of branches, each of which has its own charge.

5.4.5   Overlapping streamer current

Imagine an ideal situation, in which a gap has been overlapped by a single unbranched streamer. The streamer current measurements will contain information about the average linear channel resistivity. To find its value, it suffices to know the streamer voltage, which in our case is equal to the gap voltage. Evaluation of resistance or conductivity in such a simple way is meaningful, if the current does not vary along the streamer channel. Such a quasistationary mode takes place, after the charge redistribution along the channel is completed following the overlap. The redistribution process does manifest itself in this way or other, because while the streamer is propagating, the head potential is, by definition, larger than zero. Having contacted the opposite grounded electrode, the head 'immediately' acquires zero potential. The potentials of other points of the channel change respectively, causing charge redistribution. A neutralization wave is said to pass along the channel after the contact, although the channel charge does not, of course, vanish completely. How strong the wave will be depends on the channel conductivity. But in the vicinity of the point of contact with the grounded electrode, the current may abruptly rise for a short time even at a low channel conductivity, since the head charge goes into the ground. In case of a grounded cathode, the head positive charges will, naturally, stay where they are; they will be neutralized by electrons detached from the cathode surface.

Anode and cathode current oscillograms look differently (Figure 5.11). While a cathode-directed streamer is developing, the conduction current through the cathode remains zero, until the contact. Then it suddenly rises, sometimes greatly exceeding the anode current, and the process of head charge neutralization begins. The anode current does not change as much after the contact. It does not necessarily grow; it may continuously decrease, if the channel conductivity is very low. Following the charge redistribution, the current becomes identical in all channel cross sections. Its further evolution depends on the gap voltage and linear channel resistivity, which changes rather slowly. It is under these conditions that the evaluation of average channel conductivity $g = i/U$ becomes meaningful.

A situation like the one just described was studied experimentally in a short air gap ($l = 1.5$ cm) with an unbranched streamer [5.7]. In case of cathode-directed streamers of tens of centimeters long, current measurements on the anode side did not give the desired result: total flash current, rather than the current of an individual streamer, was being registered. So, the channel conductivity turns out to be overestimated. One has to rely only on measurements on the cathode side, but in this case the detector is to be connected to a measuring section of such a small area that a simultaneous start of several streamers from it is very unlikely [5.28]. As a result,

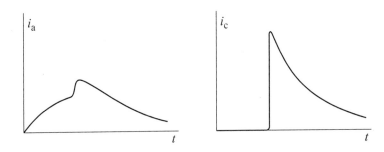

FIGURE 5.11
Schematic oscillograms of current through the anode and cathode during the
streamer propagation through an air gap.

errors in average conductivity measurements for a single streamer cannot
be avoided: the values of g will be underestimated, because the currents of
many branches are summed up in the flash stem nearer to the anode.

Still, current registration by means of small measuring sections on the
cathode can provide valuable information. By placing the sections at different distances from the gap axis, one can count the streamers 'by the
piece' and, in particular, evaluate the frequency of their occurrence during
a leader process. The result obtained is useful for the theory and simulation of long leaders. But of primary interest is the possibility to answer
the key questions in spark discharge physics—whether a long streamer possesses any essential conductivity and whether it is high enough to transform directly a streamer channel to a spark. For this, the absolute values of
parameters are not as important as the tendency for the current to change,
after the gap has been overlapped by a streamer.

Two situations are possible. If the older streamer portions have lost
nearly all their conductivity over the time of the streamer flight across
the gap (it is then an ionization wave trace rather than a channel), a
short current spike associated with the head charge neutralization can be
registered following the contact with the measuring electrode section. Then
the current drops to zero. This can be clearly seen in Figure 5.12(a) showing
an oscillogram of charge transported to the measuring section. After a
short spike, the charge value is stabilized and remains constant during
further registration. Clearly, a current spike does not propagate along the
channel, so it cannot contain quantitative information on its conductivity.
Of interest, however, is the charge amplitude—it must be of the order of
the streamer head charge, which cannot be measured, even roughly, by
other methods.

The current oscillogram for a channel with preserved conductivity was

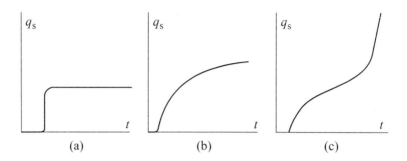

FIGURE 5.12
Typical oscillograms of charge transported onto the cathode by a single
streamer: the streamer has lost its contact with the anode (a); the streamer
conductivity decreases after the gap overlap (b); the conductivity increases (c).

described above. The charge oscillogram (current integral) has a smooth
front with a shorter, steep initial portion resulting from the neutralization
process. The growth rate of charge decreases [Figure 5.12(b)], as the channel conductivity becomes lower, or, on the contrary, increases in time [Fig-
ure 5.12(c)], if electron production is still active after the overlap. The latter course of events may sometimes lead to a streamer breakdown. The
oscillogram in Figure 5.13(a) was recorded during a breakdown in a 25 cm
pure argon gap. The voltage necessary for this was 16 kV (for comparison,
it is over 125 kV in air, but the breakdown would not be of the streamer
type all the same).

Charge measurements in single streamers moving in various gases have
been described in [5.2, 5.13, 5.20, 5.28]. The experiments used measuring
sections with radii as small as 0.5 mm. For current integration, the section was grounded through a storage capacitor. Charges over 10^{-11} C were
registered reliably. Both types of streamers were studied—from a corona
flash and from a leader head. Under normal conditions in air, the charge
pulse was always similar to that in Figures 5.12(a) and 5.13(b), rising for
about 0.1 μs and remaining nearly constant afterwards. This indicates
that a streamer that has overlapped a gap of several tens of centimeters in
length lost its galvanic connection with the anode on the way; therefore,
no conduction current flows in it. In the experiments mentioned, the current after the neutralization was less than 10 μA and the resistance per
unit channel length, $R_1 = E_c/i$, was over 5×10^8 $\Omega\,cm^{-1}$. The amount
of charge transported through the measuring section had a statistical deviation $(2-14) \times 10^{-10}$ C. The size of the section did not affect the results, if single streamers were identified reliably. The average charge value
of 5×10^{-10} C seems quite reasonable for a streamer head, whose field is

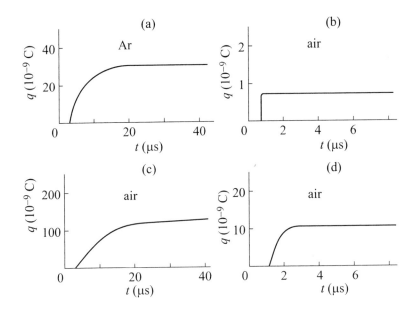

FIGURE 5.13
Oscillograms of the integral of conduction current through the cathode surface for the case of a single streamer: pure argon at $p = 1$ atm, $T = 290$ K (a); air at $p = 1$ atm, $T = 300$ K (b); air at $p = 1$ atm, $T = 900$ K (c); air at $p = 0.33$ atm, $T = 300$ K (d).

45 kV cm^{-1} for the head radius of 10^{-1} cm. The large deviation is most probably due to the radius variation caused by an extensive branching of streamers, whose head charge grows with the radius.

The situation in positive gases, like pure nitrogen and argon, in which electron loss processes are much slower [Figure 5.13(a)], is radically different. The gap current slowly decreases for several tens of microseconds, during which time the charge grows. The maximum amount of charge transported to the cathode is by two orders larger than in air, 5×10^{-8} C. This charge is, naturally, due not only to the head but mostly to the channel conduction current. Under the conditions shown in Figure 5.13(a), the streamer resistivity per unit length in argon at atmospheric pressure is 10^5 Ω cm^{-1} in a 25 cm gap, or at least by three orders less than in air.

The channel conductivity begins to increase, as soon as the electron loss in the streamer plasma slows down. But to understand the behavior of a long spark, one must take into account the heating effect. As the air gap temperature rises, the detachment is accelerated, while the electron-ion recombination rate decreases, because the proportion of complex positive ions (primarily O_4^+) becomes smaller. The electronegative properties of air are degenerated and it behaves as a positive gas. The charge oscillograms in Figure 5.13(b,c) show that, as the air is heated (at constant pressure), the charge pulse front elongates and at $T = 900$ K it becomes the same as in argon or nitrogen. The amplitude rises 200 times and becomes equal to 10^{-7} C. The average linear resistivity (measured in Ohms per centimeter) of a 0.5 m streamer at $500 < T < 1000$ K (1 μs after the gap overlap) is described by the empirical formula [5.20]:

$$R_1 \approx 1.0 \exp\left(9600/T[\mathrm{K}]\right) \; \Omega\,\mathrm{cm}^{-1}$$

It follows from here that a long streamer, practically nonconducting at normal air temperature, is capable of passing a current of about 1 A, when heated to 10^3 K, at the gap field of about 5 kV cm^{-1}.

When the gas is heated at constant pressure, its density decreases, retarding the electron attachment. To separate the effects of temperature and density, streamer currents were measured at normal temperature but lower pressure. Figure 5.13(b,d) shows characteristic oscillograms of the current integral on the cathode for a single streamer. The pulse rise time and amplitude grow with decreasing relative air density δ, but they remain an order of magnitude smaller than for the same value of δ provided by heating. This is important for analysis of the streamer-leader transition mechanism. Heating prevents a decrease in the plasma conductivity in this situation.

5.5   Numerical simulation

Numerical simulation of long streamers can be approached from different viewpoints, which will determine the requirements imposed on it. The researcher may not be satisfied with a simplified treatment, in which a streamer is artificially divided into a head and a channel, as was done in Chapter 3. There is a natural desire to consider the process of ionization wave propagation in a more comprehensive manner, taking into account all channel and head charges to evaluate the field in the wave front region, which, in turn, involves the calculation of charge distribution over the channel length. But we hardly need a cumbersome simulation to understand the ionization wave behavior. It suffices to evaluate correctly the ionization rate at the front, the electron drift exposing space charges in the head and in different channel cross sections, to calculate the resulting field, and, in case of a long-living streamer, to allow for the electron losses in recombination and attachment reactions.

Quite different should a model be, if it has to make up for a lack of experimental data. The limited potentialities of experimental studies are clear from the foregoing. They normally give only a small set of external streamer characteristics, providing no information about the plasma parameters, the distribution and dynamics of electric field in the channel, let alone in the head, or about the contribution of particular elementary events to the streamer propagation. The main task of a numerical simulation, therefore, becomes a quantitative description of discharge propagation.

The scarce data available must be used for the simulation check-up, since agreement between experiment and theory provides some grounds for confidence in the conclusions drawn exclusively from the simulation. Clearly, a numerical model must be as comprehensive as possible. Coincidence of theoretical and experimental values, usually in a narrow range, is by no means always an indication of the simulation validity. But the opposite statement is nearly always true. Therefore, if a simulation yields values and relations radically different from those obtained in experiment, it should hardly be regarded as reliable.

When designing a numerical experiment, it is not always possible to foresee the significance of this or that elementary event: many of them are manifested differently in varying electric fields. But, otherwise, field estimation may become simply impossible. So one sometimes has to consider the whole sequence of events, often with many unknowns, in order to understand the role of a particular process involved. For this reason, the most complete kinetic scheme available is preferable, but the final choice is made during numerical experiments, when the elementary events are ruled out, one by one, to register the response of the results to this procedure. Such an 'excessive' approach is quite commonly used in simulations, not only with respect to kinetics but also to particle densities, space charge, and electric fields.

Finally, a simulation may involve the computation of some engineering parameters, such as the electrical strength necessary for a discharge to overlap a gap, the charge incorporated into the gap by a branching flash, and so on. The requirements on engineering computations are very high, and the desired result can rarely be derived from an idealized model of a channel, usually straight and unbranched. Methodologically, it is reasonable to consider two basically different approaches to the solution of engineering tasks. One is associated with constructing a 'straightforward' streamer model, which is ideologically quite simple. A discharge starts with a single electron, which, by gaining energy from a known electric field, collides elastically and inelastically to produce new excited electrons and ions. Modern computers are capable of simulating statistically all possible elementary events responsible for the behavior of an electron and many generations of secondary particles, allowing one to dream about a comprehensive statistical model. Like a laboratory experiment, a numerical simulation would produce, in each run, a new streamer with an unpredictable trajectory, its own branching pattern, plasma parameters, and field variation. This should be a three-dimensional simulation, which would only require, as an input, the initial field distribution, the probabilities of electron occurrence at a given point in space, and elastic and inelastic collision cross sections. It would be applicable to research tasks and applications. But today, it is only a dream; in reality, we have to adjust ourselves to the capabilities of commercial personal computers.

Available models designed for applied tasks are semi-empirical in nature. They are based on a well calibrated and verified model of an ideal discharge element—a single streamer in our case. Such statistical parameters as the number of branches or their distribution throughout a gap are taken into account by means of averaged correction factors obtained by comparing experimental findings and calculations. Most of what we have said and will say here equally refers to the leader simulation.

5.5.1   The model dimensionality

The highest simulation standard accessible today is a two-dimensional (2D) simulation. A streamer is considered in a 2D-model as having a straight, axially symmetric structure. The computation yields the ionization wave velocity and shows the evolution of spatial distributions of all particle densities in an ionized gas, $n_k(x, r, t)$, and of electric field, $E(x, r, t)$. The requirements on time and space resolutions are very stringent: for the computation of fast events, the time steps must be about 10^{-12} s and the space steps in computing the field in the wave front vicinity about $10^{-3}-10^{-2}$ cm. Still, the amount of computations in a 2D-model is much less than in the hypothetical 3D-model mentioned above. The major simplification results from the use of a diffusion-drift approximation. It ignores the energy distribution of electrons, while drift velocities, the coefficients of diffusion, ionization frequency, excitation, attachment, recombination and others are taken to be known functions of the local field. Today, 2D-models can successfully describe the processes of avalanche-streamer transition and streamer elongation by $1-3$ cm [5.29–5.48]. Attempts to describe a long streamer in terms of a 2D-approximation have not been particularly successful because of the large amount of computations and low stability of computation algorithms. For this reason, so-called 1.5D-models are widely employed [5.49–5.56] in addition to 2D-models.

In the 1.5D-model, the channel radius is taken to be constant and *a priori* known. The plasma parameters are averaged over the channel cross section, and the longitudinal field is ascribed to the whole cross section, while the radial field is neglected. All radial processes unrelated to the field, for example diffusion, are ignored. As a result, particle densities and electric field become only dependent on the axial x-coordinate and time. In fact, the value of the channel radius is only used to calculate space charge, field and current (this restriction has given the model its name—1.5D-model). The reduced dimensionality decreases the amount of computations by an order or more, because one employs available formulas instead of solving Poisson's equation.

Although the 1.5D-model is fairly useful for finding distributions of charge, current, longitudinal field, and densities of charged and neutral plasma components, as well as for the understanding of the role of attachment, recombination, conversion and other processes, it has an evident disadvantage—the channel radius r_c is to be given artificially. But the radius is closely related to streamer velocity and head plasma density: from equation (3.9), $v_s \sim n_c r_m$; $r_m = r_c$. It also determines the maximum field in the head region from its potential, $U_t \sim r_m E_m$, and, hence, defines the ionization rate. Clearly, the ambiguity in r_m will introduce ambiguities into many results through a chain of cause-effect relationships. Of course, by varying r_c as an arbitrary characteristic, one can achieve an optimal fit to experiment. This is a tempting but insidious way, because a coincidence in details does not always mean a complete agreement. This procedure is sometimes employed, and it can provide some useful information, which, however, should be treated as qualitative rather than quantitative.

5.5.2 The set of equations

Both the 2D- and 1.5D-models use the same initial set of equations described in Chapter 2. It includes the continuity equations of the type of equation (2.24) for electrons, various positive and negative ions, electron-excited molecules (atoms); the energy balance equation for gas temperature; and Poisson's equation for electric field. In the simplest model involving only one sort of positive and negative ions and excited particles, these equations have the form:

$$\frac{\partial n_e}{\partial t} + \operatorname{div}\left(n_e \boldsymbol{v}_e\right) = \left(k_i N + k_i^* n^*\right) n_e - \left(k_a' + k_a'' N\right) N_a n_e$$
$$+ \left(k_d N + k_d^* n^*\right) n_- - \beta_{ei} n_+ n_e + S \qquad (5.9)$$

$$\frac{\partial n_+}{\partial t} + \operatorname{div}\left(n_+ \boldsymbol{v}_+\right) = \left(k_i N + k_i^* n^*\right) n_e - \beta_{ei} n_+ n_e$$
$$- \beta_{ii} n_- n_+ + S \qquad (5.10)$$

$$\frac{\partial n_-}{\partial t} + \operatorname{div}\left(n_- \boldsymbol{v}_-\right) = \left(k_a' + k_a'' N\right) N_a n_e - \left(k_d N + k_d^* n^*\right) n_-$$
$$- \beta_{ii} n_- n_+ \qquad (5.11)$$

$$\frac{\partial n^*}{\partial t} = k^* N n_e - k_i^* n^* n_e - k_q N n^* \tag{5.12}$$

$$\boldsymbol{v}_k = \boldsymbol{v}_{d\,k} - D_k \nabla \ln n_k \tag{5.13}$$

$$\triangle \varphi = e\left(n_e + n_- - n_+\right)\varepsilon_0 \qquad \boldsymbol{E} = -\nabla \varphi \tag{5.14}$$

Here, N, N_a, n_e, n_+, n_-, n^* are densities of all neutrals, negative components of the gas, electrons, positive and negative ions, and electron-excited particles, respectively; $\boldsymbol{v}_k\left(E\right)$ and $\boldsymbol{v}_{d\,k}\left(E\right)$ are vectors of average and drift velocities of particles of the k-th kind, D_k is their diffusion coefficient; k_i, k_i^* are ionization rate constants of unexcited and excited molecules; k_a', k_a'' are the rate constants of dissociative and three-body attachment; k_d, k_d^* are the rate constants of negative ion disintegration by unexcited and electron-excited particles; k^*, k_q are the rate constants of excitation and quenching of electronic states; β_{ei}, β_{ii} are electron-ion and ion-ion recombination coefficients; S is the source of photoionization; φ is electric field potential. If account is taken of various ions produced in the gas (in air, simple O_2^+, N_2^+ and complex O_4^+, N_4^+, $O_2N_2^+$ ions), specific balance equations are written allowing for the ion conversion. One can also include the balance equations for other particles, for example, for oxygen atoms rapidly destroying negative ions in air.

Without the account of thermal expansion and radial heat removal over the time of the streamer propagation, the energy balance equation looks as

$$c_v N \frac{\partial T}{\partial t} = \left(1 - \xi_V\right) jE + Q_{VT} + Q_{eT} \tag{5.15}$$

where j and E are the current density and the longitudinal channel field; $1 - \xi_V$ is the portion of electron energy expended for the translational and rotational degrees of freedom of molecules; c_v is the respective heat at constant volume; and Q_{VT} and Q_{eT} are the heat input from deactivation of vibrational and electronic states. The vibrational energy density of molecules, ε_V, affecting Q_{VT} and vibration temperature, which, in turn, determines the rates of some elementary processes, is described by the relaxation equation:

$$\frac{\partial \varepsilon_V}{\partial t} = \xi_V jE - Q_{VT} \qquad Q_{VT} = \frac{\varepsilon_V - \varepsilon_V\left(T\right)}{\tau_{VT}\left(T\right)} \tag{5.16}$$

where ξ_V is the portion of electron energy expended for vibration excitations, $\varepsilon_V\left(T\right)$ is equilibrium vibration energy, and $\tau_{VT}\left(T\right)$ is the VT relaxation time.

The set of equations (5.9)–(5.16) is closed and can equally describe processes occurring in an ionization wave and a streamer channel. Moreover, the artificial subdivision of a streamer into a channel and a head made for the sake of simplifying the ionization wave problem is hardly justifiable in a numerical experiment, which treats a streamer as a unified system. Thereby, the simulation of nonstationary modes in, say, the initial channel near the anode is simplified, when ionization processes can go on far behind the head.

5.5.3 Charge and field computations

Field computations take up most of the computation time, since they have to be performed for each section along the streamer at each time step. The accuracy of field computations determines that of all basic streamer parameters, in particular, of the ionization wave. In the most important propagation modes, the ionization frequency ν_i is strongly field-dependent, $\nu_i \sim E^k$, $k \approx 2-3$; so, the field in the ionization region should be computed as accurately as possible. Since the space resolution of the model must be able to reconstruct the near-head field, the computation step must be smaller than the channel radius.

The 2D-model always deals with a straightforward solution of Poisson's equation, and there is nothing much to discuss except the procedure to be used. The straightforward solution is too cumbersome to be applied to a long streamer simulation. Some computation time can be saved by using the iteration method employed in [5.35], but in any case, the search for field values in the 2D-model is a most sophisticated task. It can, however, be simplified by modeling a streamer with a channel of constant radius. Computation in the 1.5D-model is based on the account of electrode charges and linear charge density $\tau(x)$. What remains to be done is to find the charge distribution over a channel cross section. Many authors consider it to be uniform [5.49–5.51, 5.56]. Charge density $\rho = \tau/\pi r_c^2$ is taken to be constant within a small section Δx along the axis. The field of a uniformly charged disk of radius r and length Δx at point x counted from the far base is defined as

$$\Delta E\left(x\right) = \frac{\rho}{2\varepsilon_0}\left\{\left[\left(x-\Delta x\right)^2 + r_c^2\right]^{1/2} - \left(x^2 + r_c^2\right)^{1/2} + \Delta x\right\} \qquad (5.17)$$

The total field at this point is given by an algebraic sum of fields of all charges of elementary disks and electrodes, including those induced by electrostatic induction. This approach is known as the disk method.

Two circumstances make us question the plausibility of formula (5.17). First, a uniform charge distribution over a channel cross section contradicts the actual state of things. Charge is concentrated at the plasma surface rather than in its bulk. Indeed, space charge dissipates for the time of the order of Maxwellian time $\tau_M = \varepsilon_0/\sigma$; together with it does the radial field vanish in the plasma bulk. At $v_s \sim 10^9 \text{ cm s}^{-1}$, $n_c \sim 10^{14} \text{ cm}^{-3}$, and $\sigma \sim 4 \times 10^{-4} \, \Omega^{-1} \text{cm}^{-1}$, this happens so fast ($\tau_M \approx 0.2$ ns) that the streamer can elongate only by 0.2 cm, a distance of the same order of magnitude as its radius. The imperfection of formula (5.17) can be corrected by substituting space charge ρ by the equivalent surface charge $\tau = \pi r_c^2 \rho$. As a result, the field of an elementary ring, rather than disk, will become equal to

$$\Delta E(x) = \frac{\tau}{4\pi\varepsilon_0} \left\{ \left[(x - \Delta x)^2 + r_c^2\right]^{-1/2} - \left(x^2 + r_c^2\right)^{-1/2} \right\} \qquad (5.18)$$

For $x \gg r_c$, both formulas yield a nearly identical result.

Similar are the computed streamer characteristics. What is important, however, is the fact that the stability of the computation algorithm with surface charge distribution is much higher than with space charge. Other conditions being equal, this permits larger axial dimensions of a channel to be calculated, which is especially important for the simulation of long streamers.

Note another circumstance that follows from the simulation practice. It is always preferable to compute field strengths, rather than boundary potentials φ_1 and φ_2, which might seem to be more suitable parameters for average field evaluation. Average field values could be used to compute the rate constants for elementary processes, since the accuracy requirements here are not very high. But the use of average field to evaluate the charge gain per unit channel length is harmful because of an unacceptably large error.

Charge variation is computed from the difference in electron fluxes (primarily of drift nature) through the boundaries of the channel section to be computed. It is the field at the boundaries that is essential for their exact evaluation. The substitution of the field at a point by the average field over a channel section naturally leads to an error in the flux difference and, hence, in the charge gain. The relative value of the latter error is by a factor of $n_e/\Delta n$ (where $\Delta n = n_+ - n_e$) larger than the former; the difference is as large as two orders of magnitude for strong ionization waves with a nearly neutral plasma (Section 3.1.2).

Numerical experiments have shown that such errors tend to grow, especially in the disk method, so that the model stability is disturbed. This disadvantage can be easily removed by using formula (5.18) for boundary field computation, when one wants to find the electron fluxes. One then deals with true electron drift velocities at the boundaries rather than with velocities averaged over a channel section. As a rule, ion motion does not have to be taken into account.

Difficulties encountered in space charge computations become more serious in 2D-models because of the necessity to compute the radial field and consider electron motion not only in the axial but also in the radial direction. The solution is sought for by splitting the space coordinates to minimize the numerical dissipation, which distorts the results [5.57].

Naturally, the computational grid should not be uniform. The cell density must grow toward the wave front, where the field variation in space is especially dramatic. This leads to as fast, or even faster, density variation of various plasma components. As the wave propagates, the denser part of the computational grid must move together with it. Far behind the front, the cell size may be increased many-fold, since all parameter variations along the channel occur much slower than in the head.

The larger size of grid steps requires an additional computation time for readdressing and redistributing the cell information; so, these operations should not be excessive. In any case, the redistribution procedures should be handled with as much care as electron flux computations: even a minor disturbance in the charge balance resulting from shifting the cell boundaries may become a hazard, if it has accumulated in repeated operations. The plasma parameters near the streamer starting point also vary faster than in the rest of the channel. Therefore, the cell size here must be kept small, of the order of the channel radius, over the whole computation period.

5.5.4 Initial ionization

Initial ionization is a stumbling block in computations, nearly as serious as in theory. Clearly, a streamer cannot propagate without initial electrons before the wave front. But in laboratory experiments, we always observe a continuous movement of a streamer; therefore, it has sufficient electrons to propagate. The estimations made in Section 3.4 confirm this observation.

A rigorous simulation should, probably, involve photoionization and all other processes occurring in front of the streamer head, but this would make the problem too sophisticated, since the factual information available on photoevents is extremely inadequate. The main fact is that the initial electron density n_0 before the wave front is by many orders less than the final density n_c behind the front. In the theoretical formulas for streamer parameters (Section 3.1.2), the quantity n_0 was only used in the logarithmic ratio n_c/n_0.

The effect of the initial electron density in numerical experiments is as insignificant as in theory: the value of n_0 must be increased by 2 or 3 orders of magnitude to change the calculated streamer velocity by 10%. In most long-streamer models, in which the description of the avalanche-streamer transition is unimportant, the value of n_0 given in space outside the streamer is of the order of 10^7–10^8 cm^{-3} and for strong fields even 10^9 cm^{-3}. One cannot make a large error here. Photoionization in a streamer process operates in an on-off mode, and there is no need to worry, if it goes in the right direction.

5.5.5   Results from simple models

By simple models we mean 1.5D-models with fixed channel radii. They can give us an idea of the process onset and acceleration. A streamer starts in the external electrode field, which is not yet enhanced by the channel charges. The gap voltage does not directly affect the conditions of the start but will be manifested later during the streamer development, when its drop becomes substantial. The initial streamer velocity is determined exclusively by field strength, E_a, on the electrode (as above, we assume it to be the anode). Streamers start identically from anodes of different radii at different voltages but constant E_a. With distance from the anode, the ionization wave is affected less and less by the external field, but, at the same time, the streamer charge grows, leading to a rapid rise of the total field at the wave front. It may exceed many times the external field, if the anode radius r_a is large enough [Figure 5.14(a)]. Usually, the total field at the wave front reaches its maximum, $E_{m\ max}$, when the streamer length becomes close to r_a. At this length, the charges of the head and adjacent regions are no longer affected by the anode charge as before, and the voltage drop in the channel is still insignificant.

Increase in E_m is accompanied by an increase in the streamer velocity v_s and final electron density n_c behind the wave front, both parameters reaching their maxima at about the same moment as E_m [Figure 5.14(a)]. At constant voltage, the streamer continuously slows down; this deceleration is enhanced as the voltage in the channel decreases. As more electrons are lost to the plasma, the conductivity becomes lower, still enhancing the retardation. At velocities less than $v_{s\ min}$, the channel practically stops its elongation. Computations for atmospheric air and argon give the minimum velocities of about 10^7 cm s^{-1} and 10^6 cm s^{-1}, respectively, which is in good agreement with experiment (Section 5.2). Note that both minimum and maximum velocities are nearly independent of the initial channel radius. Therefore, unlike many other parameters, the computed values of $v_{s\ min}$ and $v_{s\ max}$ are reliable.

Consider the initial channel section at the anode. Some of its features were discussed above in connection with current measurement. When the external field E_a does not much exceed the ionization threshold E_i in air, the ionization wave front leaves behind an extremely low electron density. The longitudinal field is not driven out of the channel, and the ionization goes on, although the wave front moves on through the gap. The computations are illustrated in Figure 5.14(b). The ionization lasted for 35 ns,

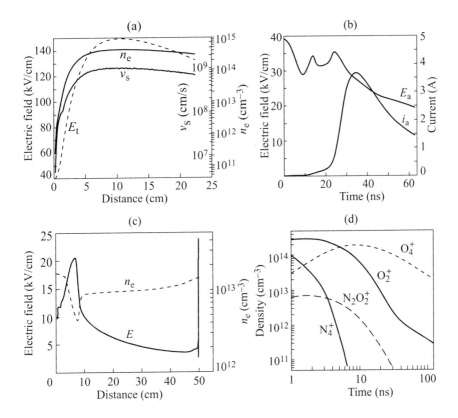

FIGURE 5.14
Calculations for a streamer of 0.1 cm radius starting from a spherical anode
($r_a = 12.5$ cm) in air at $p = 1$ atm, $T = 300$ K, and 500 kV rectangular pulse
amplitude: the velocity, head field, and plasma density behind the head versus
the channel length during initial acceleration (a); channel current and field near
the anode (b); field and plasma density distributions along the channel (c); the
composition of positive ions in the channel at a distance of 12.7 cm from the
anode (d).

and the current through the anode surface increased for nearly the same
time of 33 ns. This picture is similar to that obtained from current measurements. We should confess that the approach to the experimental data
analysis was found after we had simulated the initial streamer to reveal the
plasma density variation.

The region of enhanced field at the anode turns out to be steady. It
somewhat shifts toward the anode but persists for hundreds of nanoseconds,
affecting appreciably many elementary events. The longitudinal field in the
channel is much lower than at the anode, the more so in the wave front

region. Analytical estimation of the channel field behind the wave has yielded several kilovolts per centimeter for atmospheric air. A similar result has been obtained from a numerical simulation. The field near the anode first decreases slowly toward the head, but having passed the minimum, it goes rapidly up to the wave front value [Figure 5.14(c)].

Except for a small portion before the ionization wave, where n_e dramatically changes along the axis, $n_e(x)$ of a long streamer approximately corresponds to constant conduction current ($n_e \sim E_c^{-1}$). One can also identify an accelerated streamer section near the anode with a rapid change in the electron density [Figure 5.14(c)]. In the rest of the streamer, it is much leveled owing to two factors: (i) with distance from the anode, the head potential drops together with maximum field in the ionization region and maximum electron density; (ii) on the other hand, the electron losses due to recombination and attachment are not as great as in younger channel portions.

The streamer current in the cross section behind the head is determined by the wave front velocity: it grows with the streamer acceleration and decreases with its retardation. The current distribution along the x-axis strongly depends on the voltage pulse shape. At constant voltage and streamer velocity, the current distribution is almost uniform, except for the wave front region, where there is always a high current peak (Section 3.2.2). During a rapid voltage rise, the current increases toward the streamer head but then drops, as the voltage decreases.

The streamer dynamics analyzed in terms of 1.5D-simulation data remains basically the same for any reasonably given value of r_c. As for the details and numerical results, most of them strongly depend on r_c. One should keep this in mind, when analyzing the simulation data. The results concerning plasma parameters are the most ambiguous ones. The radius increase from 0.01 to 0.1 cm reduces the value of maximum electron density by over an order. This causes a nearly two-fold reduction in the path length, before the streamer entirely stops, and nearly as great increase in the average streamer field. The maximum field in the wave front region becomes as low as one fourth of the previous value, while the amplitude of conduction current through the anode doubles. It is quite clear that a 1.5D-model is of little value for obtaining quantitative results and of still smaller value for engineering applications. Such a model remains useful only for a qualitative description of functional relationships among some external streamer characteristics, the parameters of elementary processes and those of the voltage pulse applied to the gap. As for applied voltage, a 1.5D-model can provide practically all relationships known from experiments.

5.5.6 Requirements on a kinetic 1.5D-model

A 1.5D-model can be used to advantage to find criteria for the streamer kinetics and for reaction selection. Quantitative accuracy is not very important in this case; one should rather identify and understand the general tendencies. We will leave aside pure monatomic (inert) gases, because the possible set of elementary events in them is so small that it can hardly be simplified further. Molecular gases are more complicated, especially a gas mixture like air. An oxygen-nitrogen mixture may involve over 250 reactions [5.58], which cannot all be represented in a model, especially if many of the rate constants are poorly known. On the other hand, it is impossible to make *a priori* a rigorous selection of, say, one 'major' kind of ions to be introduced into the model as a 'permanent' parameter. The ionic and molecular composition of a gas may show a great variation in time, different in different channel cross sections. This fact will be illustrated with reference to positive simple (O_2^+, N_2^+) and complex (O_4^+, N_4^+, $N_2O_2^+$) ions in air. This is important, because the coefficients of electron recombination with various ions differ by about one order. The results below were borrowed from the computation of a streamer of 0.1 cm radius propagating in atmospheric air from an anode of $r_a = 12.5$ cm at applied voltage 500 kV [5.54].

The plasma density at the anode grows slowly (over 30 ns); as slow is the parallel increase in the densities of all sorts of ions. Most of these are positive oxygen ions, whose abundance is at least an order of magnitude larger. When the ionization at the anode ceases, N_2^+ ions disappear in conversion reactions for a few nanoseconds (the O_2 ionization potential of 12.2 eV is lower than the N_2 potential of 15.6 eV); this is profitable, because some energy is saved. Following N_2^+ ions, N_4^+ ions disappear too; complex $N_2O_2^+$ ions live longer but their proportion is less than 1%. The fate of nitrogen ions far from the anode is generally the same; only their traces can be recorded 10 ns after their production [Figure 5.14(d)].

Long-living O_2^+ and O_4^+ ions behave differently in the anode region and in the rest of the channel. Both kinds of ions are present at the anode in almost equal proportions for a long time. In about 50 ns, the amount of O_4^+ becomes three times over that of O_2^+, and this proportion remains almost constant. Outside this region, the events go on faster: 5 ns after a new channel section has been formed, the density of O_4^+ ions rises to that of O_2^+ ions; 10−20 ns later, the amount of O_4^+ ions increases by one order of magnitude. The O_4^+-to-O_2^+ ratio continues to grow with time, eventually exceeding the initial proportion by two orders of magnitude [Figure 5.14(d)].

It is quite tempting to simplify this model so as to exclude the conversion of O_2^+ to O_4^+ ions from the equations and to ascribe the electron-ion recombination coefficient for O_4^+ to all oxygen ions. This will hardly affect

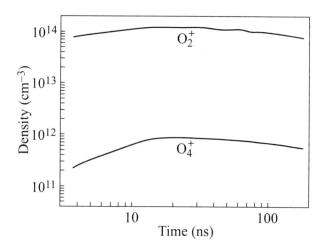

FIGURE 5.15
Variation in the positive ion composition in an air streamer at $T = 900$ K and $p = 1$ atm at a distance of 5.1 cm from the anode ($r_a = 1$ cm); the streamer radius 0.03 cm; the rectangular pulse amplitude 34.8 kV.

such channel characteristics as velocity, final length, and average field. But then, the description of the initial channel may not be quite adequate. The deceleration of electron recombination may not manifest itself, so that the computation may not show the actual effect of a higher local plasma density at the base. This effect can be neglected in long streamers but should be taken into account in the analysis of leader production. This process takes place at the bases of streamers, which start from the anode and then are triggered by the leader head (see Chapter 6).

Much care should be taken in handling a model with simplified kinetics, even though one ignores such details as the streamer behavior at the anode. Each time one wants to make computations in a new set of conditions, one has to check preliminarily if the assumptions made are acceptable. Suppose a model allows for only one sort of ions, O_4^+ ions, and is used for the computation of a streamer in air of reduced density. The density can be reduced by decreasing the pressure at fixed temperature or by heating the gas at fixed pressure. The streamer conductivities in the two situations will be different (Section 5.4.5). Computations have shown that an adaptive model (with O_4^+ ions) for cold air of reduced pressure does not lead to a major error, because the amount of simple O_2^+ ions is smaller than that of complex O_4^+ ions for most of the time. For hot air, however, the results are hopelessly faulty and disagree dramatically with experiment. Heating changes radically the plasma ion composition: the proportion of O_4^+ ions is less than 1%, and the plasma mostly contains O_2^+ ions, which we were just going to exclude from the model (Figure 5.15).

Therefore, the most reasonable way of modeling such complicated phenomena as a long streamer in air is to test various kinetic schemes in a 1.5D-model, to select the most significant components and reactions, and only after that should one employ a cumbersome 2D-model.

5.5.7 Recombination and attachment in cold air

In spite of many limitations, a 1.5D-model allows a reliable assessment of a particular reaction in the streamer process. A key problem in our understanding of spark discharge in air is the mechanism of conductivity loss in a long streamer. This problem may be solved as follows. In perfectly dry air, with which we will deal here, the dissociative recombination of complex ions $O_4^+ + e \rightarrow O_2 + O_2$ was shown above to be most significant among electron-ion recombination reactions. Let us compare the role of this mechanism of electron loss and that of attachment. Three-body attachment (Section 2.3)

$$O_2 + e + O_2 \rightarrow O_2^- + O_2$$

is the most important process in dry air of normal density for a streamer field less than $3-7 \, \text{kV cm}^{-1}$.

Figure 5.16(a) presents the simulation results for a streamer that has started from a spherical anode of 1 cm radius at a constant gap voltage of 105 kV. One can see the plasma state at 5.1 cm from the anode (beyond the acceleration region); the time is counted from the moment of the ionization wave arrival at this point. The electron density decreases by at least one order, as compared to the density n_c right behind the wave, primarily due to the recombination with O_4^+ ions produced by O_2^+ ions. Then the attachment gradually takes over. The streamer may be said to be devoid of the rest of its conductivity by the attachment, but it is the recombination that primarily limits the energy input into the channel. This conclusion is essentially different from the commonly accepted view that attachment is the principal mechanism of electron loss in air. Of course, the computations vary with the radius (as much as 3.5-fold for the electron density at 0.03 and 0.1 cm radii), but the conclusion about the major role of recombination still holds.

5.5.8 A cathode-directed streamer in hot air

Simulation of a streamer in hot air is of interest for two reasons. First, the computations can be compared with available measurements made in sim-

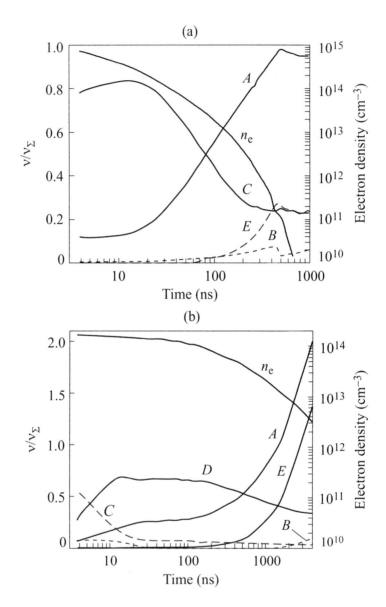

FIGURE 5.16
Relative contributions of three-body (A) and dissociative (B) attachment, of
electron recombination with complex (C) and simple (D) ions, and of
detachment (E) to the variation rate of electron density n_e in a streamer at
5.1 cm from a spherical anode ($r_a = 1$ cm). The streamer (radius 0.03 cm)
propagates through a sphere-plane air gap of 20 cm long at $p = 1$ atm:
(a) $T = 300$ K, $U_0 = 104$ kV; (b) $T = 900$ K, $U_0 = 34.8$ kV.

ilar conditions (Section 5.4.5). Second, elementary events responsible for the variation in the streamer parameters at increased temperature can be revealed. Both issues are important for treatments of the leader inception and development. According to a popular hypothesis [5.6], the streamer-leader transition is based on electron liberation in negative ion decay, whose rate abruptly rises if the air is heated even to 1000 K (see Section 6.9). The conductivity of the heated streamer section persists at a high level for a longer time. More power is contributed to it to produce greater heating, until a new leader section is formed. It was this hypothesis that gave impetus to laboratory experiments with hot air streamers. It might seem that the results obtained do not contradict the hypothesis.

Let us now look at the streamer channel. There are no elementary processes in the head other than ionization and excitation of molecules, since a given volume of air stays for too short a time in the strong field of the wave front. By varying the gas temperature T at constant pressure $p \sim NT$, the streamer velocity and plasma density behind the wave front can also be varied because of the changing gas density N, since the electron drift velocity and ionization rate constant are the functions of E/N rather than just E. No other manifestations of heating are to be expected in the ionization wave. The channel with its numerous slow processes involving heavy particles and varying with T is quite another matter. With this in mind, consider the simulation results for a streamer in air at $T < 10^3$ K.

In addition to the negative ion decay due to the interaction with excited oxygen atoms and molecules, the kinetic model has taken account of the temperature variation in the rate constants of the following reactions [5.54, 5.55]: the dissociative electron-ion recombination involving simple O_2 and N_2 ions, the ion-ion recombination and ion conversion, the formation of the first metastable state of $N_2 \left(A^3 \Sigma_u^+ \right)$ molecules, the dissociative attachment

$$O_2 + e \to O^- + O$$

in a weak channel field, and the three-body attachment

$$O_2 + e + M \to O_2^- + M$$

Conversion is worth a special note, because it can substantially affect the course of events. With increasing temperature, the rate constant of the reaction

$$O_2^+ + O_2 + M \to O_4^+ + M$$

falls, while that of the reverse process rises (Section 2.3). Heating influences other complex ions in much the same way. They are produced in smaller amounts in heated air and immediately decompose into simple ions on further heating.

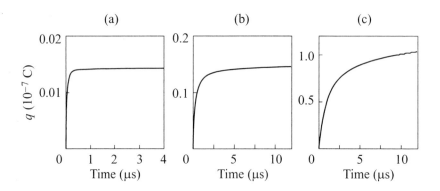

FIGURE 5.17
Calculated charge transported to the cathode by an air streamer at $T = 300$ K; the relative density $\delta = 1$ (a); $\delta = 0.33$ (b); $\delta = 0.33$ and $T = 900$ K (c). In all calculations, the average reduced field in the gap was taken to be $E/N = 2.08 \times 10^{-16}$ V cm^2 ($U_0 = 104$ kV at $\delta = 1$, and $U_0 = 34.8$ kV at $\delta = 0.33$).

Computations were made in a 1.5D-model for a sphere-plane gap of 20 cm long ($r_a = 1$ cm) at the streamer radius of 0.03 cm [5.59]. The latter value was chosen because, under normal atmospheric conditions, the streamer average field in the overlapped gap was found to be $E_{av} = 5.2$ kV cm^{-1}, which is 10% larger than in the experiment. This created, at least, one point of support for data comparison. As in the experiment, a comparison was made of the results for heated air at normal pressure and those for the same gas density attained by reducing pressure at room temperature. The computed and experimental relationships between the average field in an overlapped gap and relative air density δ are compared in Figure 5.5. The computed field decreases faster on air heating than on pressure reduction, as in experiment. The relative discrepancy between the computed and experimental curves for $E_{av}(\delta)$ corresponding to varying T at $p = $ const and to p at $T = $ const has been shown to be similar. Moreover, the values of E_{av} differ less than by 30–35%. Similar is the agreement for the charge (current integral) transported through the cathode (Figure 5.17). In air heated to 900 K, appreciable current flows through the streamer contact with the cathode, transporting a charge of 10^{-7} C for 10^{-5} s, in cold air of the same density 10^{-8} C for 10^{-6} s, and in normal atmospheric conditions approximately 10^{-9} C for 10^{-7} s. Similar results have been obtained from experiments (Section 5.4.5). Therefore, the principal features of the phenomenon concerned have been reproduced correctly by the model, which should be considered as an adequate tool for interpretations.

The processes of electron loss in air at room temperature were discussed

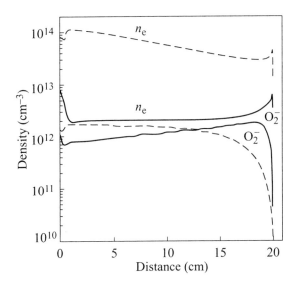

FIGURE 5.18
Distributions of electron and negative ion densities in an air streamer at the
moment of contact with the cathode at $p = 1$ atm (for conditions, see
Figure 5.16). Solid lines, $T = 300$ K; dashed lines, $T = 900$ K.

in Section 5.5.7: the electron density first drops by an order of magnitude
due to recombination with O_4^+ ions, then to attachment, which completely
damages the streamer plasma. These processes are very fast. When a
streamer reaches the cathode 250 ns after the start, the electron density in
its older sections drops by two orders [Figure 5.16(a)].

A hot air streamer retains much of its initial conductivity on arrival at
the cathode [Figure 5.16(b)]. The reason for this is not the compensation of
electron loss by their liberation from negative ions, as is commonly believed
by those following the hypothesis above [5.6]. The point is that by the
moment of contact with the cathode, electrons have just not been produced
in an adequate amount, so that their density is nearly two orders less than
along the channel (Figure 5.18). The conductivity is preserved due to a
slow recombination of electrons because of the absence of O_4^+ complexes,
which cannot be produced in large quantities at high temperature.

Thus, the maximum effect the detachment can produce is the liberation
of electrons from all negative ions. But if the amount of ions is small, the
number of liberated electrons is small too. So the reduction in conductivity
in cold air and its maintenance in hot air are primarily due to the action of
recombination. In cold air, the recombination with O_4^+ abruptly decreases
the electron density, while in hot air the electron loss is strongly retarded

by the absence of O_4^+. The attachment and detachment begin to dominate over the recombination much later.

5.6 The ionization expansion model

5.6.1 Computed channel geometry

The 1.5D-model above can be somewhat generalized to bring it closer to 2D-simulation by introducing the ionization-induced expansion of the streamer channel. The principal parameter necessary for this—linear charge density $\tau(x)$—has been defined in terms of the 1.5D-model. What remains to be done is to calculate the radial field E_r at the outer channel surface and, taking advantage of the set of equations (5.9)–(5.16), to evaluate the radial characteristics of a propagating ionization wave. If the field E_r is strong, a wave propagates without obstacles, because initial electrons are produced at the streamer lateral surface, too. Most photons emitted with a time delay by molecules excited in the head leave the channel through the lateral surface rather than in the longitudinal direction. Moreover, photons are also produced in the radial wave itself.

If the distribution $\tau(x)$ is known, the radial field at the point (x, r) can be obtained by summing the fields created by charges in the channel sections with the linear charge density τ taken to be constant:

$$\Delta E_r(r) = \frac{\tau}{4\pi\varepsilon_0 r} \left\{ \frac{x_2 - x}{\left[(x_2 - x)^2 + r^2\right]^{1/2}} - \frac{x_1 - x}{\left[(x_1 - x)^2 + r^2\right]^{1/2}} \right\} \quad (5.19)$$

Here, x_1, x_2, and x are the axial coordinates of the section boundaries and the point, respectively; r is the radial distance between the point and the axis. If $E(r, x, t)$ is known, we can find the position of the radial wave front, $r(x, t)$, and the densities of all plasma components in the new radial space: $\Delta r(x) = r(x, t + \Delta t) - r(x, t)$. The calculated densities are summed with those in the older channel cross section and averaged over the new radius $r(x, t + \Delta t)$. The latter operation is the only distinction from a true 2D-model. The ionization expansion may be assumed to be completed, when the electron density at the radial wave front drops to 5×10^{11} cm^{-3}.

A certain 'initial' streamer radius has to be introduced in the model in question. It may exist in the head region for a short time, but it does not affect the computations as much as the channel radius in a common 1.5D-model. The channel radius r_c varies along the axis at a given moment of time, being dependent on charge $\tau(x)$ that creates the radial field. The

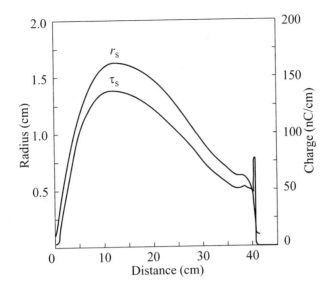

FIGURE 5.19
Linear charge distribution along a streamer, τ_s, and variation of its radius r_s with account of ionization expansion (anode radius 5 cm, initial streamer radius 0.1 cm, gap voltage 500 kV).

radius is shortest at the triggering electrode, since there is no radial field there. It grows monotonically with distance from the electrode to a maximum at $x \sim r_a$, where r_a is the electrode (anode) radius. After that, r_c first decreases slowly, then with acceleration (Figure 5.19). The channel geometry $r_c(x)$ is determined by two factors. The electrode charge repels the like charge of the channel, reducing it near the electrode; the reduction is greater for the larger electrode and for the channel cross section nearest to it. On the other hand, the potential $\varphi(x)$ and charge $\tau(x)$ in a channel with finite conductivity must decrease with distance from the electrode. Therefore, $\tau(x)$ and $r_c(x)$ reach their maximum values at a certain distance, so that a single streamer does not at all look like an ordinary cylinder but becomes pear-shaped (Figure 5.19) [5.55].

Clearly, the average radius of a single streamer increases with the gap voltage; it also varies with the electrode radius and pulse rise time at fixed voltage. The effect of the larger anode extends farther, so the maximum cross section is also located farther. At constant voltage, $\varphi(x)$ and $\tau(x) \sim \varphi(x)$ are closer to the electrode, and so is the radius. For example, at 500 kV, a streamer starting from an anode with $r_a = 5$ cm expands to a 1.5 cm radius; at $r_a = 12.5$ cm, the maximum radius is three times as small.

The effect of the pulse front is associated with the fact that a streamer initiated at a slow voltage rise has a low electron density in the channel and high voltage loss reducing the streamer charge. The dependence on the rise time t_f is only noticeable at $t_f < r_a/v_s$. At $r_a = 5$ cm, the maximum value of r_c is doubled, when t_f drops from 100 to 10 ns.

Thus, the results derived from a generalized 1.5D-model, together with some experimental data (Section 5.1.2), indicate an inadequate formulation of the problem, in which the streamer radius is considered as a constant quantity. In fact, it varies with the gap voltage, the size of the electrode having a strong field, and the field distribution in the gap. Radii like $r_c \sim 10^{-2}$ cm are valid only for short, low voltage streamers. The characteristic radius of a high voltage streamer is 1 cm. The idea of a long streamer being very thin has been greatly exaggerated. Nevertheless, the conventional cylindrical model remains fairly attractive because of its simplicity and applicability to many aspects of the streamer process.

5.6.2   The stem formation

It might seem that there is no connection between the channel expansion and the stem formation. The stem is formed at the same electrode, from which the streamer starts, and is located where the radial field is weak and the ionization expansion is practically absent. However, photographs show that the stem entirely envelopes the actually thin streamer channel inside it. Having excluded the radial field effect, one has to relate the stem formation to the longitudinal field. The fact that the field persists at the base for a long time and is high enough for ionization to occur follows from both the results obtained in the 1.5D-model (see above) and the current measurements discussed in Section 5.4.2. A model involving ionization expansion provides a still more convincing picture [5.60].

After the axial ionization wave front has left the near-anode region and the channel charge has become sufficiently large to induce the ionization expansion, the electron flux pushed by the field through the narrow neck at the channel base begins to grow. The current passes through its small cross section owing to the enhanced longitudinal field (Figure 5.20), which is strong enough to sustain the ionization. But if the streamer charge is unable to drive the external field out of the channel, it cannot do so in the neck vicinity either, because the strength here is slightly higher than in the channel (curve B in Figure 5.20). Initial electrons are provided by the ionization radiation from the existing channel. As a result, a kind of secondary electron shell with a radius of the order of the ionizing photon path length (10^{-1} cm in normal air [5.61]) is produced around the primary channel due to the avalanche ionization in the remaining anode field. The

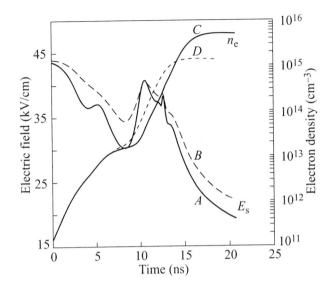

FIGURE 5.20
Longitudinal electric field in a streamer near an anode of $r_a = 12.5$ cm (A) and outside the streamer at $2r_s$ from its axis (B); the plasma density at the anode with ionization expansion (C) and without it (D) (rectangular voltage pulse amplitude 550 kV, initial streamer radius 0.03 cm).

ionization in the shell begins with a time delay, as compared to the central channel; so, its degree is lower. It is probably the external radius of the secondary shell which determines the stem boundary usually registered as a less bright region around the central channel.

If the bright stem core is considered as a trace of the primary, strongly ionized channel, conventional photographs can be used to measure the initial radius. It is then important to use an adequate exposure to make the core stand out clearly against its shell. High quality photographs were presented in [5.1] together with experimental details. They can be used to simulate the streamer development in terms of the ionization expansion model. The computations given in Figure 5.21 show that the axial length of the actively expanding portion at the anode is about 2 cm and the maximum radius is 0.15 cm. These values fit the experimental findings very well.

The expansion model generally provides correct values not only for the stem length but also for its velocity. The stem tip travels through a gap together with the longitudinal field maximum in the near-electrode region. Computations show such a displacement from a 12.5 cm anode to be 5 cm with an average velocity of 10^8 cm s^{-1}, while experiments give 7 cm and 1.2×10^8 cm s^{-1}, respectively, for the same conditions [5.11].

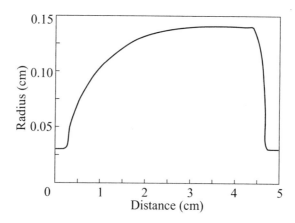

FIGURE 5.21
The calculated radius of the stem core for the experimental conditions of [5.1]
(anode radius 3.1 cm, voltage 140 kV).

5.6.3   Current through the anode

Limitations of the 1.5D-model become quite evident in evaluating streamer
current. The computed values are 3 or 4 times smaller than the experimental ones [5.53]. The computations presented in Section 5.4.2 show that the
surface of a perfectly conducting streamer cannot, in principle, hold the
measured charge, unless r_c is be very large, about 1 cm at $U = 500$ kV.
But if such a radius is introduced into the 1.5D-model, the field in the wave
front region and the computed electron density will be extremely low.

The ionization expansion model automatically removes this discrepancy.
The channel expands immediately behind the head, and the field in the
wave front region remains fairly strong. The channel expansion affects,
though indirectly, the anode current amplitude. There is no radial field near
the anode, so the channel does not expand there. Electrons enter the anode
through a narrow neck (the current through the stem cross section is likely
to be weak). But due to the channel expansion in the gap interior, more
electrons are driven to the anode to enhance the field. So the ionization
rate at the anode increases, and the plasma density grows several times
over that in the 1.5D-model (Figure 5.20).

The current amplitude in the expansion model is nearly independent of
the arbitrary initial radius r_0. The channel radii are, however, established
automatically under the action of the radial ionization wave. For the experimental conditions of [5.5, 5.25] (spherical anode 12.5 cm in radius, external field at the anode 40–45 kV cm^{-1}), the computed amplitude of total
current through the anode was found to be 9 A for $r_0 = 0.02$ cm and 11 A

for $r_0 = 0.1$ cm, which is in agreement with the measured value of 10 A.

5.6.4 Streamer development in a uniform field

This problem should be treated in the framework of the ionization expansion model, since conventional 1.5D-simulation is meaningless in this case because of the strong dependence of the results on the channel radius. For example, for the external field of 10 kV cm^{-1}, the computed streamer velocity changes by one order of magnitude, if the radius is changed from 0.02 to 0.1 cm.

The problem is formulated as follows. Let us find a critical external field E_S high enough for a steady streamer development and then the field dependence of the streamer velocity v_s at $E > E_S$. By a steady development we mean the streamer propagation at a velocity, which would not vary along an arbitrary path of considerable length. Numerous computations made so far for short streamers in uniform fields [5.33–5.36] cannot give even approximate characteristics of this process. The velocity cannot become stabilized along a path of several centimeters in length. It is impossible to predict whether it will ever be stabilized or will start decreasing after it has reached a peak. Unlimited acceleration, obtained in the model of a perfectly conducting channel [5.62], has little to do with the reality. It was demonstrated in Section 3.2.3 that the longitudinal field does not decrease with higher streamer conductivity; the conductivity growth is accompanied by a greater rise of streamer current and, hence, of the field. When the channel field becomes as strong as the external field, there is no channel polarization and no charge is pumped into the head; as a result, its acceleration drops to zero.

In the numerical experiments [5.55], a streamer started from a spherical anode of 0.2 cm radius at $U = 10$ kV. Having covered a 1 cm distance, it left the region of anode-enhanced field and entered a region of unlimited size with a given external uniform field E_0. The streamer velocity in a weak field reached a peak and then dropped to zero. In a fairly strong field, it stabilized at a level which rose with E_0. The stabilized velocity strongly depended on E_0; for example, when the external field changed from 10 to 20 kV cm^{-1}, the velocity of a streamer in atmospheric air increased by an order (Figure 5.22).

The onset of a stabilized mode is always accompanied by field stabilization in the head region. This occurs when the longitudinal field becomes equal, on the average, to the external field E_0 [Figure 5.23(a)] when it is no longer pushed out of the channel plasma. The electron flux toward the anode becomes weak, although the streamer velocity remains constant. Such a situation is possible only when the extended positive region behind

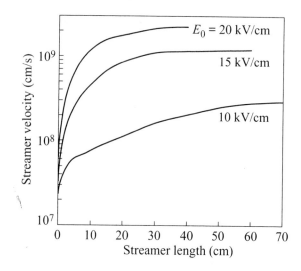

FIGURE 5.22
Initial streamer acceleration in a uniform electric field E_0.

the head moves together with it, while the total charge variation decreases along the streamer. The linear charge falls towards the channel base, then drops to zero and changes its sign [Figure 5.23(b)]. The distribution of charge along the channel is similar to that of polarization charge along a conductor of finite length having no contact with electrodes. Nonuniform is also the distribution of current along the channel [Figure 5.23(b)].

An unexpected result of the computation was the high value of critical field E_S, at which the velocity stabilizes. Instead of $E_S \approx 5 \text{ kV cm}^{-1}$ obtained in some experiments [5.5, 5.15, 5.19], the computed values exceeded 8 kV cm^{-1}. All attempts to bring them into agreement with experiment by varying the rate constants of basic elementary processes failed. However, there are no reasons to question the model validity, because it gives a good fit with experiment for nonuniform fields [5.55]. The question arises whether the experimental data on E_S were interpreted adequately and whether they can be employed at all for the evaluation of the critical field in the sense we are using the term here.

Few measurements have been made in a uniform field, that is, in a field with local nonuniformities at the streamer start. All of them have been performed without registration of the streamer velocity dynamics. It is only the overlapping time that is measured and then used to find average velocity. The gap length is usually less than 20 cm. It is possible that the streamer velocity did not stabilize in the experiment, so that the streamer

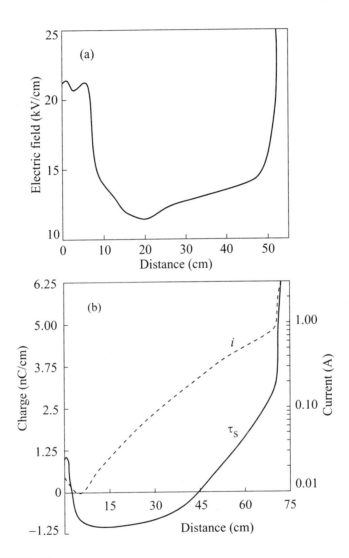

FIGURE 5.23
Steady-state streamer parameters in a uniform electric field: field distribution
along the streamer at $E_0 = 15 \text{ kV cm}^{-1}$ (a); linear charge and current at
$E_0 = 10 \text{ kV cm}^{-1}$ (b).

may have overlapped the gap in the retardation regime. It is quite clear that the problem of critical uniform field is still far from being solved and requires further theoretical and experimental studies.

6

The Leader Process

6.1 General description

6.1.1 The leader structure

In a conventional photograph taken with 10^{-7} s exposure, a spark leader looks like a thin channel slightly expanding towards its base at the electrode. The average radius of a long laboratory spark is usually a few tenths of a centimeter. The leader channel begins with a bright head, or tip, of 1 cm radius, in which its inception actually occurs. A leader often branches, so there may be more than one head, each attached to a short stem of 3–10 cm long. All initial stems, except one, die away during the leader development, being replaced by new ones. In front of the head is a streamer zone looking like a diverging column of diffuse glow. The streamer zone length is determined by the gap voltage and may be several meters in a megavolt experiment, while its maximum width is usually two or three times less than its length. When photographed through a narrow longitudinal slit, a streamer zone may show numerous individual streamers, with one streamer head per 1–100 cm^3 of the streamer zone [6.1]. Streamers starting from the head of a leader transporting positive space charge to the gap are of the cathode-directed type.

Continuous streak photography performed by an electron-optical image converter shows clearly the three components of a leader—the channel, the head, and the streamer zone (Figure 6.1). The head is the brightest structure looking in a streak picture as a narrow stripe, whose tilt can be conveniently used to measure the leader axial velocity (Section 4.7.2), or the projection of its real velocity onto the gap axis. The brightness ratio of the channel and the streamer zone varies with the photocathode used, since the channel primarily emits in the visible (red) region and the streamer zone in the near-ultraviolet. For this reason, a channel may appear like a

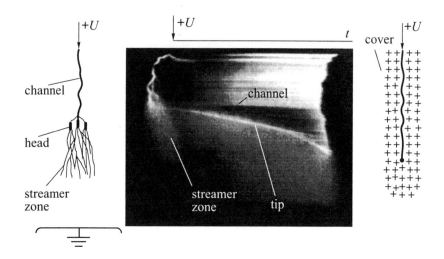

FIGURE 6.1
A positive leader in air: a schematic diagram, continuous streak photography,
and the charge cover.

dark wedge following a bright trace of the head, when a highly sensitive
cathode is used in the ultraviolet region. The channel brightness may vary
with time. One can sometimes observe its spontaneous short-lived flashes at
voltage pulses longer than several hundred microseconds; their brightness
considerably exceeds that of the general background glow.

Another important leader structure is indistinguishable but present in
a photograph—it is a space charge cover around the channel. The charge
transported by streamers into the streamer zone is accumulated in the
cover. As the leader propagates, its head moves through the gap space together with the point of inception of new streamers and with the streamer
zone boundary. The earlier streamers and their space charges do not move
forward; as a result, the elongating leader enters the space filled with
charge. It looks as if a channel pulls over an available charged jacket, making it its own cover (Figure 6.1). The characteristic radius of the cover is
close to that of the streamer zone. The charge cover is not at all rigid, but
there is a continuous charge redistribution going on in it, which may be accompanied by ionization events in the radial field at the channel surface.
As a result, the current varies along the leader axis.

The radial field direction does not always correspond to the leader polarity, and the amount of charge in the cover does not always grow. The
radial field may reverse its sign, so that some of the charge returns to the
channel. A corona of opposite sign may appear at the channel surface in a

strong reverse radial field, with opposite charge being incorporated into the cover, thus decreasing its total charge. Of course, the ionization rate at the lateral channel surface is much lower than at the head, but then the lateral surface is much larger than the head surface. Therefore, their contributions to the leader current may be quite comparable. The leader current and energy supply cannot be described adequately without analysis of charge flows through its lateral surface. In Chapter 1, we mentioned a steady-state corona as a form of discharge in gaps with a strongly nonuniform field. The corona theory may come in quite handy for the analysis of processes occurring in the leader cover, since all basic corona features remain the same, although we deal here with a plasma channel instead of an electrode.

6.1.2 Two stages of the leader process

The initial stage starts with the leader inception and lasts until the streamer zone comes in contact with the electrode of opposite sign or with a counterpropagating spark [Figure 6.2(a)]. Since none of the discharge elements has closed the gap, the leader current is closed by displacement current on the other electrode. This means a continuous redistribution of the electric field, which grows in time in front of the streamer zone and decreases respectively in the leader zone. All the charge transported by the streamers is accumulated within the gap.

The leader ability to propagate through a gap is determined by the fields around the head and in the streamer zone in front of it. At constant voltage,

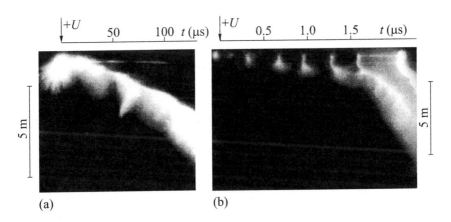

(a)   (b)

FIGURE 6.2
Typical streak photographs of a positive leader at the initial stage: a continuous mode (a), a flash mode (b).

the field at the head decreases, as the leader elongates; this slows down the leader process and may even stop it. There are two reasons for the field getting lower. One is the voltage drop along the elongating channel, due to which the head potential and charge become smaller. The other reason is the growth of charge of opposite sign, induced in the head by space charge in the gap, namely in the streamer zone and the cover. We will discuss this effect once more, because it inevitably appears, in this way or another, in a leader process characterized by the presence of a conductor under high voltage (channel and head) and of space charge isolated from the conductor.

Imagine a metallic sphere (channel head, or tip) of radius r_t, connected to a voltage source U_0 by a metallic thread (channel). The sphere potential relative to the ground will be identical to U_0 created by all the system charges: q_t concentrated on the sphere, Q distributed in the gap, and the thread charge, which we will ignore for the time being. Suppose Q is located at distance R from the sphere center, creating the potential $U = Q/4\pi\varepsilon_0 R$, so that we have $U_0 = (4\pi\varepsilon_0)^{-1}(q_t/r_t + Q/R)$. The sphere charge

$$q_t = 4\pi\varepsilon_0 r_t (U_0 - U) = Q_0 + Q_i \qquad (6.1)$$

can be assumed to be the sum of the charge $Q_0 = 4\pi\varepsilon_0 r_t U_0$ of an isolated sphere and the induced charge $Q_i = -r_t Q/R$, whose counterpart $-Q_i$ is 'pushed out' into the ground through the thread and voltage source. If Q has the same sign as U_0 and is so large that U is close to U_0, the actual sphere charge q_t turns out to be small relative to an isolated sphere charge at the same U_0. This is also true of a leader (see Section 6.7.1). The average field on the sphere, $E_t = q_t/4\pi\varepsilon_0 r_t^2 = (U_0 - U)/r_t$, decreases with larger Q and smaller R, that is, with a greater electrostatic induction effect.

Generally, charge Q creates its own field $\Delta E \approx Q/4\pi\varepsilon_0 R^2$ around the sphere, which can be either added to or subtracted from the sphere charge, depending on the position of charge Q relative to the point on the sphere where the field is to be found. In any case, this squared effect at $R \gg r_t$ $[\Delta E/E_t \sim (r_t/R)^2]$ will give way to the effect of electrostatic induction proportional to $U/r_t E_t \sim r_t/R$.

Therefore, as a leader elongates and its cover and streamer charges grow, the head field tends to decrease. In order to make a leader propagate continuously, the voltage applied to the gap must be initially high enough or be raised during the leader development. If the voltage rises too slowly, there occurs so-called flash-like discharge: a short-lived discharge is followed by a pause of a few hundred microseconds, during which time all ionization processes in the gap cease [Figure 6.2(b)]. But sooner or later, flashes give way to a continuous discharge.

We would like to emphasize the arbitrariness of the notion of continuous discharge. High speed streak photography with a good spatial resolu-

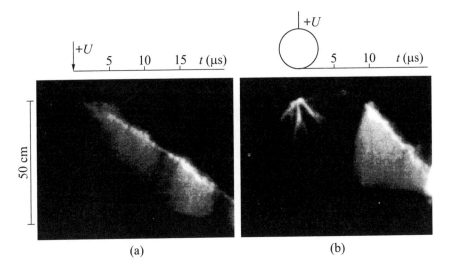

FIGURE 6.3
Streak photographs of the initial leader from a needle-shaped anode (a) and
from a spherical anode (b).

tion clearly shows that, in actual reality, a leader propagates by means of
consecutive flashes, each of which extends the leader by a length of about
its head radius [Figure 6.3(a)]. One can conceive the following sequence
of events: a streamer flash starting from the leader head is followed by
a pause, which results from space charge having decreased the head field,
after which the head transforms to a new portion of the leader channel,
and so on. In this sense, the leader start is not an exception—the initial
streamer flash just occurs on the electrode surface in a yet undistorted external field. If the electrode radius is large, the flash turns out to be quite
powerful, incorporating much charge into the gap, and the first pause in
the leader process also proves to be long [Figure 6.3(b)]. But if we deal
with an electrode of small radius, for example, with a rod having a sharp
tip, the first flash does not differ from the others at all [Figure 6.3(a)].

In the second stage, streamers contact the opposite electrode (or some
structural elements of a counterpropagating discharge). The conduction
current can now close the gap. This stage is termed as the through stage in
the Russian literature, while in English it is called the final jump because
of a great rise in the leader current and velocity. The leader is accelerated
immediately after the streamer contact with the opposite electrode. This
is clear because the streamers do not leave all their charge in the gap space
but some of it is transported onto the electrode. The current through the

channel may be several orders of magnitude higher than in the first stage, sometimes as high as 10^2–10^3 A even in a laboratory experiment. The current rise is accompanied by a great energy input into the spark channel, so that its temperature and conductivity rapidly rise while the voltage drop decreases. Unless the gap voltage is deliberately lowered, the leader in this stage will not be able to stop but will necessarily close the gap. For this reason, the conditions, under which the transition to the final jump occurs, may be identified in this case with the breakdown conditions.

The leader contact with the opposite electrode initiates the main spark stage. This is the process of channel neutralization discussed in Section 5.4.5, but here it is more intensive because of the higher leader conductivity.

6.1.3   The leader origin

The problem of the leader origin appears to be more ambiguous than that of the streamer origin, since a leader arises in a field distorted by the initial corona charge, creating a reverse field at the active electrode. The amount of incorporated charge varies widely, together with the value of ΔE denoting the field decrease at the electrode surface (Section 5.4.4). A weaker field inhibits the ionization, producing a discharge pause. To revive the leader, the field decrease must be compensated by a higher gap voltage. If the voltage is raised rapidly (within a few microseconds), the leader channel reappears from the corona stem and propagates as far as the stem half length, then it turns away nearly at the right angle to go out into free space. It has been pointed out [6.2] that a leader revives when the field at the electrode is raised to a fixed level, close to 28–30 kV cm^{-1} in atmospheric air. Evidently, the voltage rise must be greater for larger q_c that has decreased the external field. Hence, we observe an extremely wide range of leader inception voltages U_{iL}. The value of U_{iL} nearly doubles in a 2 m rod-plane gap (rod radius 1.5 cm), when q_c varies from 0.4 to 1.2 μC (Figure 6.4). Other conditions being equal, the U_{iL} range becomes smaller with larger active electrode radius, since the field is increasingly dependent on the charge of the electrode, whose capacitance rises. Besides, the excess voltage U_{iL} over the initial corona voltage U_{iC} also decreases. The value of $\Delta U = U_{iL} - U_{iC}$ did not exceed an instrumental error in studies with anodes of 50 cm radius [6.3]. A similar result was obtained for anodes with a 30–50 cm radius [6.4], while at $R_a = 5$ cm, ΔU was over 200 kV, or 40% of U_{iC}.

A leader may not develop at all, if the field grows too slowly, for hundreds of microseconds after the initial flash. Then the pause is followed by another, more powerful corona flash rather than by a leader. There may be

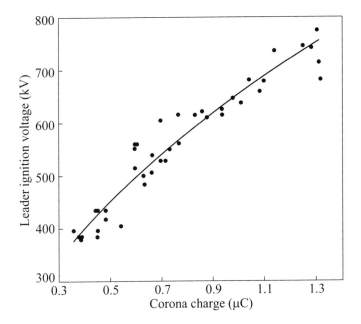

FIGURE 6.4
Ignition voltage for a positive leader in an air rod-plane gap of 2 m in length
($r_a = 1.5$ cm).

a series of flashes before a leader appears, if the voltage rises too slowly. No
leader may arise at all at a constant, but relatively low voltage. Instead,
an infinite number of flashes of nearly equal charge would appear between
pauses of 10–20 milliseconds. Evidently, the incorporated charge vanishes
over this time, restoring the near-electrode field to a value necessary for a
new flash.

6.2 Leader velocity measurement

Leader velocity is registered in nearly all experimental studies of spark behavior; nevertheless, the velocity data available are still insufficient for designing computational models for a long leader. This is due to several reasons. First, leader velocity data are usually averaged over a long period
of time, often over the whole initial leader stage. It is very difficult to follow the velocity evolution; so, the results obtained characterize the velocity
projection onto the gap axis (axial velocity) rather than the actual propa-

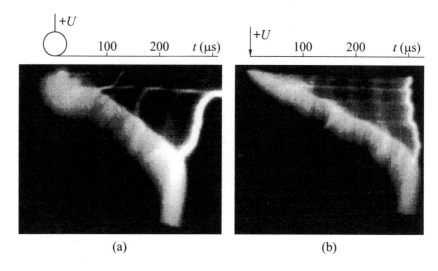

(a)   (b)

FIGURE 6.5
Streak photographs of a positive leader in a sphere-plane gap with
$r_a = 50$ cm (a) and in a rod-plane gap with $r_a = 2.5$ cm (b); $d = 9$ m.

gation rate of a leader. These values vary considerably for very long gaps,
in which a spark makes a lot of bends.

Second, it is very hard to relate the measured velocity to the leader parameter of interest, for example, to its current or head potential. Such a
relation would be of great value for the understanding of the leader mechanism. Unfortunately, a leader always develops in an electric field much
distorted by space charge. The experimenter knows exactly the gap voltage but has only a general idea of the field around the leader head. Similar is the situation with leader current. It can be readily registered in the
channel base, but this value will be very different from that in the head not
only because the current varies along the channel axis, but also because
there may be several heads and their exact number is unknown. So in the
majority of cases, the velocity data turn out to be unrelated to any other
data, especially in the initial leader stage.

More representative and reliable are the leader velocity data for air,
where gaps of several dozen meters long have been studied in detail [6.5].
Let us compare the streak pictures characteristic of the leader evolution in
a sphere-plane gap [Figure 6.5(a)] and in a rod-plane gap [Figure 6.5(b)]
of 9 m long. The respective radii of the sphere and the rod differed much,
50 and 2.5 cm, whereas the voltage in the first gap was only 15% larger, so
that the initial external field distributions were quite different. However,

the leaders had nearly the same average axial velocities in the initial stage: 1.25×10^6 cm s^{-1} for the rod and 1.35×10^6 cm s^{-1} for the sphere. These velocities did not practically vary: the leader head shows in the picture almost a straight line. A similar result was obtained for longer gaps [6.6]. At positive voltage pulses of 2/7500 μs duration and 4.9 MV amplitude, a leader jumped off a high voltage screen, which served as a roof for a pulse voltage generator, and moved toward the ground from a 40 m height, exhibiting a much curved trajectory of 70 m in length. Its vertical velocity component remained nearly constant and equal to 2×10^6 cm s^{-1}. The same velocity was obtained from other measurements [6.7] for a 45 m leader, which was initiated from a short rod mounted horizontally at 28 m height on a tower generator roof. The generator produced positive voltage pulses of 300/10000 μs. At the pulse amplitude of 3.4 MV, the spark moved horizontally without touching the ground, sporadically going up a little.

This evidence suggests that neither the gap length, nor the anode radius, nor the voltage rise time can change radically the axial leader velocity in the initial stage, whose minimum value for a continuous propagation in atmospheric air is $(0.5-1) \times 10^6$ cm s^{-1}. The average velocity increases but slightly with the closed gap length. If the gap voltage amplitude is made larger, the leader velocity is supposed to increase, but there is no possibility to check the relationship $v_L\,(U)$ in laboratory conditions because of the lack of space. Streamers from an initial corona flash are known to cross a gap even in an average external field of $4.5-5$ kV cm^{-1}, so that the discharge goes to its final jump from the very beginning. One can state with certainty that the external field dependence of velocity for a leader is much less pronounced than for a streamer. A 2–2.5-fold increase in gap voltage accelerates a streamer by one order of magnitude or more, whereas a leader increases its average axial velocity in the initial stage only from 1.3×10^6 to 2.1×10^6 cm s^{-1} at a 2.2-fold voltage rise [6.4]. A weak leader velocity response to overvoltages is characteristic of all long gaps.

Different velocity responses to external field variation in a discharge gap is a manifestation of different mechanisms for the streamer and leader processes. Indeed, a streamer propagates through a cold gas as an ionization wave, whose velocity is determined by the ionization frequency in the wave front, that is, by the maximum local field near the streamer head. In contrast, the slow propagation of a leader would be impossible without a strong gas heating. But heating is quite a different process, time- and energy-consuming, whose rate cannot be defined only by the local field enhancement.

Because of numerous bends, the real velocity of a leader is always higher than its axial velocity, approximately by 25% for a 10 m gap [6.4]. There are reasons to believe that this value may be larger for longer gaps. In any

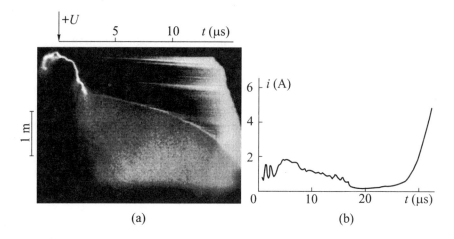

(a)   (b)

FIGURE 6.6
A streak photograph (a) and a current oscillogram (b) of a leader in the final
jump in a rod-plane gap with $d = 3$ m and $r_a = 1.5$ cm.

case, sparks of 100 m long were observed in a 30 m gap between a high
voltage electrode and the ground at voltages above 3.4 MV [6.5]. In these
conditions, a comparison of the axial and actual velocities becomes meaningless. The actual velocity is independent of the local leader trajectory: a
leader may move along the external field vector or normal to it with nearly
the same velocity. Its variation along the discharge trace is also slight.
The average velocity at minimum breakdown voltage in a 10 m gap varies
from one discharge to another by about 3% of that averaged over many experiments, while the axial velocity shows a 20% variation. Like the axial
velocity, the actual velocity responds weakly to gap overvoltage. For the
above 2.2-fold rise, the actual average velocity in a 10 m gap increases only
from 1.8 to 2.6 cm μs^{-1} [6.4]. According to other data [6.8], the velocity
of 1.8×10^6 cm s^{-1} should be taken to be optimal for long air gaps with
strongly nonuniform fields: a gap then requires a minimum voltage for its
breakdown.

The leader velocity may vary much greater in the final jump stage (Figure 6.6). It is equal to $(2–5) \times 10^6$ cm s^{-1} at the moment of contact of the
streamer zone with the opposite electrode, but it rapidly increases, as the
leader head comes closer and the streamers become shorter. If the gap voltage is not lowered, the maximum velocity may be as high as 10^8 cm s^{-1}.
According to some data [6.3], this requires an average streamer zone field
of about 20 kV cm^{-1} in air, which is four times greater than at the moment the leader feels ready for its final jump.

There have been several attempts to find the value of v_L (i_L) from current measurements made during the final jump. Such measurements are relatively easy to perform, because the current can also be registered from the grounded electrode side. The data are commonly approximated by the power function $v_L = k_v i_L^a$. For instance, some workers [6.9] used $a = 0.66$ and the coefficient $k_v = 0.858$ cm $(\mu s\,A^a)^{-1}$ for $1 < i_L < 22.6$ A and $k_v = 2.65$ cm $(\mu s\,A^a)^{-1}$ for $85 < i_L < 1100$ A. The expression suggested appears to be quite rough—it gives lower velocities than those measured in [6.4] for currents of $1-2$ A. The calculated values for $i_L = 50-200$ A, on the contrary, prove to be 1.5–2 times greater than the experimental values from [6.2], which give a weaker current dependence of the leader velocity:

$$v_L = k_v i_L^{0.5} \qquad k_v = 4 \text{ cm } \left(\mu s\,A^{0.5}\right)^{-1}$$

Note that i_L is the total leader current, as for the initial stage, rather than the current only through the head from which the channel propagates.

Obviously, the final jump current can be easily limited by a series connection of large resistances. Resistance of about 10^6 Ω will suppress the current to several amperes or to a few fractions of an ampere at the maximum supply voltage. In this case, the gap voltage will decrease appreciably with increasing spark length and respective shrinkage of the streamer zone (the sum of their lengths coincides with the gap length in the final stage). By choosing appropriate resistances, one can reduce the voltage so much that the leader velocity during the final jump will be as low as in the initial stage. The leader deceleration to 10^6 cm s^{-1} over the entire period of the spark propagation permitted the first streak pictures to be taken with a Boys camera [6.10, 6.11]. Their analysis led to the discovery of a leader in the laboratory spark.

6.3 Current and charge measurement

The minimum current capable of providing the propagation of a continuous positive leader is as small as a few tenths of an ampere. A similar current provides the leader development at a minimum critical voltage for air gap breakdown. Minimum current slightly rises with the gap length d from 0.6 A at $d = 5$ m to 0.8 A at $d = 10$ m [6.4]. These data were obtained from measurements of the leader base current, using a detector placed from the anode side. There are no data on the leader head current, although this current is directly related to the conditions for the streamer-leader transition and to the leader velocity.

The base current grows appreciably with the gap voltage: its average value changes from 0.8 to 9 A at 2.2-fold overvoltage in a 10 m gap [6.4]. It is clear that the current in the channel head must vary less, since the actual leader velocity under the same conditions has increased only 1.45 times, while the relation between the head velocity and current is at least $v_L \sim i^{0.5}$ (Section 6.1.4). One reason for a slight head current rise might be a larger number of heads existing simultaneously, and another reason might be associated with a nonuniform current distribution along the channel.

There is much evidence indicating current variation along a leader channel, for instance, the leader acceleration accompanied by current drop (to zero!) in the base (Figure 6.6). This can be observed during the final jump, when the gap voltage rapidly decreases. Acceleration necessarily requires a higher current in the head, the seat of a new leader portion, but this process cannot go on at zero head current. So we have to assume that the currents in the channel head and base do not differ merely quantitatively but also in the tendencies for a change. We will explain this with a simple model similar to the streamer model discussed above. Let us assume a leader channel to be a perfect conductor with a potential equal to that of the anode, U_a. We will make use of average linear capacitance C_1 as the sum of the capacitances of the channel and its cover (the latter is, in fact, larger than the former). If i_0 and i_1 are the currents in the base and behind the head, then

$$i_0 = \frac{d\left(C_1 l U_a\right)}{dt} = C_1 U_a v_L + C_1 L \frac{dU_a}{dt} = i_1 + C_1 L \frac{dU_a}{dt} \qquad (6.2)$$

The base current turns to zero when the rate of the voltage drop is $|A| = -dU_a/dt = i_1/C_1 L = U_a v_L/L$. At $i_1 \approx 1$ A and total leader capacitance $C_1 L = 100$ pF characteristic of channel length L of several meters, this gives the limiting value $|A| = 10$ kV μs^{-1}, which is quite realistic for long air gap experiments using so-called standard thunderstorm pulses with the rise time of 1 μs and duration of 50 μs (0.5 level).

The opposite tendency is also clear from expression (6.2)—the leader current must increase from the head towards the base, if a rising voltage is applied to the gap. The practical significance of this phenomenon is beyond doubt. Owing to the current rise, the voltage supply contributes additional energy to the channel, resulting in a higher channel temperature and conductivity. Other conditions being equal, the channel acquires the ability to develop at lower voltage, and the gap electrical strength becomes lower. We will return to this problem in Chapter 7, but here we only emphasize that breakdown of a long gap is possible at a minimum voltage if a rising, rather than rectangular or decreasing, pulse is applied to it.

The leader development in the initial stage is accompanied by a continuous charge incorporation into the gap. Its amount may exceed several times the active electrode charge; hence, the initial field distribution in the gap is greatly violated. In particular, the field in front of the leader head, where the streamer zone is formed, becomes lower (Section 6.1.2). Naturally, the leader tends to propagate in the direction of the largest total field vector, bypassing the already existing streamer zone. This phenomenon was described in [6.7] for a positive leader of 45 m long. A shot-by-shot registration by an electron-optical image converter shows the streamer zone rotation—it circumscribes a wide cone in front of the leader head. The rotation is not smooth but discrete. At the moment of turning, when the streamers move in space yet unfilled by space charge, the leader current and velocity appreciably grow. The head circumscribes a kind of spiral about its axis, determining the general direction of propagation.

Numerous experiments have shown that minimum linear charge $\tau_{\min}$ of a leader corresponds to the breakdown conditions at minimum voltage, and its average value is $\tau_{\min} \approx 0.3\text{--}0.4 \ \mu\text{C cm}^{-1}$ [6.4, 6.7, 6.12]; it somewhat increases [6.4] at higher air humidity, from $0.2\text{--}0.3 \ \mu\text{C cm}^{-1}$ at $(0.4\text{--}0.7) \times 10^{-5} \ \text{g cm}^{-3}$ to $0.4\text{--}0.5 \ \mu\text{C cm}^{-1}$ at $(1.3\text{--}1.5) \times 10^{-5} \ \text{g cm}^{-3}$. When the conditions do not meet the requirement of minimum electrical strength, the linear leader charge may exceed its minimum value several times, or even by an order of magnitude for extremely long leaders like lightning. The average channel current $i \approx \tau v_{\text{L}}$ rises respectively. For lightning with velocity about $10^7 \ \text{cm s}^{-1}$, it is as high as 100 A instead of a typical value of 1 A in a laboratory spark under the optimal conditions.

The leader current rises abruptly after the process enters the final jump stage. If the voltage source possesses a sufficiently high power and low resistance to maintain the gap voltage, the current may exceed $10^2\text{--}10^3$ A. The current is higher for smaller gaps and higher head potential. The latter dependence is very strong, which is clear from the following data:

Head potential (kV)	400	750	1200
Leader current (A)	2	50	200

They were obtained for a sphere-plane gap of 1 m long and anode radius of 12.5–50 cm [6.3]. Because the channel was short during the measurement, its head potential was nearly the same as the anode potential.

6.4 Channel and streamer zone fields

If the average fields in the leader channel E_{L} and the streamer zone E_{s} (or leader head potential U_{t}) are known, the breakdown voltage in a gap of length d can be expressed as

$$U_{\text{b}} = E_{\text{s}} L_{\text{s}} + E_{\text{L}} \left(d - L_{\text{s}} \right) \qquad U_{\text{t}} \sim E_{\text{s}} L_{\text{s}} \tag{6.3}$$

since the streamer zone length L_s at minimum breakdown voltage does not vary much in long gaps and is equal to two or three meters in normal air.

6.4.1 The streamer zone field

The conditions for streamer propagation through a discharge gap were discussed in Chapter 5. The average field in a streamer zone can be found by applying a voltage pulse with the rise time of a few microseconds to a gap of length d and by measuring the amplitude U_a, at which the initial corona streamers contact the opposite electrode. A leader may also start from the high voltage electrode during the streamer flight, but its length L will not be large by the moment the gap is bridged. The range of values $U_a/d < E_s < U_a/(d - L)$ defines the average field in the streamer zone, E_s, with a satisfactory accuracy; as mentioned above, it is $E_s \approx 4.5$–5 kV cm^{-1} in normal air for gaps as long as 10 m. At lower air density, the average field E_s decreases more rapidly than the density. Experiments made in the mountains at relative air density $\delta = 0.7$ in gaps up to 5 m long yielded $E_s = 2.8$–3.0 kV cm^{-1} [6.13], a value close to that obtained in the laboratory conditions at $d = 0.5$ m and the same δ. It may be suggested that the independence of E_s of the gap length is a common feature of streamer zones. This greatly facilitates the evaluation of air insulation strength.

Indirect evidence for a uniform field distribution in the streamer zone is provided by streamer velocity measurements—they all move uniformly. An attempt was made [6.14] to register the field directly, and the result obtained was similar. The experiment was performed in a 12 m rod-plane gap at positive voltage pulses with a rise time from 15 to 300 μs. The field detector was based on the Pockels effect and used a 0.86 μm laser as a light source. The laser radiation illuminated a bismuth silicate crystal of 8 mm in length and 1.5 mm in radius and was then transmitted to a light detector through fiber lightguides. The detector was precalibrated with a known field and had a sensitivity better than 50 V cm^{-1}.

Unfortunately, the authors did not mention the upper limit of the registered field, which was to be distorted by local corona charge from the crystal faces. An axial field oscillogram from a detector placed on the axis of a 4 m gap at 1.2 m from the anode shows that the field rises until the moment t_1, at which time the streamer zone approaches the detector, and reaches 5 kV cm^{-1}. Between the moments t_1 and t_2, while the detector is still within the streamer zone in front of the leader head, the field remains nearly constant, but it begins to decrease after the leader head has passed the detector. The same value of 5 kV cm^{-1} was registered from the initial corona region. Here, too, the field was stabilized during the streamer flight and did not vary with the detector position relative to the anode.

The field decrease after the leader head has passed the detector is a direct indication of the channel field being lower than the streamer zone field. Clearly, the field near the channel is not equal to the longitudinal field in the channel, but their difference can hardly be large when the detector is placed close to the leader. The time of the field decrease Δt and the leader velocity can provide information about the length of the transition region, in which weakly conducting streamers are transformed to a well developed leader channel. The average leader velocity in the initial region of 1 m long was found to be $v_L \approx 4$ cm μs^{-1} and the time $\Delta t \approx 15$ μs [6.14]. This means that the transition region length is $\Delta l = v_L \Delta t \geq 60$ cm. The inequality sign indicates that the leader rapidly passes to the final jump under the conditions described; therefore, its velocity rises as it approaches the cathode. The values of v_L and Δl were probably considerably underestimated.

It is possible that the transition region extends as far as a few meters behind the head. This should be borne in mind, when analyzing breakdown of air insulation gaps in high voltage transmission lines, where insulator strings are five meters or less. The leader length in such gaps does not exceed 3 m prior to the final jump, which means that the leader channel has no time to become stabilized. This creates additional difficulties in breakdown voltage calculations, because the rapid fall of the longitudinal field from the head toward the base is to be taken into account.

6.4.2 Channel field and head potential

There are no direct data on field distribution in a spark channel, and scarce are its indirect experimental evaluations. Average channel field has been a subject of numerous and, in many respects, fruitless debates. The existing opinions differ so much that there seems to be no hope to make them compatible. For instance, the classical experimental study [6.15] was the first one to give a clear concept of the leader structure and mechanism, but it ascribed arc gradients of 55 V cm^{-1} to an air spark channel. This was due to an erroneous interpretation of voltage oscillograms from closed gaps.

In another work [6.16, 6.17], the values deduced from experimental data were by two orders of magnitude larger. The authors observed the leader development at voltage pulses of $1/40$ μs, at which the streamer zone crossed the gap, and partially cut the voltage at a fixed channel length L. The voltage cut was chosen to be minimal for arresting the discharge. Voltage U_L and average longitudinal field $E_L = U_L/L$ were calculated from the difference between the residual gap voltage and the voltage in the streamer zone, whose average field was taken to be 5 kV cm^{-1}. The obtained value was $E_L \approx 2-4$ kV cm^{-1}, decreasing with the channel length and lifetime.

These measurements were made at the leader current of about 100 A and discharge lifetime less than 20 μs. Such conditions are characteristic of the final jump but not of the initial stage important for the calculation of electrical strength in multimeter gaps. Moreover, the method of partial voltage cut is unsuitable for the initial stage, since, in addition to the channel and the streamer zone, the voltage also drops in the gap space free from discharge. The latter circumstance introduces an ambiguity in data interpretation in all cases except the transition to the final jump, when the streamer zone just touches the grounded plane.

Similar measurements made at the moment of contact have been described [6.18] for 3–5 m gaps at positive, slowly rising voltage pulses of 150–300 μs; they yielded $E_L \approx 0.3$ kV cm^{-1}. The measurement accuracy, however, was very low. Indeed, the leader length in the initial stage was close to 2.5 m even for $d = 5$ m; the channel voltage was $0.3 \times 250 = 75$ kV, whereas the gap voltage was 1300–1500 kV, or 20 times greater. The desired quantity was sought for as the difference between two close values, so that even a small error in the channel and streamer zone lengths or in the choice of field E_s could be fatal.

Another indirect method for field evaluation was employed in [6.4]. Voltages U_a lower than the breakdown voltage, applied to a gap of length d, produced discontinuities at the channel length $L < d$. The values of L and the charge incorporated into the gap over the entire leader lifetime were measured experimentally. This was followed by theoretical calculations, for which the charge was assumed to be distributed uniformly over the surface of a cylinder of length L and arbitrary radius and over the surface of a hemisphere of the same radius with the center in the leader head. This was the way to model the leader charge cover and the streamer zone. The head potential U_t was calculated as the sum of potentials created by the charges of the electrode, cover, and streamer zone. The average channel field was estimated as $E_L \approx (U_a - U_t)/L$.

The values obtained for a leader starting from a spherical anode of 30 cm radius in a gap with $d = 10$ m at voltage pulses of 600/10000 μs and a variable amplitude were:

L (m)	1.0	2.0	3.0	4.0	5.0
E_L (kV cm^{-1})	9.0	5.0	3.3	2.3	1.7

The main error source in this calculation seems to be the component U_t associated with the cover and streamer zone space charge. The calculations made in [6.4] give an incredibly small error, less than 400 kV. The total head potential calculated by the same authors does not exceed 800 kV and somewhat decreases with leader elongation; $U_t \approx 600$ kV at $L = 5$ m. This value contradicts optical registrations, in which the streamer zone just before the final jump had the length $L_s = 3$ m; therefore, its voltage and head potential are higher than $E_s L_s = 1400$ kV, where $E_s \approx 4.65$ kV cm^{-1} is the average streamer zone field. It is hard to find the physical reason for this great increase in the gap charge along the last 1.5–2 meters of the leader path prior to the final jump that could reasonably account for the doubled head potential. This circumstance makes us question the E_L estimates in [6.4], in spite of their being frequently quoted in the literature. Lower values of E_L appear to be more probable.

Less arbitrary assumptions were made in the work [6.19], which describes parameter measurements with stabilized leader current. For this purpose, voltage was applied to a rod-plane gap with $d = 1$ m through a large (up to 1 MΩ) load resistance, such that the leader current would rise as slowly as possible after the final jump. Discharges providing stabilized leader current in the initial stage and most of the final jump were selected for further data analysis. Stabilized current is an indirect indication of steady-state conditions of the streamer propagation in front of the head. Therefore, it may be suggested that the average field in the streamer zone did not vary in the experiment and remained practically the same during the final jump, as at the moment of the streamer contact with the grounded plane.

The gap voltage $U(t)$, streamer zone length $L_s(t)$, and the leader length projection on the vertical gap axis $L(t)$ for a fixed moment of time t were registered. At the moment of the streamer zone contact with the cathode, the average channel field was evaluated as

$$E_L(t) = \frac{U(t) - E_s L_s(t)}{L(t)} \tag{6.4}$$

with the average streamer field taken to be $E_s = 4.65$ kV cm^{-1}. The estimation of E_L is accurate enough, since the voltages in the channel and streamer zone prove to be comparable. At the moment of the channel contact with the cathode, all the gap voltage was assumed to be applied to the channel: $E_L(t) = U(t)/d$. This estimate is also fairly good, because the load resistance did not permit the leader acceleration. The time of its contact with the plane was registered by an electron-optical converter with an accuracy to the head size in a streak picture.

The other two evaluations were less accurate. In the first one, the average channel field during the final jump was found from the same formula of (6.4) and the same value of E_s. The other procedure provided the average field near the anode. Suppose the channel has increased its length by ΔL over the time $t_2 - t_1$.

It may be suggested that the whole leader channel with stabilized current does not differ in its parameters at the time t_1 from a channel section of the same length $L(t_1)$ (counting from the head) at the time t_2. Then the change

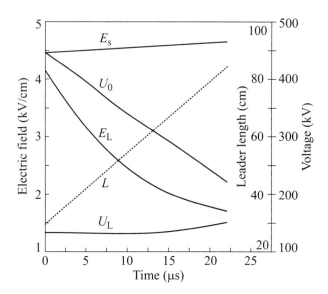

FIGURE 6.7
Leader parameters in a 1 m gap at $t_f = 150$ μs: U_0, gap voltage; L, leader length; U_L, voltage drop in the channel; E_L and E_s, average fields in the channel and streamer zone, respectively.

in the gap voltage $\Delta U = U(t_2) - U(t_1)$ is determined by the streamer zone shrinkage in the final jump by the value ΔL plus the voltage in the near-anode channel section of the same length: $\Delta U = -E_s \Delta L + E_L \Delta L$. Hence, we have $E_L = E_s + \Delta U / \Delta L$.

All measurements were made at positive voltage pulses of 150 μs rise time. After its start, the leader existed in a flash form and only during $10-15$ μs prior to the jump did it propagate continuously. At the moment of the streamer contact with the cathode, the leader length was 30 cm. The average duration of the final jump before the gap overlap was 25 μs. The channel field estimates were made for the leader currents of $1-1.5$ A. The current was destabilized $2-3$ μs before the breakdown and could rise to 4 A. At the moment of transition to the final jump (after $10-15$ μs of continuous propagation), the field was $E_L \approx 2.5-4.5$ kV cm^{-1}; the smaller values refer to larger times. At the end of the jump, the value of E_L dropped to 1.2 kV cm^{-1} and was approximately 7 times less than that estimated in [6.4] for a leader of the same length, 1 m. The field E_L decreased in time with deceleration, the voltage drop in the channel being nearly constant (Figure 6.7), although the channel length increased by a factor of 3.

At the moment of transition to the final jump, the longitudinal field in a

TABLE 6.1
Leader parameters derived from experiment

d (m)	5	10	15
U_0 (kV)	1.3	1.9	2.2
L_s (m)	2.3	3.2	3.6
L (m)	2.7	6.8	11.4
U_t (kV)	1.1	1.5	1.7
E_L (kV cm^{-1})	0.75	0.59	0.44

section of 10−15 cm near the anode was 1.2 kV cm^{-1} at 1 A, which is three times less than the average value for the channel. This indicates a great field variation along the channel. The field behind the head may be even larger than $E_s \approx 4.5$–5 kV cm^{-1} in the streamer zone; otherwise, the experimental value of 3.5 kV cm^{-1} could not be obtained for a 30 cm channel.

Probably, the most reliable experimental value for the head potential was obtained from measurements made at the moment of the streamer zone contact with a plane cathode. The streamer zone length was measured using an electron-optical converter. Such measurements give quite reliable and informative data with an accuracy of about 10% [6.20–6.22, 6.18, 6.4] for gaps with $d = 5$−15 m, whose U_0 increases at an average rate of 5 kV μs^{-1} over the leader lifetime. They can be used to obtain some leader parameters (Table 6.1). The length of the streamer zone L_s is given for the moment of its contact with the cathode, and the length of the axial channel projection is $L = d - L_s$. The head potential U_t is calculated as $U_t = E_s L_s$ with $E_s = 4.65$ kV cm^{-1}. The potential difference between the head and the plane during the final jump falls on the streamer zone, and its average field is above E_s. At the beginning of the jump, the field is probably close to this value. The average field in the channel can be found from the known gap voltage as $E_L = (U_0 - U_t)/L$, but the accuracy of this estimate will not be high, since $U_0 - U_t$ represents a small difference of large values.

However, the above result seems quite convincing qualitatively: the average field monotonically decreases with the channel elongation. This is natural, because the field becomes weaker in the older, better heated channel sections with a tendency to turn to the arc type. For the leader velocity $v_L \approx 1.5$–2 cm μs^{-1}, its lifetimes in the longest gaps are 0.5−0.7 ms, which is sufficient to reduce the average field to 0.5 kV cm^{-1}. The actual average field in the channel may be found to be weaker because of the numerous bends giving a 25–35% elongation.

6.4.3   Short spark experiments

Studies of superlong leaders require the knowledge of field strength in a channel with a large lifetime (several milliseconds) and of its response to slow and fast current variations. This problem has not yet found an experimental solution. Some information can, however, be obtained from voltage measurements in relatively short sparks after the gap overlap. The experiment must produce such a spark that would differ very little in its final parameters from a long leader section of the same length as the spark and the same lifetime of several microseconds. When supplied by a typical leader current, the spark plasma may be suggested to behave in the same way as in the leader section. Indeed, the principal reasons for possible differences are the near-electrode effects and heat removal conditions. However, the effects are not very essential in a channel exceeding several centimeters in length, and the differences in heat removal, primarily at the cost of the convective component, would affect large lifetimes but would hardly be significant in a millisecond experiment.

The complexity of such an experiment is associated with the necessity to observe strictly the experimental requirements. A generator with a large internal resistance can provide the necessary power supply mode during the spark propagation through a short gap and stabilize the current at a fixed level after the overlap, but at the moment of overlapping, the capacitances of the electrodes, voltage divider, and connective wires will discharge into the spark channel. Their energy will be released and will heat the channel, leading to a considerable increase in its conductivity. In order to avoid this effect and to preserve the channel in the condition preceding the overlap, all the capacitances should be made as small as possible or be separated from the gap by large resistances.

A spark discharge in a 7 cm gap between thin rod electrodes was studied in [6.12]. Although the voltage was measured by a high resistance divider, its response time was less than 10 μs because of the analog electrical-to-light signal conversion and optical information transmission. During the registrations, the channel was supplied by stabilized current varied from 0.05 to 10 A in different experiments, and the registration time was several milliseconds. The initial, reliably registered field in the channel at low currents was 1–2 kV cm^{-1}, close to that in a 'young' leader. Typical average field oscillograms obtained at various currents are presented in Figure 6.8(a). The field decreases to a current-dependent steady-state for 1.5–2 ms or a little faster at higher currents. This dependence is approximated by the expression:

$$E = 32 + \frac{52}{i\,[\mathrm{A}]} \ \mathrm{V\,cm^{-1}} \qquad\qquad (6.5)$$

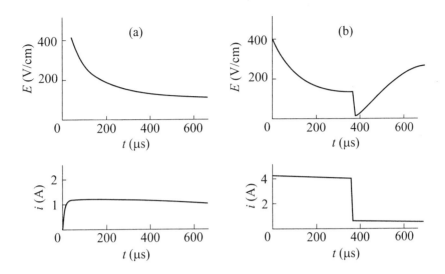

FIGURE 6.8
Leader simulation by a short spark: average field oscillograms in the channel at stabilized current (a) and at abruptly decreased current (b).

This is close to what is normally observed in a steady-state arc.

For evaluations of the leader state in the flash stage, it is necessary to know the channel response to a fast current drop from the initial to the new value. This issue was also studied in the experiment described above. The channel field first dropped as much, since the plasma temperature and conductivity measurements took time [Figure 6.8(b)]. But as the gas became cooler, the field gradually rose to a new steady-state level described by formula (6.5). The time for the transition to the new state is of the order of 100 μs and rises with the energy input into the channel, until the current drops. The channel response to the jump-like current rise is 5–10 times faster.

6.5 Leader channel expansion

What undoubtedly follows from experimental studies is the fact of gas expansion in a spark channel. The first information about the expansion rate was obtained from streak photographs taken through a narrow transverse slit. It was found half a century ago that a leader with a current rise of 10^6 A s^{-1} expanded with velocity $u_r \sim 10$ m s^{-1}, starting with

TABLE 6.2
Channel expansion measurements [6.24]

Current rise $(A\,s^{-1})$	3×10^6	2×10^7	8×10^7	5.5×10^8
Leader current (A)	4–30	12–120	40–360	80–900
Channel lifetime (μs)	12	7	4.5	1.7
Expansion rate $(m\,s^{-1})$	~ 30	~ 60	~ 100	~ 300

the initial visible diameter of about 10^{-2} cm [6.23]. Subsonic velocities at weak currents were later registered by an electron-optical image converter with a high time and spatial resolution (about 10^{-2} cm). For example, at a slowly varying current of $1-4$ A, a leader expands at an average velocity of 60 m s^{-1}, and its visible diameter reaches $d_L \approx 0.25$ cm over $25-30$ μs [6.19]. Channel expansion for 18 m s^{-1} at a positive voltage pulse of $130/3000$ μs and amplitude 885 kV was registered in [6.8]. The largest visible diameters of $0.4-0.8$ cm were obtained in [6.4] from still photographs of a leader in a 10 m gap (the photographs were analyzed for the image microscopic structure to identify the region of maximum density). No relation was found between d_L and leader current or transported charge. Unfortunately, the authors did not mention the spatial resolution of the optics used.

The expansion rate increases with the rate of leader current rise. This is clear from Table 6.2 showing experimental data borrowed from [6.24] for a leader with continuously rising current. Only at $di/dt > 10^8$ A s^{-1} and $i > 100$ A does the expansion rate approach the sound velocity. Under the 'normal' conditions of slowly varying currents, $1-10$ A, the expansion is slow, which means that the pressure in the channel is close to that in the ambient gas. This conclusion is important for the understanding of the electron production and loss processes in the leader plasma. Constant pressure during the heating indicates that the gas density in the channel decreases. On moderate heating, the field necessary for impact ionization decreases in proportion with the density.

The visible diameter gives little information on the current distribution over the leader cross section. More informative in this respect are the data on so-called 'thermal' diameter usually registered by the shadow method [6.4], which can provide some knowledge about the channel core, in which most energy is released. The measurement error was estimated to be 0.01 cm. At positive voltage pulses with $t_f = 30$ μs in a 1.5 m gap, the thermal diameter of a near-anode channel increased from 0.03 to 0.18 cm over the leader lifetime [6.4]; it turned out to be smaller than the visible diameter. Direct measurements of gas temperature in the most heated region of the channel would be very helpful.

6.6 Comparison of leader and streamer parameters

A comparison of the leader and streamer parameters is, at least, useful to question the popular illusion about a typically weak streamer with low current and a powerful leader. The base current in the initial leader stage is close to 1 A, sometimes reaching several amperes. Such streamer currents are easily attainable without applying high voltages. For example, simulation of a cathode-directed streamer initiated from a sphere of 1 cm radius by a rectangular voltage pulse of 100 kV amplitude has yielded $i_{max} \approx 5$ A. Experimental studies [6.26] leave no doubt as for the validity of this result. The existence of streamers with currents of 10 or even 100 A has been confirmed by both numerical simulation and experiment. Note that such currents can be obtained easier in a streamer than in an initial leader, where the value of 100 A is only characteristic of lightning with megavolt leader head potential.

   Therefore, it is the initial leader, rather than the streamer, that should be referred to as a weak gas discharge structure. On the other hand, a weak streamer is not at all an exception, because its current drops to 10^{-3}–10^{-2} A at velocities close to the minimum velocity. Such streamers, for example, were observed in air gaps with a uniform field of about 5 kV cm^{-1} [6.27] and may, probably, occur in the streamer zone of a leader at a low average field.

   As for the plasma density, it has never been measured directly in a leader or streamer channel, but the value of n_e can be found indirectly. For this, it is necessary to know the current, longitudinal field in the channel, and its radius. The 'thermal' radius is preferable, since it is closer to the current cross section radius than the 'optical' leader radius. Following the authors of the work [6.4], let us take $r_L = 0.1$ cm, a value also characteristic of streamers at moderate voltages. The leader current $i = 1$ A in the near-anode base of a long spark corresponds to $E_L \approx 1 \text{ kV cm}^{-1}$ [6.19]. It is no use trying to check its accuracy now, because the channel gas density N is unknown, while the electron drift velocity is $v_e \sim E/N$. If the gas density is $\delta \approx 0.1$ relative to the normal density corresponding to the moderate leader temperature $T = 3000$ K (simulations deal with twice as large values), then we have $E/N \approx 4 \times 10^{-16} \text{ V cm}^2$ and $v_e \approx 6 \times 10^6 \text{ cm s}^{-1}$; hence, the density is $n_e = i/\left(\pi r_L^2 e v_e\right) = 3.3 \times 10^{13} \text{ cm}^{-3}$.

   The plasma density value will be smaller, if we take a higher temperature and field and replace the thermal radius by the 'optical' one. This means that the plasma density has been overestimated rather than underestimated. But the value of $n_e \approx 3 \times 10^{13}$ cm^{-3} is quite ordinary for a streamer plasma. Calculations give even an order of magnitude higher estimates for streamers.

It is seen that the streamer is not inferior to the leader in the plasma density either. The average longitudinal field in a long air streamer is close to 5 kV cm^{-1}, being larger than the leader longitudinal field. Considering that the streamer current is not lower than the leader current, the energy release per unit channel length of a streamer may also be equal to or higher than that of a leader.

Thus, the leader has no advantages as for extremal parameter values—at short lifetimes, it is inferior to the streamer in practically all characteristics. Its advantages are manifested only at large distances, when a longer time is necessary to sustain the current, to confine electrons to the plasma, and to provide the channel with a continuous energy flow.

The streamer is unsuitable for such purposes. A fast ionization wave that forms a streamer may acquire a great instantaneous power and leave behind it a fairly dense plasma. However, the energy released into the channel is insufficient to heat the gas; so, the streamer remains cold. Electrons cannot be reproduced in a cold channel, and their density in air rapidly decreases due to attachment and recombination with cluster ions.

In the above example of a streamer from a spherical anode of 1 cm radius, the base current drops by an order of magnitude over a period of 0.02 μs and the dissipated power over 0.01 μs. The plasma density n_e decreases by two orders in 0.1 μs after the start. For a leader, however, this time interval is just an instant, for which it elongates by 1 mm, while the path it has covered may be dozens of meters long.

The advantages of a leader are unrelated to the values of its plasma density, current, and power—in their limiting values, a leader may even lag behind a streamer. Of crucial importance here is the leader ability to preserve its current and power at the initial levels for a long time—for tens of microseconds and even milliseconds. Owing to the low rate of the leader process, the gas is heated to several thousands of degrees, so that the energy stored in molecular vibrations rapidly transforms to heat, the recombination is decelerated, and the electron attachment is compensated by detachment. Then, the electron losses can be compensated owing to the ionization in a relatively weak longitudinal field, stimulated by a low density hot gas, and then to thermal ionization. The reason for all this lies in the greater effective linear capacitance: a leader receives much more energy than a streamer (cf. Section 3.3.1). We will now discuss the origin of this effective capacitance.

6.7 Streamer zone and cover charges

Experimental data on charges in the streamer zone and in the leader cover are very scarce. The total cover charge is usually measured experimentally, and the linear charge, that is the charge per unit channel length, is calculated. Sometimes, experiments can also provide information on the length and maximum radius of the streamer zone; the latter can give an idea of the cover radius. Charge spatial distribution can be inferred only from rough calculations made from the results of field registration on the grounded plane (Section 4.6.2). It is hard to say, however, how much of the charge is concentrated in the streamer zone in front of the leader head and in its cover, behind the head. These are important parameters related not only to leader current but also to breakdown voltage.

6.7.1 The streamer zone

We will start the discussion of charge with the streamer zone, because it is here that the leader cover charge originates. When the leader head moves on together with the streamer zone, the charge it leaves behind envelopes the leader channel to form its cover. Streamers incorporating charge into the gap fly out of the head all the time. Experimenters believe that the streamer velocity in the initial leader stage is probably constant and close to the minimum possible velocity $v_s \sim 10^7$ cm s^{-1}. Constant velocity indicates that the field is distributed more or less uniformly within the zone space and that it is not strong, about 5 kV cm^{-1}. This is confirmed by indirect experimental findings (Section 6.4).

The streamer flight within a streamer zone of 1–3 m long takes several tens of microseconds. During this time, most of the plasma in the older streamer sections decays, so the current in the streamer base becomes very low. Streamer charge stops growing long before a streamer stops flying. During further streamer elongation, the available charge just flows out of the older but still conducting section to the new streamer section. Under these conditions, the positive charge of a slow cathode-directed streamer is concentrated only in the head and a short region behind it, $v_s \Delta t \sim$ 1–10 cm, where $\Delta t \lesssim 10^{-6}$ s is the plasma decay time in cold air and $v_s \sim 10^7$ cm s^{-1}. There is practically no positive charge in the rest of the channel; on the contrary, there may appear even a negative charge of low linear density resulting from polarization of the partly conducting plasma. Nevertheless, space charge is distributed nearly uniformly along the streamer zone because of the presence of numerous streamers in various developmental stages.

In addition to the plasma decay, there is another reason for an appreciable decrease in the streamer linear charge—the electrostatic effect of charged streamers on one another, which is enhanced, as the number of streamers becomes larger and the distance between them becomes smaller.[1] This effect can be illustrated with a set of N thin conducting parallel rods of radius r and length $l \gg r$. Suppose all of them are connected to a common voltage source U (something like this happens in a streamer zone, with the leader head as a source). If the rods are separated by 'infinite' distances (in reality, much larger than l), there is no interaction, and the average linear charge τ of each rod is the same as that of a separate rod (Section 3.2.1). It is related to the potential as

$$U = \frac{\tau}{2\pi\varepsilon_0} \ln\left(\frac{l}{r}\right) \qquad (6.6)$$

Let us now reduce the distance between the rods by placing them into a cylinder of the same length l and radius R such that $r \ll R < l$. The potential imposed by the voltage source of each rod, say, of the first one, is determined not only by its own linear charge τ_1 but also by charges τ_k of the other rods. Calculations yield for the first rod center:

$$U = \frac{1}{2\pi\varepsilon_0} \left[\tau_1 \ln\left(\frac{l}{r}\right) + \sum_{k=2}^{N} \tau_k \ln\left(\frac{l}{z_{1k}}\right) \right] \qquad (6.7)$$

where z_{1k} is the shortest distance between the first and the k-th rods.

If, for better clarity, we distribute all rod axes uniformly over a circle of radius R instead of the cylinder base, the charges of all the rods will, by virtue of identical conditions, be identical, τ_N. It is clear from expression (6.7) that τ_N is smaller than τ and decreases with the number of rods. For instance, at $l = 200$ cm, $R = 50$ cm, and $r = 0.1$ cm, we have $\tau_{10} = 0.4\tau$ and $\tau_{100} = 0.05\tau$. At $1 \ll N < N_{\max} = 2\pi R/r$, the value of τ_N approaches a $1/N$ fraction of the linear charge of a separate rod with length l and radius R, $\tau_N' \approx 2\pi\varepsilon_0 U / [N \ln(l/R)]$; therefore, we obtain $\tau_{10}' = 0.55\tau$ and $\tau_{100}' = 0.055\tau$. Figure 6.9 presents calculations for the case when the rod axes are, for simplicity, arranged uniformly and concentrically over the cylinder area. There is a rapid charge reduction from the outer surface inward.

Besides, the mutual electrostatic effect reduces the head charges distributed in the streamer zone, but it is much weaker, since the mutual effect of two point charges is characterized by the geometrical factor $1/z_{1k}$,

---

[1] Being of purely electrostatic nature, this effect is also present in a branched corona flash, where streamers may be more powerful and faster and where the plasma cannot decay so much in a shorter time.

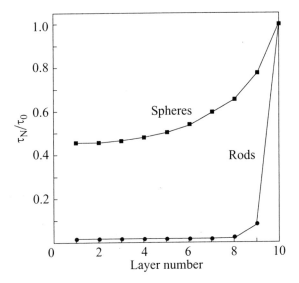

FIGURE 6.9
Charges of metallic rods and spheres under the same potential, located at the
distance $r = 2N$ cm from the central axis.

which diminishes faster than in the case of linear charge, $\ln(l/z_{1k})$. The
charges of the heads located inside the streamer space are reduced several
times, as compared to those on the outer surface, but not by orders of magnitude like in the channels (Figure 6.9).

Thus, both reasons—the lower conductivity due to the plasma decay
and the electrostatic interaction—lead to the streamer charge being concentrated in the streamer heads rather than in their channels. This conclusion has been convincingly confirmed by experiment. Optical registration of an initial flash has shown that its streamer branches have about
the same length. In other words, their heads are located in a narrow layer
at the external corona boundary, where most charge is concentrated. This
could not have happened, if charge were transported not only by the heads
but also by the channels.

Let us see now what determines the integral characteristics of a streamer
zone—its size and charge Q—in the initial leader stage. These are important because the transverse dimension of a streamer zone and its space
charge density ρ are the primary parameters for the channel cover. Represent a leader as a cylinder of radius R with an adjacent hemispherical
streamer zone of the same radius. A thin channel of radius r_L extends
along the cylinder axis. At the end of the channel is a head (tip) of radius

$r_t \ll R$ with the hemisphere center in it ($r_L \ll r_t \ll R$); the head has potential U_t and certain charge q_t, which will be shown to be small relative to Q. Suppose the system has already stabilized, and the structure and size of the streamer zone do not change, or only quasistationarily. This model is similar to the streamer model discussed in Chapter 3. While the leader is in the initial stage and has a length comparable with the streamer zone radius R, the model can be simplified by replacing the channel cover by another hemisphere, thereby surrounding the leader head with a 'streamer' sphere of radius R.

Streamers fly out of the leader head all the time, which means that the field E_t on its surface is sufficient for ionization and is over $E_i \approx 30 \text{ kV cm}^{-1}$ in atmospheric air. On the other hand, E_t cannot greatly exceed E_i, because the streamer flux would become more powerful and its space charge would decrease the field (Section 6.2.1). The fact of $E \sim E_i$ has been both theoretically and experimentally established for electrodes displaying continuous coronas [6.28]. In the case concerned, the head displays a quasistationary corona. The head charge in such a low field, $q_t = 4\pi\varepsilon_0 r_t^2 E_i \sim 3 \times 10^{-8} \text{ C}$ ($r_t \sim 1$ cm), is capable of providing only a small portion ($r_t E_i \sim 30$ kV) of the potential which, in actual reality, is measured in megavolts in long leaders. So we can ignore the head charge on the assumption that its potential is entirely created by the gap space charge.

For evaluating the potential at the center of a 'streamer' sphere, the distribution of space charge ρ along the radius might be taken to be $\rho = $ const, but this would contradict the experimental fact of field E being uniform in most of the streamer zone. The Gaussian theorem with $E(r) = $ const is, however, consistent with the charge distribution $\rho = 2\varepsilon_0 E/r$, which we will accept. If the leader head is far from the ground, its potential is

$$U_t \approx \frac{1}{4\pi\varepsilon_0} \int_0^R \frac{4\pi r^2 \rho \, dr}{r} = 2ER \tag{6.8}$$

while on the outer sphere surface $U_R = Q/(4\pi\varepsilon_0 R) = U_t/2$, where $Q = 4\pi\varepsilon_0 E R^2$ is the total sphere charge and $E = (U_t - U_R)/R$ is the average field in the streamer zone.

Streamers are known to move slowly in the initial leader stage, because they are 'weak'; therefore, the field is close to the minimum capable of sustaining streamers, $E \approx E_s \approx 4.65 \text{ kV cm}^{-1}$ for atmospheric air. If the gap voltage and channel length are such that $U_t = 1.5$ MV, the streamer zone has, according to (6.8), size $R = 160$ cm; this value is in reasonable agreement with experiment. The total sphere charge is $Q = 1.26 \times 10^{-4}$ C and the streamer zone charge is $Q_s = Q/2 = 6.3 \times 10^{-5}$ C. If average charge of a streamer is equal to $q_s \approx 5 \times 10^{-10}$ C, a value derived from calculations and experiments (Sections 5.4.5 and 6.8), there are 10^5 streamers in a streamer zone, or one streamer head per 100 cm^3. The head charge $q_t \approx Q\,(E_t/E)\,(r_t/R)^2 \sim 10^{-3}Q$ is small enough to be neglected. Incidentally, its value cannot be found exactly from electrostatic considerations, since it is determined by processes in the gas around the head.

The length of a streamer zone changes, as it approaches a grounded plane. At the moment of contact with the plane, all voltage U_t, but not half of it as before, drops in the streamer zone, since there is no other space between the head and the plane, so the zone length doubles, $L_s = U_t/E_s$. This is clearly seen in streak pictures of a leader, if an electronic shutter screened the image exactly at the moment of transition to the final jump [Figure 6.5(a,b)]. The greatly extended streamer zone does not look like a hemisphere at all, and it should be approximated by a more complex cylinder model.

6.7.2   Ionization in the leader cover

We have mentioned that radius R and charge density ρ in the initial cover cannot differ much from those in the streamer zone interior. This was a key idea of evaluations in the previous model, in which the streamer zone and initial cover were replaced by a sphere with space charge. The cover charge is produced by numerous charged inclusions—the former streamer heads. These charges are in their initial form, since the streamer channels have already lost their conductivity and connection with the conducting leader channel. It might seem that there is nothing that could disturb the cover state during 10^{-3} s, while the leader is elongating to dozens of meters, but this is not the case, and the main reason is partly of electrostatic nature.

Assume for simplicity that the channel possesses a perfect conductivity and that constant positive voltage U_0 is applied to the gap. Then the channel potential will vary neither axially nor in time, coinciding with the head potential, $U_t = U_0$. But as the leader length L increases, the constant space charge density of the 'invariable' cover is supposed to stimulate the growth of potential φ, because its total charge grows with L. Beginning with a certain length, potential φ will inevitably exceed the actual value, U_t, leading to charge removal from the channel to the source. The channel charge will alter its sign, and its linear density τ_1 must become negative to provide electrostatic equilibrium, $\varphi + \varphi_1 = U_t$, where φ_1 is the potential created in the channel by its charge τ_1. At a certain length, the charge τ_1 necessary for the equilibrium would become so large that a strong reverse field would arise at the channel surface, initiating a powerful corona. The corona would incorporate negative charge into the cover, reducing its resulting charge. This is what actually happens in reality but in a quieter, quasistationary way, without strong reverse fields. The token of the quasistationary mode is the relative low rate of leader development, at which a corona of opposite sign quenches the reverse field. Our hyperbolization of the effects was aimed at making the causative relations stand out clearer.

The above reasoning will now be illustrated by the calculation of linear leader charge corresponding to the quasistationary state. Irrespective of the law for the radial charge density $\rho(r)$ in a cylinder cover, the potential φ induced by this charge in the channel is expressed by the same formula (3.17) or (6.6) through the linear charge density:

$$\varphi = \frac{\tau}{2\pi\varepsilon_0} \ln\left(\frac{L}{R_{\text{eff}}}\right) \qquad \tau = 2\pi \int_0^R \rho(r)\, r\, dr \tag{6.9}$$

Here, R is the actual cover radius coinciding, in our model, with the radius of a hemispherical streamer zone and R_{eff} is the effective cover radius, on which the linear charge is concentrated. The calculation yields $R_{\text{eff}} = R/\bar{e}^{1/2}$ ($\bar{e} = 2.718\ldots$) for $\rho(r) = \text{const}$ and $R_{\text{eff}} = R/\bar{e}$ for $\rho = \text{const}/r$, as in the streamer zone. Since φ varies with R_{eff} only logarithmically, the choice of $\rho(r)$ does not matter much, especially at $L \gg R$.

If the cover state were invariable, its characteristic density and linear charge would be close to those in the streamer zone, where $\tau_{\text{s}} = Q_{\text{s}}/R \approx \pi\varepsilon_0 U_{\text{t}}$, according to the results of Section 6.7.1. By substituting $\tau = \tau_{\text{s}}$ into expression (6.9), we find $\varphi \approx (U_{\text{t}}/2)\ln(L/R_{\text{eff}})$. The potential φ becomes larger than U_{t} with $L > \bar{e}R$ (at $\rho \sim 1/r$). Since the potential created by the channel charge is equal to $\varphi_1 = (\tau_1/\pi\varepsilon_0)\ln(L/r_{\text{L}})$, the condition of electrostatic equilibrium $\varphi + \varphi_1 = U_{\text{t}}$ is fulfilled at

$$\tau_1 = 2\pi\varepsilon_0 U_{\text{t}} \frac{[1 - (1/2)\ln(L/R_{\text{eff}})]}{\ln(L/r_{\text{L}})} \tag{6.10}$$

For example, for $U_{\text{t}} = 1.5$ MV, $L = 10$ m, $R_{\text{eff}} = R/\bar{e} = 0.6$ m, and $r_{\text{L}} = 0.3$ cm, we get $\tau_1 \approx -3.8 \times 10^{-8}$ C cm^{-1}. The field at the channel surface $E(r_{\text{L}}) = \tau_1/2\pi\varepsilon_0 r_{\text{L}} \approx -245$ kV cm^{-1} could indeed stimulate a very powerful negative corona.

But in reality, no catastrophe occurs: the leader propagates slowly, increasing its length up to $L = 10$ m over $t \sim 1$ ms at $v_{\text{L}} \sim 10^6$ cm s^{-1}. When the reverse field exceeds the corona ignition threshold E_{ic}, which is 50 kV cm^{-1} for a 'wire' of radius $r = r_{\text{L}} = 0.3$ cm from the empirical formula

$$E_{\text{ic}} = 31\delta \left[1 + \frac{0.308}{\sqrt{\delta r\,[\text{cm}]}}\right] \text{ kV cm}^{-1}$$

then a reverse corona of moderate intensity is ignited at the channel surface. It contributes to the cover a charge opposite in sign to that already present there. The total cover charge decreases gradually. As a result, the reverse field at the surface of a slowly developing leader is maintained at a level close to the corona threshold E_{ic}. This is the way the quasistationary nature of the leader process manifests itself.

Owing to the reverse corona, the linear charge density in the channel $\tau_1 \approx 2\pi\varepsilon_0 r_{L} E_{ic}$ does not prove to be too large. The respective potential $\varphi_1 = r_{L} E_{ic} \ln (L/r_{L})$ is only a small fraction of the actual average potential in a long leader. Its value $\varphi_1 \approx 120$ kV is 8% of the average potential value, $U \approx 1.5$ MV at $L = 10$ m, $r_{L} = 0.3$ cm, and $E_{ic} \approx 50$ kV cm^{-1}. Therefore, we can neglect the component φ_1 in the electrostatic equilibrium equation

$$\varphi_1 + \varphi = \varphi_1 + \frac{\tau_{L}}{2\pi\varepsilon_0} \ln \left(\frac{L}{R_{\text{eff}}} \right) = U \qquad \tau_{L} \approx \frac{2\pi\varepsilon_0 U}{\ln (L/R_{\text{eff}})} \qquad (6.11)$$

and evaluate the average linear charge τ_{L} from average potential U.

One can see that we return to the conventional formula (6.6) for the linear charge (and capacitance) of a long linear conductor, meaning by its radius the effective radius of a charged cylinder R_{eff} and by τ_{L} the actual linear leader charge, or the cover charge. With $L = 10$ m, $R_{\text{eff}} = 0.6$ m, and $U = 1.5$ MV typical of laboratory discharges, we obtain $\tau_{L} \approx 0.3$ μC cm^{-1}, a value 1.3 times lower than the average linear charge of a streamer zone, which is the initial cover charge. It is worth noting that the problem of cover charge origin associated not only with the streamer zone but, probably, also with ionization processes in the cover itself is of little importance. This conclusion will permit estimation of the initial leader current in a way similar to that for a streamer (Section 3.2.2).

Thus, a leader cover is a dynamic structure, and its linear charge varies with the leader evolution. Only in case of a rapid voltage rise in the gap can the cover charge be unipolar and its linear density can increase in time (owing to the incorporation of charge of the same sign, when the near-surface field is positive and exceeds E_{ic}). If the voltage decreases, remains constant or rises slowly, the cover becomes two-layered: the outer layer charge has the same sign as the streamer zone and the inner layer charge is opposite in sign, with the total leader charge decreasing in time.

6.8   The origin of leader current

The origin of currents in the initial leader stage and in the final jump is somewhat different. Until the streamer zone has crossed the gap and there is no charged particle flow through the grounded electrode, the current in the leader base at the high voltage electrode is determined by how fast the gap charge $i_0 = dQ_{L}/dt$ is accumulated. This is the charge accumulated in the cover, on the leader surface and in the streamer zone. After the overlap, the charges of streamer heads arrive at the opposite electrode, making the current convective. It is now determined by charge transport by streamers from the leader head to the grounded electrode. Let us analyze both mechanisms.

6.8.1   The initial leader stage

The current through the leader base can be written as $i_0 = \mathrm{d}\,(CU)\,/\mathrm{d}t$, where C is the total capacitance of the leader system and U is its average potential. In case of a long leader, we have to deal with distributed linear capacitance $C_1\,(x)$ and potential $U\,(x)$:

$$i_0 = \frac{\mathrm{d}}{\mathrm{d}t} \int_0^L C_1\,(x)\,U\,(x)\,\mathrm{d}x \qquad (6.12)$$

In a simple situation of a leader developing in a steady-state mode at constant voltage, when the streamer zone does not vary in size, we have

$$i_0 = C_1 U v_\mathrm{L} = \tau_\mathrm{L} v_\mathrm{L} \qquad C_1 \approx \frac{2\pi\varepsilon_0}{\ln\,(L/R_\mathrm{eff})} \qquad (6.13)$$

where $C_1 = C/L$ is average linear capacitance and R_eff is an equivalent radius of the charge cover. Experimentally, we obtain $L/R_\mathrm{eff} \sim 10\text{--}20$ and $v_\mathrm{L} \approx 1\text{--}2$ cm $\mu\mathrm{s}^{-1}$; from formula (6.13) we have $i_0 \approx 0.5\text{--}1.5$ A. This is the range for many experimental data on current in the initial leader stage. If the gap voltage varies in time, a second term $C\mathrm{d}U/\mathrm{d}t$ appears in the expression for current i_0. With a rapidly varying current, its absolute value may be comparable with $\tau_\mathrm{L} v_\mathrm{L}$ and even exceed it.

In a long leader with a large linear resistance, the current markedly varies along its length. This variation is also stimulated by changes in the gap voltage U_0 (Section 6.3). With a rapid voltage rise, the base current turns out to be stronger than at the head, because some of the current input into the channel is spent for raising the linear leader charge. If U_0 drops in time, the charge leaves the leader to enter the electrode, and the system becomes discharged. The current near the electrode decreases, increasing along the channel towards the head.

Even in the case of constant voltage and small leader resistance, when the potential is nearly constant along the channel, the head current exceeds the average leader current or the base current (6.13). A similar situation was considered for the streamer in Section 3.2.2. Average linear capacitance in a streamer zone is larger than that in a channel, and it receives a larger charge per unit time. Indeed, charge Q_s of the streamer zone is transported to a new site for the time $\Delta t = R/v_\mathrm{L}$—the zone is displaced forward by its radius R. The head current at this moment is

$$i_t = \frac{Q_s}{\Delta t} = Q_s v_L R = \tau_s v_L = \pi \varepsilon_0 U_t v_L \qquad (6.14)$$

The latter equality uses the result of Section 6.7.1. The ratio of the head current and the average leader current is

$$\frac{i_t}{i_0} \approx \frac{\tau_s}{\tau_L} \approx \frac{1}{2} \ln \left(\frac{L}{R_{\text{eff}}} \right) \qquad (6.15)$$

This formula is similar to (3.24), except that here the radius r_c of the charge region is replaced by the effective cover radius R_{eff}. The difference in the current values is largely due to the different logarithms. For a meter-scale streamer, $L/r_c \sim 10^3$ and i_t/i_0 may be as large as an order of magnitude, while for a laboratory leader $L/R_{\text{eff}} \sim 10-20$ and the current ratio is only 1.5.

However, the origin of leader head current is somewhat specific. Charge is incorporated into a streamer zone by macroscopic structures—streamer heads which act as current carriers. The phenomenological expression (6.14) for head current coinciding with streamer zone current in a steady-state can be written in the canonical form of a charge carrier flow:

$$i_t = \frac{Q_s v_L}{R} \approx \pi R^2 q_s n_s v_L \qquad n_s = \frac{3 Q_s}{2 \pi R^3 q_s} \qquad (6.16)$$

where n_s is an average carrier density, q_s is carrier charge, v_L is an average velocity, and πR^2 is the current cross section.

Note that the average carrier velocity is different from the actual streamer velocity v_s but is equal to the leader velocity v_L. This is not surprising, because streamers cross the streamer zone at velocity $v_s \gg v_L$ and stop in the weak field region. The leader head slowly passes them by at velocity v_L. But the current in it is determined by the average relative velocity of the head and the charge carriers, which coincides with v_L because the time of flight for a streamer $\Delta t' \approx R/v_s$ is small, as compared to the total time of relative motion of the head and the carriers, $\Delta t = R/v_L$. Thus, the head current is maintained owing to a continuous arrival of streamer charges, which enter the cover, as the leader elongates.

Although the direct source of leader current is a continuous emission of streamers by the leader head, the emission frequency ν_s is, in a sense, a secondary quantity. It is not the emission frequency that gives rise to current, but the frequency is established in such a way as to provide current [see expresson (6.16)]. If we neglect streamer branching in the gap space and ignore the contribution of streamer conduction current, which is small, then the frequency ν_s will be defined by the equality $i_t = \nu_s q_s$. For the charge of a streamer head equal to $q_s \approx 5 \times 10^{-10}$ C (Section 5.4.5), the current of 1 A will result from the frequency $\nu_s = 2 \times 10^9$ s^{-1}. This is a very large value (it will be shown below to be supported by experiment). On the average, a streamer is emitted every 0.5 ns. This time is too short for the leader to cover even a distance of its head length. The emission rate varies with the leader head field. As a result of a complicated self-regulating process involving both the streamer production and the electro-static effect of the ever varying streamer charge on the leader head (every charge appearing in front of the head, sooner or later, finds its way to the cover behind it), the established field is such that the removal of charge carriers from the streamer zone is compensated by the production of new ones. A comprehensive quantitative description of this process, playing a key role in leader development, is a matter of future research efforts.

6.8.2 The final jump

During the final jump, streamers reach the opposite electrode, taking their head charges away from the gap. If streamers were conductors, the conduction current would flow through the leader channel and streamers from the anode directly to the cathode. Nothing of the kind happens in air. Because of the plasma decay, the conduction current through a single streamer is too small to be detected. Electron charge comes only from a short section behind the streamer head, which has not yet lost conductivity, and its average value of 5×10^{-10} C has been estimated to be close to the charge of a streamer head. This probably is the case, because a streamer channel practically has no charge, when it is in a tight bunch of similar streamers.

The streamer heads arriving at a grounded electrode can be counted 'by the piece'. Such an experiment was performed using small measuring sections on a grounded cathode (Section 5.4.5) to detect every individual streamer [6.29]. The sections were placed at different radial distances from the gap axis, and the number of streamers that arrived at each section over a fixed time was used to find their radial distribution function. Integration over the cathode area gave the total number of streamers. The number of streamer branches in a 1 m rod-plane gap was found to be $N = (2.4–2.7) \times 10^5$ at an average current of 5 A for 25 μs of a positive leader propagation in the final jump. Their average frequency of $\nu_s \approx 10^{10}$ s^{-1} did not coincide with the initial streamer frequency at the leader head because of their extensive branching. If the average charge of a single streamer head was $q_s = 5 \times 10^{-10}$ C, the transported charge $Q = q_t N = 1.25 \times 10^{-4}$ C coincided, within a 5% accuracy, with integral measurements of the cathode current. This again supports the convective nature of leader current in the final jump.

If streamer heads are current carriers and if they cover the distance between the leader head and the cathode like elementary charges, the current in the final jump can be found from an expression similar to (6.16):

$$i_f \approx \pi R_f^2 q_s n_f v_s = \tau_f v_s \qquad (6.17)$$

where R_f is the cross section radius of a streamer zone, n_f is the average streamer head density over the zone space, v_s is the average streamer velocity, and τ_f is the average linear charge of the streamer zone in the final jump. The linear charge τ_f can be found in a way similar to that for the streamer zone in the initial stage, taking into account its extension prior to the contact with the cathode (Section 6.7.1); it is reasonable to represent it as a cylinder rather than a hemisphere. The leader head will be located in the center of one cylinder base, the other base coinciding with the cathode plane. The cylinder length, or the streamer zone length in the final jump, L_f, decreases as the head approaches the cathode, but right after the moment of contact, L_f is several times larger than the diameter $2R_f$.

As in the initial stage, the potential at the leader head center, U_t, practically coincides with that of the streamer zone charge plus the image charge in the cathode plane. If $\rho(x,t)$ is space charge density, we have

$$U_t = \frac{2\pi}{4\pi\varepsilon_0} \int_0^{R_f} r \left[\int_0^{L_f} \frac{\rho dx}{(r^2+x^2)^{1/2}} - \int_{L_f}^{2L_f} \frac{\rho dx}{(r^2+x^2)^{1/2}} \right] dr$$

To take the integral, we assume $\rho = $ const and restrict ourselves to an extended streamer zone with L_f several times over R_f. By expressing the average space charge density ρ through the average linear density $\tau_f = \rho \pi R_f^2$, we obtain

$$\tau_f = \frac{4\pi\varepsilon_0 U_t}{\ln\left(\bar{e}^{1/2} L_f / R_f\right)} \qquad (6.18)$$

With a reasonable choice of L_f/R_f, the value of τ_f from expression (6.18) coincides, in order of magnitude, with the average linear charge of a streamer zone in the initial stage, $\tau_s \approx \pi\varepsilon_0 U_t$,[2] at the same head potential U_t. It follows from expressions (6.17) and (6.14) that leader currents in the two stages are related as streamer and leader velocities, that is, the leader current increases after the transition to the jump, at least, by one order, as compared to the initial stage. This result has been supported experimentally.

As the leader head approaches the grounded electrode and the streamer zone becomes shorter, its average field $E_f = U_t/L_f$ rises, leading to a higher streamer velocity. It is as high as $10^9\,\mathrm{cm\,s^{-1}}$ at $E_f \approx 20\,\mathrm{kV\,cm^{-1}}$, and the

---

[2]This is also valid for other reasonable distributions of density ρ; for example, $\bar{e}$ appears instead of $\bar{e}^{1/2}$ in expressions (6.18) at $\rho \sim 1/r$.

leader current becomes equal to 10^3 A, or by three orders of magnitude over the characteristic initial values. Since the time of streamer flight becomes much shorter, streamers retain a high conductivity, thereby increasing the current input. Due to their residual conductivity, streamers connect, like thin wires, the leader head to the grounded electrode. The higher the streamer conductivity, the lower is their production frequency providing the same current. This was confirmed in [6.30–6.32] by counting streamers arriving at the cathode in hot air, nitrogen, and in argon with a small proportion of oxygen: in these gases streamers preserve their conductivity longer. For example, the streamer frequency dropped to 10^8 s^{-1} at $T = 650$ K and to 10^7 s^{-1} at $T = 900$ K. It was only one hundredth the value in cold air at the same current. Nitrogen streamers appeared as individual flashes, and the streamer zone between them did not glow. A similar picture was registered for argon.

6.9 Streamer-leader transition

The term 'streamer-leader transition' is often applied to both the leader inception at the electrode in the initial corona stem and to the formation of a new leader section within the streamer zone. The latter is similar to a corona flash with the leader head playing the role of the electrode. The streamer-leader transition is the least understood and, theoretically, the most difficult stage in the leader evolution. The transition function, its phenomenological description, and the final result are generally clear. A new leader section is produced by numerous streamers starting from the leader head. Their total current is concentrated in a narrow channel, heating the air and inhibiting electron losses. Owing to the conductivity maintenance and considerable heat release (Section 2.2), the air is eventually heated up to 5000–6000 K, as evidenced by spectroscopic measurements. The plasma state in a hot channel is such that it passes leader current of about $i \sim 1$ A at a relatively low longitudinal field $E \sim 1$ kV cm^{-1} and less (reduced field is $E/N < 10^{-15}$ V cm^2; see Section 6.10).

An important role in the understanding of the streamer-leader transition in air was played by I. Gallimberti's hypothesis, but today it can be accepted only with reservations. In this hypothesis, air in initially cold streamers is to be heated to about 1500 K to make this process irreversible. This temperature is capable of liberating electrons from a large number of negative ions produced in the streamer plasma; the electron attachment is later compensated by their accelerated detachment. The heating goes on due to the restored conductivity. Further, during the initial current flow through cold air, about $\xi_V \approx 95\%$ of the energy gained by electrons from the field is transferred to vibrations of nitrogen molecules. The vibration relaxation time τ_{VT} abruptly decreases with increasing gas temperature, reaching $\tau_{VT} \approx 10^{-5}$ s at $T \approx 2500$ K. At this temperature, the vibrational energy transforms to translational energy for $10^{-5}-10^{-4}$ s, and the channel temperature approaches the 'final' value of 5000 K.

This scheme is basically valid, but some points need to be refined. Calculations show (Section 5.5) that the detachment accelerated at $T > 10^3$ K is unable to increase the number of electrons by several orders of magnitude for the simple reason that there are not as many negative ions in a streamer plasma as is necessary for producing so many electrons. At high electron density, the attachment is dominated by the dissociative recombination with complex O_4^+ ions (Figure 5.16). For the attachment time $\tau_a = \nu_a^{-1} \sim 10^{-7}$ s at the recombination coefficient $\beta_{ei} \approx 3 \times 10^{-7}$ cm^3 s^{-1}, corresponding to field $E \approx 5$ kV cm^{-1} typical of a streamer, the recombination 'cuts' any high electron density down to $n_e \approx (\beta_{ei}\tau_a)^{-1} \approx 3 \times 10^{13}$ cm^{-3}, as is clear from expression (2.14). For this reason, the densities n_- and n_e are comparable, but more often $n_- < n_e$ (Figure 5.18). At $T \approx 1500$ K, the attachment is indeed mostly compensated by the detachment due to the appearance of oxygen atoms and active molecules capable of destroying negative ions. The plasma decay is decelerated not only for that reason but also because of the disappearance of complex ions, primarily O_4^+ recombining faster than O_2^+.

But the principal problem we are faced with when analyzing the streamer-leader transition is quite different, and judging from the available literature, it has not been discussed in detail. We mean a specific mechanism for the accumulation of currents from numerous streamers. Leader current of $i \sim 1$ A is able to heat only a limited air volume up to the necessary 1500 K (for the estimations of streamers, see Section 3.3). Therefore, there must be a mechanism for the current accumulation within a region of small radius inside the head, which is the only structure that becomes heated to transform to a new leader section. Let us evaluate the upper initial radius r_0 corresponding to cold air density ρ_0. According to the results of Sections 3.3.1 and 6.7.1, the unit elongation of a leader in a well-developed streamer zone requires the energy $C_{1s}U_t^2/2$, where $C_{1s} \approx \pi\varepsilon_0$ is the linear leader capacitance in the nearby region. The maximum volume of air to be heated from $T_0 = 300$ K to $T = 1500$ K is defined as

$$\pi r_{0\,\text{max}}^2 \rho_0 c_p (T - T_0) = \frac{\pi\varepsilon_0 U_t^2}{2}(1 - \xi_V) \tag{6.19}$$

where c_p is specific heat at constant pressure (a leader moves slowly enough for the heated gas to expand). At $U_t \approx 1$ MV, we have $r_0 = 0.03$ cm.

It is likely that the current accumulation is due to instabilities similar to those causing the contraction of a glow discharge [6.35]. Such announcements have already been made in the literature, but that is the only thing that has been done. We are suggesting a mechanism that seems quite probable. Whatever happens to the head, where a new portion of the leader is formed, its current remains constant, because it results from processes in the streamer zone and is fixed by its huge equivalent resistance $U_t/i \sim 10^6 \ \Omega$ ($U_t \sim 10^6$ V is the leader head potential). The leader head is a source of mild streamers produced at frequency $\nu \sim 10^9 \ \mathrm{s}^{-1}$ and propagating with velocity $v_s \sim 10^7 \ \mathrm{cm \, s}^{-1}$. Their channels lose their conductivity for the time $\tau_a \sim 10^{-7}$ s over the length $l \sim v_s \tau_a \sim 1$ cm. In other words, a bunch of short ($l \sim 1$ cm) conductors—young streamers numbering $N_s \sim \nu \tau_a \sim 100$—start from the leader head at any moment of time. The streamer current i is transported through a large streamer zone by the streamer heads separated from the leader by a nonconducting space. The current near the bases of these conductors is still conduction current. All N_s conducting streamers almost touch one another near the starting point to form a continuous conductor. Let us represent it as a cylinder of radius $r_{\mathrm{sum}} \sim r_s N_s^{1/2}$, where r_s is a streamer radius. The cylinder possesses initial electron density $n_e = n_0$ and conductivity σ and is in a longitudinal field $E = i/\left(\pi r_{\mathrm{sum}}^2 \sigma\right) \sim n_e^{-1}$, which can vary, in contrast to the current.

Suppose the temperature of a thin column of radius r_0 in a conducting cylinder has exceeded the common initial temperature T_0 by ΔT. In terms of the instability theory [6.35], we are considering the development of perturbations transverse to current with a wavelength r_0. Electron losses in the overheated column are smaller due to accelerated detachment and a smaller number of O_4^+ ions, while the ionization rate is higher because of a higher reduced field E/N resulting from thermal expansion. The conductivity of the overheated region will increase, increasing the current through it. As a result of enhanced heat release, the temperature will become still higher, and so on.

This is the way an instability develops, resulting in current transport from the whole conductor with radius r_{sum} to the column with radius r_0. Note that field rather than current is a constant parameter, when an overheating instability develops in a glow discharge.

The time of instability development, τ_{ins}, determines the leader velocity $v_L \sim l/\tau_{\mathrm{ins}} \sim v_s \tau_a/\tau_{\mathrm{ins}}$, while the length $l \sim v_s \tau_a$ determines the size of the leader head, r_t. The latter assertion has been supported experimentally [6.31]. The leader radius in air heated to 900 K, when the streamer plasma decay rate is lower, is $r_t \sim 10$ cm instead of the usual value of about 1 cm (Figure 6.10). Incidentally, a leader head does not seem to us to be a continuous structure but rather a fan of short, still conducting 'young'

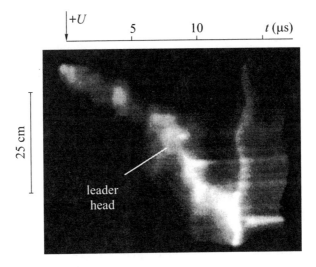

FIGURE 6.10
A streak photograph of the leader head trace in a rod-plane gap of 0.5 m long
at air temperature 900 K and $p = 1$ atm.

streamers. Their tight packing makes them look like a continuous glowing
background enhanced by light-emitting, excited and ionized molecules.

The instability development has been calculated numerically with the
equations presented in Section 5.5, taking into account the true kinetics of ionization, attachment, detachment, ion conversion, recombination,
and heating but without spatial derivatives. It is assumed that $i = 1$ A,
$r_{sum} = 0.1$ cm, $\Delta T = 100$ K, and $n_0 \sim 10^{15}$ cm^{-3}; the initial electron density has a slight effect, because it is effectively reduced by recombination to
the value of 10^{13} cm^{-3}. It turns out that the main contribution to the instability development is made by ionization rather than by electron loss in
a cold plasma or by electron liberation from negative ions, as was supposed
in [6.33, 6.34]. The reduced field in the overheated region rises abruptly
to $E/N \approx 1.2 \times 10^{-15}$ V cm^2 for 0.1 μs and continues to grow, until most
of the current is pooled into a narrow channel [Figure 6.11(a)]. This again
indicates that the role of attachment-detachment processes in the leader
formation has been exaggerated. In heated air, detachment does compensate for attachment, but even if it were not the case, the leader channel
would elongate all the same. A stronger field would arise at its head but
that is all. We emphasize that the current is imposed on the leader head
from the 'outside', while the field is adjusted to it to allow the ionization
to compensate for any electron losses.

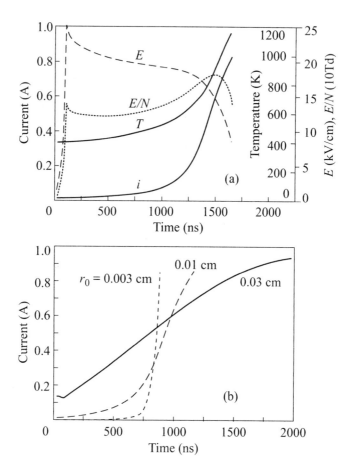

FIGURE 6.11
Numerical simulation of overheating instability in the leader head at the leader
current 1 A and current column radius 0.1 cm. One can see variations in the gas
temperature T, current i, longitudinal field E, reduced field E/N in an
overheated filament of radius $r_0 = 0.01$ cm (a); at various filament radii r_0 (b).

The evolution of initial perturbation varies with the radius r_0 of the
perturbed region, or the perturbation 'wavelength'. A smaller radius induces an explosion-like transport of the current to the narrow column [Fig-
ure 6.11(b)]; therefore shortwave perturbations prove to be more viable, although all the times are ~ 1 μs in the range of r_0 values considered. This
value has a minimum limited by perturbation attenuation due to heat conduction and ambipolar charge diffusion. Over the time $\tau_{ins} \sim 1$ μs of the
instability evolution, the heat and charges move across the current at a dis-

tance $\Delta r \sim (4\chi\tau_{\text{ins}})^{1/2}$, where χ is thermal conductivity, or the ambipolar diffusion coefficient. Both processes provide radius $r_0 \sim 10^{-2}$ cm for the lower limit and for the most viable perturbations. Thus, we have arrived at identical estimates of the initial leader radius from different (energetic and kinetic) considerations. This is natural, since the time τ_{ins} coincides in order of magnitude with the time necessary for heating a perturbed column up to temperature T at linear power iE corresponding to the leader current: $\tau_{\text{ins}} \sim \pi r_0^2 c_p \rho_0 T / iE \sim 1 \, \mu s$ at $i \sim 1$ A and $E \sim 20 \, \text{kV cm}^{-1}$. The leader velocity defined as $v_L \sim l/\tau_{\text{ins}} \sim 1 \, \text{cm} \, \mu s^{-1}$ ($l \sim 1$ cm) has a correct order of magnitude. No doubt, the above reasoning needs to be supported by a more substantial theory and calculations.

As far as the leader inception is concerned, the evaluations can be made using experimental data. A leader is initiated inside the stem of the initial corona flash (Section 6.1.3). Spectroscopic measurements give $T \approx 900$ K there. A corona flash is followed by a pause of about 10 μs. The gas in the stem does not cool during this time, but the plasma density drops by two orders and hardly exceeds $10^{12} \, \text{cm}^{-3}$. The field rises with increasing voltage and penetrates into the stem owing to the low n_e (Section 5.4.1). As a result, a secondary ionization wave passes along the stem, giving rise to a leader. One of the present authors registered the near-anode field of $47 \, \text{kV cm}^{-1}$ at gap voltage $U_0 = 270$ kV in experiments with a spherical anode of 3.12 cm radius at the moment of corona ignition. The incorporated charge $q = 1.5 \, \mu C$ decreased it to $15 \, \text{kV cm}^{-1}$. By choosing $E = 30 \, \text{kV cm}^{-1}$ as an average field, we can evaluate the energy released per unit stem length in the following two ways: as $W_1 = qE = 0.045 \, \text{J cm}^{-1}$ and as $W_1 \approx qU_0/2l \approx 0.1 \, \text{J cm}^{-1}$ (the stem length was found [6.36] to be $l \approx 1.8$ cm for the same anode and similar conditions). Both evaluations have yielded close results, coinciding, in order of magnitude, with the above energy release value in the head of an existing leader, $C_{1s}U_t^2/2 \approx 0.14 \, \text{J cm}^{-1}$. Hence, the maximum possible radius of a hot stem region is of the same order of 10^{-2} cm, as inside the leader head. Thus, the processes of leader inception and channel reproduction in the leader head occur more or less similarly.

6.10 Leader support voltage

Voltage U_0 applied to a gap partly drops in the leader channel, adding potential U_t to the head. In the initial stage, nearly half of the latter drops in the streamer zone and the rest in the unperturbed gap region as far as the cathode, if the leader is positive. A quasistationary state with a more

or less constant longitudinal potential gradient is established in most of a
multimeter channel. It is then reasonable to write the voltage balance as:

$$U_0 = U_{\mathrm{L}} + U_{\mathrm{t}} \equiv E_{\mathrm{L}} L + U_{\mathrm{t}} \tag{6.20}$$

where E_{L} is an average channel field. If the channel is less than a few meters long and the plasma lifetime is $L/v_{\mathrm{L}} < (1\text{--}3) \times 10^{-4}$ s, the quasistationary state is still being established, and the field may appreciably vary
along the channel, so that the average value of E_{L} from expression (6.20)
cannot characterize adequately the plasma state. But for a long leader, the
evaluation of the necessary voltage U_0 reduces to finding two parameters:
the head potential U_{t} and the average channel field E_{L}.

6.10.1   Leader head potential

The leader head potential must be high enough to provide the formation
of a new leader section, adequate current, and the leader propagation. The
channel of a long leader is rather passive—it only transmits the high anode
potential to the leader head, like a positive column of glow discharge or
arc, where the cathode and its sheath are responsible for current support.
Of course, voltage losses due to the finite resistance are inevitable. One
may suggest that the gas temperature of about 5000 K satisfies a certain
'preferable' condition, when the resistance is not too high to consume most
of the applied voltage or too low to require an excessively high ionization
degree and head potential. Various leader propagation modes, more or
less close to the optimal conditions, are possible. Normally, a successful
leader development can be supported by a regular electrode voltage rise
to compensate for its drop due to elongation, unless there was an initial
excess voltage applied.

At present, there is no adequate physical theory to define analytically the
values of head potential and current necessary to promote leader propagation at a prescribed rate. Numerical models of the type described in [6.34]
involve such a large number of assumptions and additional parameters that
it is hard to see the principal tendencies behind the suggested equations
and calculations. There is a chance that no simple relations exist just because the process is too complicated; therefore, we will restrict ourselves to
commenting on some key points and to making some simple estimations,
using experimental findings wherever possible.

As was mentioned, the formation of a unit leader length requires energy
of about $C_{1\mathrm{s}} U_{\mathrm{t}}^2 / 2$, where $C_{1\mathrm{s}} \approx \pi \varepsilon_{0\mathrm{m}} = 2.78 \times 10^{-13}$ F cm^{-1}. This value
takes into account everything: the ionization and gas heating in the leader
channel and around it, the ionization along every streamer, molecular excitations (the most energy-consuming in the leader cover), etc. The energy

expenditures can be subdivided into two groups: those caused by the channel and those associated with its cover. It is hard to say how large they are; if the components are comparable, they should be of the order of $C_{1s}U_t^2/2$.

As for the channel, we will use the reasoning that yielded formula (6.19). The amount of energy released by molecular vibrations is quite sufficient to bring the air temperature in the initial channel from 1500 to 5000 K. Indeed, the vibrational degrees of freedom release $(1 - \xi_V)^{-1} \approx 20$ times greater energy during the relaxation than that necessary for the gas heating to 1500 K. However, specific enthalpy $h(5) = 12$ kJ g^{-1} of actual air at $p = 1$ atm and $T = 5$ kK is only 8 times greater, $h(1.5) \approx 1.5$ kJ g^{-1}. Therefore, the relaxation and increasing translational temperature have excess energy for heating the gas around the initial channel. The upper limit of the initial radius of a cold air column, which can be heated to the final temperature T, is described by a relation similar to equation (6.19):

$$\pi r_{0\ max}^2 \rho_0 h(T) \approx \frac{\pi \varepsilon_0 U_t^2}{2} \qquad (6.21)$$

At $U_t = 1$ MV and $T = 5000$ K, the initial radius is $r_{0\ max} = 0.054$ cm. As a result of thermal expansion, the column eventually has the radius $r_{L\ max} = r_{0\ max} [\rho_0/\rho(5)]^{1/2} = 0.26$ cm. Channels of this scale have been observed experimentally. The initial radius has a minimum $r_{0\ min} \approx 0.01$ cm defined by thermal conduction or diffusional extension of the initially small radius of a plasma column (Section 6.9). The value of $r_{0\ min}$ describes the minimum energy required for the creation of a leader channel. By substituting $r_{0\ min}$ for $r_{0\ max}$ into expression (6.21), we find the minimum potential capable of providing such energy expenditures: $U_{t\ min} \approx 200$ kV.

Let us now calculate the energy necessary for the creation of a cover. The velocity of streamers within the streamer zone is low, $v_s \sim 10^7$ cm s^{-1}, and comparable with electron drift velocity, when nearly all electrons leave the streamer head, exposing its ions (Section 3.1). Then the head charge $q_s \approx 5 \times 10^{-10}$ C determines the actual number of ions in it, q_s/e, and the number of ionization events along a streamer channel of the length of its head diameter, $2r_s$. Hence, the number of electrons and ions produced in a streamer of length l is $(q_s/e)(l/2r_s)$. The value of l coincides, in order of magnitude, with the streamer zone radius and initial cover radius R (Section 6.7.1). Thus, if $\tau_s = C_{1s}U_t$ is the linear cover charge near the leader head, the respective number of charge pairs produced is $(\tau_s/e)(R/2r_s)$.

Suppose the production of one pair of charges requires average energy w. In addition to the ionization potential, the quantity w includes energy for electronic excitation of molecules: $w \approx eE/\alpha(E)$, where α is the Townsend ionization coefficient in field E (Section 2.2). The field is close to maximum field in the streamer head, E_{max}, where the ionization largely occurred (Section 3.1). For a weak streamer with, say, $E_{max} = 80 \text{ kV cm}^{-1}$, we have $\alpha = 500 \text{ cm}^{-1}$ at $p = 1$ atm and $w = 160$ eV. By expressing the cover radius in formula (6.8) through U_t and average field E_s in the streamer zone, we obtain the energy for the formation of the cover unit length:

$$W_{1cov} \approx \left(\frac{\tau_s}{e}\right)\left(\frac{R}{2r_s}\right) w \approx \left(\frac{C_{1s}U_t^2}{2}\right)\left(\frac{w}{2eE_s r_s}\right) \tag{6.22}$$

This energy makes up a more or less definite portion of the total energy required, $C_{1s}U_t^2/2$. This energy portion is independent of the head potential but is determined exclusively by the streamer parameters varying with the streamer zone field rather than potential. The value W_{1cov} has much weight: with $E_s \approx 5 \text{ kV cm}^{-1}$, $w = 160$ eV, and $r_s = 0.03$ cm the energy necessary for the cover formation will make up 50% of the total energy. In other words, the minimum head potential necessary for a viable leader propagation estimated from expression (6.21) should be increased $\sqrt{2}$ times to give $U_{t \min} \approx 280$ kV. This may account for the experimental fact that a leader cannot survive in its initial stage in normal air at a gap voltage below 300–400 kV. It has been observed only in very short gaps crossed by streamers of an initial corona flash. But in that case, the process goes over to the final jump from the very beginning, and its energy supply mode becomes quite different.

6.10.2 Long leader plasma and field

The problem of the plasma state and field in a long leader channel, like many other problems related to the leader process, raises more questions than gives answers. This problem is of interest from the fundamental and practical points of view, for example, for prediction of gap electrical strength. The longer the leader, the more the applied voltage U_0 drops in the leader channel and the stronger is the dependence of electrical strength on the channel field, as is clear from formula (6.20).

Besides, the longer the leader lifetime, the closer is its state to that of a low current arc, which can be supported by a relatively low field (Sections 2.8 and 6.4.3). For instance, a leader of 100 m long can be sustained by 3.5 MV, even if the voltage rise mode is not optimal. If the head potential is as large as 1–1.5 MV of that value, the average channel field turns out to be 200–250 V cm^{-1}, which is only by a factor of 2–2.5 larger than the field in an air arc at the same current of 1 A as in the leader. However, a laboratory leader of a few meters long and a young section of the same length in a long leader may have 5–20 times stronger fields.

In order to find the field in a laboratory leader and to calculate the breakdown voltage in a gap of medium length $d = 5-15$ m, some workers assumed the reduced channel field to be constant and equal to $E/N \approx 8 \times 10^{-16}$ V cm^2 [6.8, 6.34], and others even refined this value to 7.8×10^{-16} V cm^2. That assumption was based on the following arguments [6.4, 6.33]. Electrons are produced in electron impact ionization of unexcited molecules and are lost in recombination (because the attachment is entirely compensated by the detachment in hot air). Under the conditions of ionization equilibrium, the rates of ionization, $\nu_i n_e$, and of recombination, βn_e^2, are identical, $\nu_i = \beta n_e$. But since the reduced ionization frequency ν_i/N is a rapidly growing function of E/N, the latter is more or less constant, being only slightly sensitive to the ionization degree n_e/N and the recombination coefficient β. This assumption can be conveniently employed for experimental data analysis. Consider, as an illustration, an experiment of Les Renardiers Group mentioned in [6.33].

The spectroscopic temperature in a leader that had reached the middle of a gap of $d = 10$ m between a conical anode and a plane at voltage pulses of $500/10000$ μs and amplitude of $1.6-1.8$ MV was about $5000-6000$ K. At the average current $i \approx 1$ A, the leader propagated with velocity $v_L \approx 2$ cm μs^{-1}. The initial channel expanded with a radial velocity of about 10^2 m s^{-1}, which decreased to 2 m s^{-1} over a period of 100 μs. Judging by the absence of expansion, the pressure in the channel was 1 atm. The shadow technique showed the average thermal radius of the expanded channel to be $r \approx 0.1$ cm. The vibration relaxation had stopped by that time, so that equilibrium was established in the heavy particle gas. From the postulated value of $E/N \approx 7.8 \times 10^{-16}$ V cm^2 and from $N = 1.48 \times 10^{18}$ cm^{-3} corresponding to $T = 5000$ K, the channel field was $E = 1.15$ kV cm^{-1}.

Suppose now that all the current flows through a column of radius r. The electron mobility in the considered range of E/N values is $\mu_e \approx 1.5 \times 10^{22}/N$ cm^2 V^{-1} s^{-1}. Hence, we have the drift velocity $v_e \approx 1.2 \times 10^7$ cm s^{-1}, conductivity $\sigma \approx i/\left(\pi r^2 E\right) \approx 2.8 \times 10^{-2}$ Ω^{-1} cm^{-1}, and electron density $n_e \approx \sigma E/\left(e v_e\right) \approx 1.7 \times 10^{13}$ cm^{-3}. These values look quite reasonable, and it was probably such estimates that made the postulate of $E/N = \text{const} = 8 \times 10^{-16}$ V cm^2 in a leader channel popular.

However, the above postulate should be used with caution. It may be quite justifiable for initial, cold channel sections or for short leaders, but it becomes unacceptable for sections heated over $3000-3500$ V. The latter refers to the above case of a 5 m leader at 5000 K. This temperature is likely to be close to the maximum gas heating in leaders with the typical current $i \approx 1$ A. For example, at $T = 10000$ K, specific enthalpy of air at 1 atm, $h(10) = 48$ kJ g^{-1}, is four times greater than at 5000 K. Where does all this energy come from?

To explain this situation, we would have to assume a current contraction

TABLE 6.3
The composition of equilibrium air at normal pressure

T (K)	4000	4500	5000	5500	6000
N (10^{18} cm^{-3})	1.79	1.60	1.48	1.35	1.27
n_e (10^{13} cm^{-3})	0.63	1.70	4.90	11.2	21.4
N_O (10^{17} cm^{-3})	4.70	4.90	4.60	4.35	3.81
N_N (10^{16} cm^{-3})	0.25	1.15	3.67	9.92	20.6
N_{NO} (10^{16} cm^{-3})	7.62	4.54	2.73	1.67	1.03

mechanism operating in a heated gas, which could make a current column still thinner. But at an average field $E = 1.1$ kV cm^{-1}, which follows from the postulate with $T = 5000$ K, the voltage in a 100 m leader would have to rise to an incredibly high value of 11–12 MV (at $U_t \sim 1$ MV). Even for a laboratory leader, the postulated E/N value leads to a 2.5–3 times greater breakdown voltage, which is not accidental either.

There is an unspoken argument behind the postulate of $E/N = $ const: the only result of gas heating that actually affects the ionization processes is a decreased gas density. The accepted high level of E/N corresponds to the necessity of electron impact ionization of nitrogen and oxygen molecules having a fairly high ionization potential. This is what happens in cold air. On heating, however, the nitrogen and oxygen molecules dissociate, making the associative ionization $N + O \rightarrow NO^+ + e$ feasible (see Section 2.2.1). Due to the low activation energy, 2.8 eV, associated primarily with the anomalously lower ionization potential of NO (9.3 eV) than that of O_2, O, N_2, and N, the reaction goes on at a rather high rate constant [see expression (2.13)]. The result of this process is clear from Table 6.3. One can see that the electron densities at 4500−5000 K are of the same order of magnitude as for the above estimates for a leader channel, while at 6000 K they are even higher by an order of magnitude. At these temperatures, the equilibrium dissociation degrees for N_2 and O_2, the NO concentration and the ionization are established over the time 10−100 μs [6.37, 6.38] taken by the leader for its elongation by 10−100 cm.

Direct electron impact ionization of NO can compete with associative ionization $N + O$ only at electron temperature $T_e \sim 20000$ K, that is, in strong fields. However, an actual plasma column with the gas temperature 5000 K and current 1 A does not need high fields. It was shown in Section 2.8 that under the heat balance conditions with the Joule heat removed from the current channel by heat conduction, like in an arc, the gas temperature T can be maintained by the linear power

$$W_1 = iE = 8\pi\lambda T \frac{kT}{I_{\text{eff}}} \tag{6.23}$$

Here, $\lambda(T)$ is heat conductivity in a current channel and I_{eff} is an effective 'ionization potential' of the gas, which characterizes the slope of the function $\sigma(T)$, when approximated by the function $\sigma \sim n_e \sim \exp(-I_{\text{eff}}/2kT)$, as for the equilibrium ionization of a monoparticle gas (Section 2.7). From Table 6.3, we have $I_{\text{eff}} \approx 8.1$ eV $= 94000$ K for the temperatures considered.

All parameters of a stabilized arc (i, current channel radius r, E, and T) are interrelated and determined by a quantity, normally by current, given by the external source. Something like this applies to a leader channel, so that the temperature of 5000 K can be regarded as being inherent in 1 A leader current defined by processes in the streamer zone. Expression (6.23), whose meaning was explained in Section 2.8, relates the temperature to the linear power, with no other parameters entering it explicitly. This facilitates the problem solution. For $T = 5000$ K, we have $\lambda = 0.02$ W cm^{-1} [Figure 2.6(c)] and $W_1 = 134$ W cm^{-1}. With $i = 1$ A, we obtain $E = 134$ V cm^{-1} and $E/N = 0.9 \times 10^{-16}$ V cm^2, which is by an order of magnitude less than the postulated value. At this E/N value, the electron temperature is twice as large as T, $T_e \approx 10000$ K (Section 2.1), but this will not affect the ionization rate. The difference between T_e and T is also great in 1 A arcs [6.35].

A field calculated with less than a two-fold error agrees with both the data for an arc described in Section 6.4.3 and the average field estimate for a long leader, $E = 200$–250 V cm^{-1}. Generally, the field necessary to support the leader current $i = 1$ A could be found from the equilibrium electron density corresponding to 5000 K as $E = i/\left(\pi r^2 e\mu_e n_e\right)$. But this kind of evaluation is not particularly informative because of an arbitrary choice of current cross section radius r greatly affecting the result. It is better to find r from an 'experimental' field, say, $E = 250$ V cm^{-1} for a long leader and $T = 5000$ K, which correspond to $E/N = 1.7 \times 10^{-16}$ V cm^2 and $n_e = 4.9 \times 10^{13}$ cm^{-3}. For $i = 1$ A, we obtain $r = 0.13$ cm, a value close to the thermal radius [6.4].

We will quote another experimental fact indicating that 'unconventional' ionization mechanisms may result in an abrupt reduction in E/N. A positive glow discharge column in purified nitrogen at elevated pressures has $E/N \approx (2$–$4) \times 10^{-16}$ V cm^2, although electron impact ionization of unexcited molecules would require the same hypothetical value of 8×10^{-16} V cm^2 or more. Special purpose experiments have shown that there is probably a more effective way of ionization, which should not be excluded for an air leader. Electrons are produced from the reaction $N_2 + N_2 \rightarrow N_4^+ + e$ involving molecules with a high vibration excitation. Such molecules result from the exchange of vibrational quanta in collisions of vibrationally excited molecules rather than from the electron impact requiring high E/N values [6.35].

Although it is clear from the foregoing why the reduced field in a long mature leader is much less than in cold channel sections, there are still questions to be answered. The field in a steady-state arc with identical current is half that in a long leader. Of course, the estimation of the leader field may contain an error, because more voltage may drop in the transition region of the leader or in front of its head than is generally believed. But there are also physical reasons to question the complete identity of conditions in a leader and an arc. The point is that the heat balance in a long arc is stabilized either by the cooled walls, to which the released heat is removed, or by a flow, as in a plasmatron. When an arc is burning in an open air, it is cooled by a convective air flow. As for the leader process, its lifetime is too short for convection to develop. Therefore, the state of the leader channel remains unbalanced. The heat spreads radially away from the channel, heating the surrounding air. This may require additional energy, thereby increasing the linear power and field. Although the heat spread after the initial channel expansion is a slow process, it should not be entirely discarded. The solution to this problem is yet to be found.

As far as the transitional section of a leader channel is concerned, its ionization degree and conductivity may be lower and the field essentially higher. Here, we must take into account the leader evolution, the heat balance, expansion, and electron-molecular kinetics, including all actual ionization mechanisms. The calculation may yield the transition duration, or the distance from the leader head, at which the field drops to the 'arc scale'. Clearly, most of the delay time and length will be associated with colder sections of the transition region, where atoms and excited molecules are accumulated slowly, the vibration relaxation is slow, etc.

It is likely that the leader state occupies an intermediate position between the arc and contracted glow discharge states. Like the leader, the latter arises from ionization-overheating instabilities. Today, however, there is no good theory to describe a contracted glow discharge.

6.10.3 Optimal regime for a long leader

An optimal regime to support a leader is the one in which the channel elongates to a prescribed length L at a minimum gap voltage. The conditions favorable for a leader are very dangerous, since the electrical strength of a gap of length d will be minimal. Why an optimal regime exists at all may become clear from equality (6.20) combined with some empirical relations. Experimental findings show that leader velocity increases with leader current and head potential U_t, which seems to be quite natural. Current also grows with U_t. On the other hand, average channel field decreases with higher current. This relationship is characteristic of an arc, and the reason for it in case of a leader is probably similar (Sections 2.8 and 6.10.2). Why the current-voltage characteristic $E(i)$ in an arc is of the descending type can be explained as follows.

In a steady-state leader, the power per unit length, $W_1 = iE$, must largely be removed from the channel by heat conduction. The heat flow grows with temperature but not very fast, while the degree of plasma ionization and conductivity σ are strongly temperature-dependent (see the $n_e(T)$ data in Table 6.3). If the field remained constant, even a slight increase in T would greatly raise σ, the current density $j = \sigma E$ and current $i = \pi r^2 j$. The linear power iE would rise, violating the heat balance. The balance is preserved, because the field drops and the power varies only a little, that is, $E \sim i^{-1}$. This determines the shape of the current-voltage characteristic for an arc or leader channel at low arc temperatures. In reality, the relationship $E(i)$ is weaker, but it approaches $E \sim i^{-1}$ at low currents $i \sim 1$ A (Section 2.8).

To return to the optimal regime, let us suppose that the relationship between the channel average field and the current is, say, $E_L = b/i$ and $v_L = aU_t^{1/2}$ supported, to some extent, by experimental data. Bearing in mind that the current is $i = C_{1s} U_t v_L$, we can find the relation $U_t = Ai^{2/3}$ with $A = (C_{1s}a)^{-2/3}$ and $v_L \sim i^{1/3}$. By substituting the relationships $v_L(U_t)$ and $U_t(i)$ into expression (6.20), we obtain $U_0 = Lb/i + Ai^{2/3}$. The function $U_0(i)$ has a minimum at $i = i_m$:

$$i_m = \left(\frac{3bL}{2A}\right)^{3/5} \qquad U_{0\,\min} = \frac{5}{3}A^{3/5}\left(\frac{3bL}{2}\right)^{2/5} = \frac{5}{3}U_{t\,m} \qquad (6.24)$$

where $U_{t\,m}$ is the head potential in the optimal regime.

It follows from expressions (6.24) that the condition for leader development at a low head potential, close in the energy criteria to the minimum possible potential $U_t \approx 280$ kV, is far from being optimal. Breakdown of a long gap at minimum voltage requires the head potential level of $3/5U_0$, approximately $U_0/2$; nearly half of the voltage supply potential is lost in the channel. This condition of half voltage loss on the way to the streamer zone has something in common with the condition, typical of electric circuits, for maximum power transfer from the source to the load, which, in our case, is the streamer zone. Another result following from expressions (6.24) is a weak nonlinear dependence $U_{0\,\min} \sim L^{2/5}$. To increase the length of a multimeter gap is not a particularly effective means of raising its electrical strength. This is one of the principal difficulties encountered in high voltage engineering.

As an illustration, we take $C_{1s} \approx \pi\varepsilon_0$, $a \approx 1.5 \times 10^3$ $V^{-1/2}$ $cm\,s^{-1}$, and $b \approx 300$ $W\,cm^{-1}$. For $L = 50$ m, we obtain $i_m \approx 1.14$ A, $U_{t\,min} \approx 1.96$ MV, $U_{0\,min} \approx 3.27$ MV, $E_L \approx 263$ $V\,cm^{-1}$, and $v_L \approx 2.1$ $cm\,\mu s^{-1}$. For a leader of $L = 100$ m, we have $i_m \approx 1.74$ A, $U_{t\,min} \approx 2.59$ MV, $U_{0\,min} \approx 4.3$ MV, $E_L \approx 172$ $V\,cm^{-1}$, and $v_L \approx 2.4$ $cm\,\mu s^{-1}$. These are reasonable values corresponding to an optimal regime for long artificial leaders.

The distributions of the longitudinal field and current along the leader depend on the time variation of gap voltage. Appropriate calculations can be made using equations (3.29)–(3.31) and taking the leader linear capacitance and resistance to be known (they may vary with current or voltage). The results show that the current increases but the field decreases from the head to the base, as U_0 is raised. These changes are accompanied by reduction in the total channel voltage. By choosing the pulse rise time t_f, the necessary leader length can be achieved at lower voltage than in the case with $U_0 = $ const. Experiments give optimal values of $t_{f\,m}$, at which gaps of length d have minimum electrical strength $U_{b\,min}$ (see Chapter 7).

In the opposite case when the voltage pulse decreases in time, which normally occurs in long spark experiments, the current is lowered towards the base and can even reverse the direction near the electrode (the charge earlier incorporated into the gap flows back to the anode). The voltage source does not supply energy to the leader, which continues to propagate for some time owing to its own resources—the electric field due to the cover space charge. Where the current reverses its direction, the energy supply is minimum, the gas is cooled, and its conductivity drops, abruptly increasing the longitudinal field. The channel voltage grows, resulting in a much higher electrical strength of a long gap, as shorter voltage pulses are applied.

6.11   The leader family

So far, we have discussed a classical leader process—the formation and propagation of a positive leader. With an adequate support by a voltage source, a leader can survive for a long time and cover large distances, like lightning. Its principal property making this possible is the ability to form a hot channel, so hot that the ionization in it can be supported by a relatively weak longitudinal field. How energy is supplied to the channel is of no principal importance. What is important is that the channel could heat itself through 'its own' effort. An ordinary leader does this by combining and accumulating currents from numerous streamers emitted by the leader head. It will be shown here that such a mechanism is not at all compulsory and that there are other ways of supplying the channel with energy. Yet, we will first discuss a version of the leader process, which has some specificity due to a different polarity but possesses the same energy supply mechanism.

6.11.1   A negative leader

A negative leader starts from the smaller electrode (cathode); a grounded plane may act as the anode. As far as a mature leader channel is concerned, its state and parameters far from the head differ only a little from those in a positive leader. Slight differences may be associated with the opposite sign of the corona from the channel lateral surface, which may alter the cover charge under some conditions (Section 6.7.2). The main specificity is associated with processes in the streamer zone and with movement of the channel front section in general. Both processes are much more complicated than those in a positive leader.

The streamer zone formation in a negative leader requires a high voltage for the same reason as a single anode-directed streamer propagation does. An electron involved in avalanche ionization in a cathode-directed streamer is driven by the field towards the leader head, so it sooner finds itself in a strong field than in an anode-directed streamer. In the latter, an electron tends to 'run away' from the head, so that the avalanche develops in a weaker field, that is, in less favorable conditions. Of course, a strong streamer with velocity v_s much greater than electron drift velocity v_e does not show much difference between $v_s + v_e$ and $v_s - v_e$. But we know that streamers are weak in the streamer zone, having $v_s \approx v_e$, so that the sum of these values and their difference differ much. The first variant has evident advantages.

Due to this, gap breakdown by negative voltage is more difficult to achieve, since the voltage must be higher. But the effect of polarity becomes weaker with larger gap length, because the head potential U_t of a long leader makes a smaller contribution to the total voltage $U_0 = U_t + U_L$. Channel voltage U_L is practically insensitive to polarity: gaps with $d \sim 100$ m can be closed by a leader at positive and negative voltage pulses of nearly identical amplitude. This must be especially true of lightning. However, the structure of a streamer zone in a negative leader has little in common with that of a positive one. Let us examine the former structure phenomenologically, using streak pictures and our own imagination.

A streamer zone of a long negative leader always consists of two types of streamers—anode- and cathode-directed streamers. Anode-directed streamers start from a high voltage electrode as a negative corona flash. It

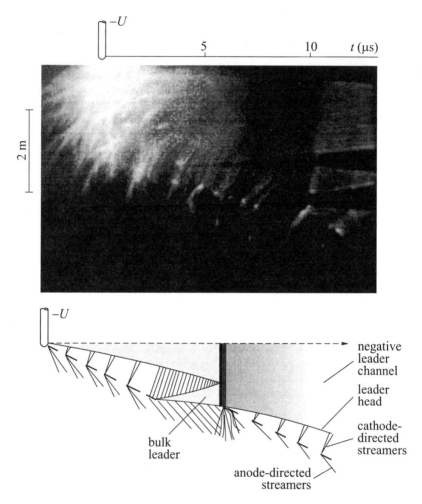

FIGURE 6.12
A streak photograph and schematic diagrams of a negative leader.

looks like a positive corona flash with a similar stem-like element near the electrode, a 1 cm visible radius and numerous streamer branches having a radius less than 0.1 cm each [6.39–6.41, 6.25]. The stem of a negative discharge looks as if it lived its own life. Streak photographs and a diagram based on them (Figure 6.12) show a stem that has separated from the cathode and is moving through the gap. The streamer zone of a negative leader was first studied in detail, using an electron-optical image converter to obtain streak photographs [6.41]; the authors termed this bright structure a *second-order anode-directed streamer*. We will call it for brevity a *plasma stem.* Omitting the description of leader inception at the cathode, we will consider the formation of new sections of an existing channel. The situation is generally similar to that with a positive voltage: the process of leader head reproduction is basically the same as its inception, and the streamer zone is equivalent to the initial corona flash at the electrode.

Suppose we have a long leader that has started from a cathode. At some distance in front of the head is a new plasma stem (Figure 6.12). 'Regular' streamers start in both directions from the stem ends—anode-directed streamers move in the same direction as the leader, while cathode-directed streamers counterpropagate. We are interested in the latter type of streamer. Its behavior is generally similar to that of a metallic rod placed in an electric field and aligned with it. The rod becomes polarized, and charges accumulating at its ends induce strong fields and emit respective streamers. Return to the plasma stem. Cathode-directed streamers start from the stem end facing the head of a negative leader and enter this head. This is a reverse picture as compared to a positive leader, in which cathode-directed streamers are produced in the head and then run ahead of it. In a negative leader, they start from the plasma stem and counterpropagate to flow into the leader head. It turns out that both types of leader are supplied by current from cathode-directed streamers.

Cathode-directed streamers from the plasma stem may be followed by a positive leader.[3] It counterpropagates with the negative leader head, generating streamers from its own head. When the heads come in contact, their charges are neutralized, leaving uncompensated positive charge at the stem end facing the anode. The charging of the leader channel begins. The leader acquires a potential close to the head potential of the negative leader, because the latter is supplied by the voltage source. This is accompanied by an abrupt current rise and energy release, temporarily increasing the radiation intensity from the basic negative leader and the additional leader, which now form a unified channel. The stem end most remote from the cathode transforms to a new leader head, and the process is repeated; namely, a new plasma stem separates from the new head, and so on.

If a discharge is registered by a device with a low time resolution and light sensitivity, as was actually done by B. Schonland in his famous studies of lightning in the 1930's [6.42], streak photographs will show only a series of flashes representing an elongating leader. These were called *negative leader*

---

[3]This kind of leader is referred to as a gap-space leader, because it originates in the gap space. Such leaders can be initiated by placing metallic rods of several centimeters long in a gap. When the field is enhanced by the space charge of the approaching streamer zone of a negative leader, positive gap-space leaders counterpropagate from the rods. Negative leaders may start from the opposite rod ends and move slowly. They have also been observed starting from plasma stems.

steps and the leader a *stepwise leader*. Every step elongates a laboratory leader by tens of centimeters, or by several meters in superlong gaps. A negative lightning discharge has been found to have steps of one hundred meters long.

In any case, the length of a step makes up a small fraction of the gap length. The current overshoot from a step in the external circuit has a short lifetime. Owing to thermal gas inertia, the consequences of pulsed energy incorporation into a negative leader are, on the average, similar to those of a continuous energy supply of a positive leader. Other conditions being equal, the total energy transferred from a high voltage source to a discharge gap does not practically depend on the voltage sign.

6.11.2 A creeping leader

A creeping leader may be considered as having no streamer zone. All the ionization process, from the avalanche multiplication of seed electrons to the current channel heating, is confined to the leader head, which may be regarded, like a streamer head, as an ionization wave. A principal diagram of a creeping discharge is shown in Figure 6.13. An ionization wave starts at a tip and slides through the gas along a thin dielectric film on a grounded metallic plane. With a film thickness measured in dozens of microns and a relative dielectric permittivity of 3–5 typical of polymers, the resultant linear capacitance of the plasma channel may be 2–3 orders of magnitude larger than that of a similar channel in a gas; therefore, the necessary energy input into the discharge can be provided by a lower voltage. Creeping sparks of several meters long produced in laboratory conditions can be excited at $U_0 < 10-20$ kV in air. Note that an air discharge develops without involving the dielectric material into the plasma production.

Figure 6.13 shows a streak photograph of a creeping discharge in a gap with $d = 100$ cm. The spark started from a rod and propagated along an organic glass sheet of 4 mm thick placed on a grounded plane that served as a cathode. A 70 μs positive voltage pulse was applied to the gap. At the moment of the overlap, the voltage was $U_0 = 100$ kV, or one fourth to one fifth of what is necessary to support a leader in a free air gap of the same length. The photograph clearly shows the discharge propagating at nearly constant velocity, $(2-3) \times 10^6$ cm s^{-1}. Such velocity is characteristic of a leader in free air. The current measured from the cathode side gradually rose from 0.3−0.4 A to 1.5 A immediately before the overlap; these values are also typical. At a gap voltage of 100 kV, the average field in a 30 μs channel did not exceed $E_L = U/d = 1$ kV cm^{-1}, and a leader propagating through free air had approximately the same field. There is no indication of a creeping spark being different from a 'classical' leader in air.

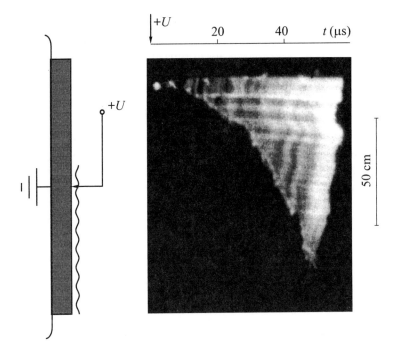

FIGURE 6.13
The basic diagram and a streak photograph of a creeping discharge in air.

Its optical pattern, however, differs much—the streak picture of a creeping spark shows no streamer zone, or rather it is so short that it cannot be identified in a photograph with an optical resolution less than 1 cm.

This kind of discharge should be termed as a streamer-free leader. The energy supply mechanism in it is principally the same as in a classical leader. If the linear channel capacitance is C_1, the energy input into its unit length is $W_1 \lesssim C_1 U_0^2/2$, which should be sufficient to heat up the channel. A leader in air attains such a value owing to the streamer zone that creates a charged cover around the channel, thus increasing the linear capacitance several times. A creeping spark does not need a streamer zone, because its capacitance is large due to the small distance between the grounded plane and the channel, comparable with the channel radius, as well as to a high dielectric permittivity. The possibility of increasing the capacitance is nearly unlimited. Successful experiments have been performed with films of micron thickness, in which the channel linear capacitance was raised up to $(2\text{--}5) \times 10^{-11}$ F cm^{-1}, which is 2 orders of magnitude higher than in a spark in free air. The conditions for gas heating to arc temperatures at such linear capacitance could be created at 10−20 kV. If the radius of a high voltage electrode is so small that this voltage is sufficient to initiate a discharge, a creeping leader propagates for many meters, with its longitudinal field being close to that in an open quasistationary arc [6.43].

6.11.3   A leader along a conducting surface

There is a special form of leader, which is of great importance for technical applications. The surface of any dirty, moist dielectric possesses conductive properties; such is the earth's surface. When a lightning discharge strikes the ground, spark leaders may run along its surface for 100 m and more. In this way, lightning current reaches protected objects, bypassing their lightning arresters. Such a discharge is of interest, because its energy supply is directly unrelated to the ionization processes in the leader head. We will analyze this mechanism with reference to a spark propagating along tap water surface with $\sigma \approx 4 \times 10^{-4}\ \Omega^{-1}\,cm^{-1}$.

It is seen from typical streak photographs taken at positive and negative voltage (Figure 6.14) that we again deal with a streamer-free leader. The general appearance of this discharge is the same as that of a spark creeping along a thin dielectric film. The channel is clearly seen to contain no streamer zone and to expand from the head towards the base. The head is not particularly bright; its intensity is even lower than in the rest of the channel. The few branches near the head can hardly affect the characteristic scale of discharge current, which is always measured at the base and may

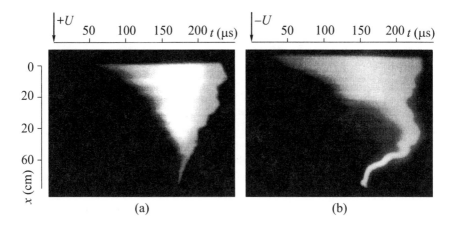

(a)   (b)

FIGURE 6.14
Streak photographs of a leader propagating along water surface at a positive (a) and a negative (b) voltage pulse of 20/1000 μs.

be so strong that one has to take special measures to keep it at a level of 100 A; otherwise an experiment with a high voltage source of limited power capacity may fail. A shallow cuvette of 100 cm long with 25×3.5 cm^2 water layer cross section was used to limit the current [6.44]. The spark started from a pointed rod normal to the water surface and propagated along it towards a plane electrode of opposite sign, which occupied all the lateral side of the cuvette. The image nonlinearity in the photographs taken along the gap axis resulted from filming by an electron-optical converter horizontally mounted at an acute angle to the water surface. The photographs were analyzed using a correction calculated from specially calibrated pictures.

The channel velocity and base current rose from the initial values $v_L \approx 3 \times 10^5$ cm s^{-1} and $i \approx 20$ A to 9×10^5 cm s^{-1} and 45 A; these values were registered 5 μs before the gap overlap and the registration time was 225 μs. The voltage necessary for a leader to close a 100 cm gap was about 45 kV for the positive rod and 10–15% higher for the negative rod. This is by an order less than in free air. Judging by the low values of v_L, the current at the head was many times less than the base current. It is quite natural because the channel current is the sum of leakage currents through the contact surface between the leader and water. It is determined by the contact surface area and resistance of the water column which is not closed by the discharge. This is the source of a continuous current rise and respective velocity increase accompanying the channel elongation. Here, we observe the principal difference from a discharge creeping along a dielectric, when i and v_L vary slowly because of a nearly unvarying linear capacitance, which, in turn, determines the current.

When the leader current is determined by linear channel capacitance, the velocity and current are interrelated, since the velocity, being current-dependent, determines this current. This relationship is violated in a discharge along a conducting surface. Of course, some current is related to the charging of new channel sections, but it is negligible as compared with leakage current, which plays the key role in providing the discharge with energy, so that the energy is supplied irrespective of leader velocity. In other words, a channel may even stop, but it will remain hot and support its conductivity on a sufficiently high level. There is no minimum leader velocity similar to 10^6 cm s^{-1} in air. In the experiment described, the discharge could not bridge the gap at 43 kV. For 0.5 ms before the leader stopped, it propagated at 2×10^4 cm s^{-1}, or 50 times slower than the minimum possible velocity in free air. Due to the voltage drop in the gap, the field at the head dropped below the value necessary for air ionization, so the leader stopped. But even in this case, current continued to flow through the leader, supporting the plasma conductivity at almost the same level.

To conclude, a long leader requires a high ionization degree of the head gas and an energy supply necessary for heating it to 'arc temperatures' and for the compensation of heat losses. How the energy is supplied is of little importance.

The streamer mechanism of energy supply, inherent in a discharge in free gas, reduces to incorporation of electrical charge into the leader cover. The cover radius is by 3 orders of magnitude larger than the channel radius. As a result, the larger linear capacitance, determining the dissipated energy, increases. The streamer mechanism is power-consuming, since streamer development requires a strong electric field over the whole streamer zone of several meters long. However, a leader process involving a streamer zone has no limitations as for the gap length and the kind of gas used.

The streamer zone shrinks, when the leader is formed along a dielectric with a grounded conductor behind it. In that case, the channel capacitance grows, while the head potential providing the necessary energy input decreases. The streamer zone of a leader creeping along a thin film deposited on a metallic grounded surface completely degenerates, but this does not change the leader parameters, because the channel is supplied by energy from its own, multiply increased capacitance. The voltage necessary for a leader to bridge a gap decreases with increasing leader capacitance. In case of a degenerate streamer zone, the voltage appears to be close to the voltage drop in the channel and depends just slightly on the kind of gas.

The mechanism of energy supply changes, when a leader develops along a conducting surface. The energy is supplied to the channel by leakage currents through the contact surface between the plasma and the conductor, the energy input being independent of the leader velocity, which may become quite low. The energy flow into the leader channel is determined by linear resistance of the conducting medium rather than by gas or plasma characteristics. Besides, linear resistance largely determines the value of overlap voltage.

Various energy supply mechanisms may coexist in particular engineering circumstances. But in any situation, the electrical insulation strength must be reduced, when some of the energy is supplied by leakage currents or by high capacitance, in case of a leader creeping along a dielectric.

7

Electrical Strength of Long Gaps

Breakdown of a long gap occurs when its voltage is sufficient for a leader to bridge the gap. A leader starts from an electrode following a corona flash and voltage rise to compensate for the decreased electrode field. The fact that a leader has started propagating does not necessarily mean that it will reach the opposite electrode. The leader process may stop at the initial stage, if the head potential drops below a threshold value due to voltage drop in the channel or to a small pulse duration. The process becomes irreversible only after the streamer zone has contacted the opposite electrode.

When evaluating the insulation properties of a long air gap, one often has to examine the whole sequence of events that have led to a breakdown, but this is not an easy thing to do. The requirements on the accuracy of engineering insulation parameters are very high: even a 5% error is often considered unacceptable. As a rule, the engineer is not satisfied with the knowledge of 50% breakdown voltage $U_{50\%}$.[1] Depending on a particular technical task, one has to determine either the withstand voltage U_{ws} corresponding to a nearly zero breakdown probability or, on the contrary, the voltage providing a 100% breakdown probability. This is done by finding, from test results, the relationship between the breakdown probability and the voltage pulse amplitude, $\Psi(U)$. This task, however, requires a large number of elaborate and costly experiments. As is clear from the foregoing, one cannot rely on theory; so, all available engineering approaches are based on experimental findings [7.1–7.7]. These problems will not be discussed here, but we rather will consider the insulation properties of long gaps in terms of the general concepts of spark discharge discussed above. Our consideration will be based on the notion of nonuniform distribution of the initial electrical field in a gap. The nonuniformity degree m is known to be the ratio of maximum field $E_{\max}$ to average field $E_{\mathrm{av}} = U/d$ in a gap

---

[1]This is a voltage, at which breakdown occurs with a 50% probability.

of length d, $m = E_{max}/E_{av}$. The value of m may serve as a convenient engineering parameter for the insulation design.

7.1 A weakly nonuniform gap

A weakly nonuniform field will be assumed to be a field, whose distribution in a gap permits the initial corona streamers arising at the smaller electrode (anode) to bridge the gap. For this, the average gap field $E_{av} = E_{max}/m$ must exceed the average field E_s providing the streamer propagation. But a corona flash arises only if the electrode field is higher than the ignition threshold E_{ic}; therefore, a gap with $m \leq m_c = E_{ic}/E_s$ may be termed as weakly nonuniform. For the electrode curvature radius $r_a \gg 1$ cm in atmospheric air we have $E_{ic} \approx 30$ kV cm^{-1}, $E_s \approx 4.5$–5 kV cm^{-1}, and $m_c \approx 6.5$–7. Streamers propagating through a weakly nonuniform gap carry their charge out of it. The anode field does not practically change after the initial corona flash, and the leader immediately follows the flash without a pause. It develops in the final jump mode from the moment of its inception, because the streamer zone of a leader, like the initial corona flash, bridges the gap. If the gap voltage is lowered relatively slowly, a leader developing in the final jump mode cannot stop (Section 6.1). Therefore, the breakdown voltage coincides with the voltage necessary for the corona ignition. Similar also are the probable voltage distributions for the breakdown and corona ignition. The lower limit of E_{ic} for a corona flash is the strength in a steady-state corona ignition, E_{i0}; hence, the steady state voltage is equal to the withstand voltage of the gap, $U_{ws} = E_{i0}d/m$. The value of E_{i0} for an electrode of a prescribed shape and radius r_a can be found, for example, from Peak's empirical formulas. For a cylindrical anode, this formula is (3.27), while for a spherical anode the coefficients 31.0 and 0.308 in this formula should be replaced by 27.8 and 0.54, respectively. For evaluation of the nonuniformity degree m, one can take advantage of the semi-empirical formulas

$$m = \left(1 + \frac{2d}{r_a}\right) + \left[8 + \left(1 + \frac{2d}{r_a}\right)^2\right]^{1/2} \tag{7.1}$$

$$m \approx \frac{\left(d^2/r_a^2 + 2d/r_a\right)^{1/2}}{\ln\left[(1 + d/r_a) + (d^2/r_a^2 + 2d/r_a)^{1/2}\right]} \tag{7.2}$$

for a sphere-plane and a wire-plane gap, respectively.

Of practical importance are gaps with anode radii $r_a > 10$ cm, in which

E_{i0} is actually independent of r_a. In this case, the validity of withstand voltage evaluation is determined exclusively by the calculation accuracy of the electrode field defining m. Note that formulas (7.1), (7.2), and others like them refer to ideal gaps, ignoring the effects of electrode suspension elements. How large these effects may be was shown in Section 4.9: the neglect of suspension elements usually leads to a greatly underestimated value of U_{ws}.

Formulas (7.1) and (7.2) clearly show the nonlinearity of the relationship $U_{ws}(d)$. Equalizing the field, or decreasing m, is an effective means to raise the gap insulation strength. However, rain drops, dust, and other contamination from the open air enhance the near-electrode field, thereby reducing the breakdown voltage, sometimes considerably, as compared to the calculated values. This effect is more frequent, as the time of voltage action becomes longer. Sporadic 'anomalous' spark discharges from screens of high voltage sources are known to occur at voltages much lower than the calculated withstand value. Such sparks are stimulated by coronas arising from accidental field nonuniformities, for example, from web threads touching the electrode.

Breakdown voltage of a weakly nonuniform gap is independent of pulse rise time, since the field does not change after the flash and there is no need to raise the voltage—the discharge is already developing beyond the pulse front. The pulse duration t_i could affect the gap strength only if it were less than the time $t_L = d/v_L$ necessary for the leader to cross the gap. But even at $d = 5$ m, the time $t_L < 50$ μs is too short ($v_L \geq 10^7$ cm s^{-1} at the final jump); longer gaps with weakly nonuniform fields are rare in practice. Therefore, for the shortest standard pulse of 1.2/50 μs, the gap strength is nearly identical to that for a long pulse or for constant voltage.

The fact that the threshold breakdown voltage is equal to the corona ignition voltage is due to streamers transporting their charge to the opposite electrode, so that the voltage does not have to be raised to initiate a leader. The role of a trigger is played by the corona flash. The situation might be quite different, if the streamers were conductive and if their number were small enough to eliminate the electrostatic effect. Then the charge would accumulate in the streamer channels but not only in their heads (Section 6.7.1), and the streamer contact with the opposite electrode would not liberate the gap from space charge, so that the leader inception would not follow immediately the corona flash, resulting in a breakdown. Indeed, in nitrogen [7.8] and hot air [7.9], where streamers appear at a lower frequency and do not lose their conductivity as quickly as in cold air, leader excitation would require twice as much voltage as would be necessary for corona initiation.

Our evaluations concern only withstand voltages but not 50% or 100%

breakdown voltages. But the difference is not large. The deviation of breakdown voltage in weakly nonuniform gaps is small, $\sigma_u < 0.01 U_{50\%}$; so, in the normal distribution law, we have $U_{50\%} \approx U_{ws} + 3\sigma_u < 1.03 U_{ws}$, with U_{ws} corresponding to a 10^{-3} breakdown probability. The discrepancy between $U_{50\%}$ and U_{ws} is close to a measurement error.

Thus, a leader process occurring in a weakly nonuniform field starts with the final jump, and the breakdown voltage coincides with the corona ignition voltage. This makes U_b dependent on the anode radius, since corona ignition requires a definite electrode field. The rate of voltage rise does not affect the gap electrical strength.

7.2 A strongly nonuniform gap

Formulas (7.1) and (7.2) yield $m = m(r_a/d)$. Besides, E_{ic} is nearly constant, except for small radii r_a. Hence, the criterion of weak nonuniformity, $m \leq m_c$, satisfies the similarity condition in r_a/d. Consider now a gap with a strongly nonuniform field and follow the effect of the anode radius r_a, rather than the r_a/d ratio, on its electrical strength. Experimental data indicate the existence of a critical radius r_{cr}, below which variation in r_a does not affect the electrical strength [7.10, 7.11]. At $r_a > r_{cr}$, the insulation strength increases with r_a. Let us evaluate r_{cr} and explain these facts.

7.2.1 Supercritical anode radius

The anode radius determines the initial voltage of a corona flash, U_{ic}; for a sphere, for example, we have $U_{ic} \approx r_a E_{ic}$. Corona streamers excited in a strongly nonuniform field do not reach the opposite electrode, nor do they carry their charge out of the gap. This charge decreases the field at the anode, and the gap voltage is to be raised to excite a leader. However, electrodes with a radius of several tens of centimeters, which are of special practical interest, require only a slight voltage rise (Section 6.1.3). The leader inception does not necessarily mean that the gap will be broken down: a leader may stop at a distance $L < d$. Breakdown is ensured when the initial corona voltage U_{ic} exceeds the minimum voltage $U_{0\min}$ necessary for a leader to cover the distance $L = d - L_s$, so that the streamer zone can contact the opposite electrode (Section 6.10.3). The voltage then does not have to be raised, because the voltage for breakdown will all the same be equal to that for corona ignition, with the withstand voltage being equal to a minimum value, at which a steady-state corona can be ignited.

By equating $U_{ic} \approx E_{ic} r_a$ for a sphere to $U_{0\min}(L)$, we can find the critical radius $r_{cr}(d)$ from formula (6.24). At $r_a > r_{cr}$, the above will be valid:

$$r_{cr} \approx \frac{U_{0\,min}}{E_{i\,0}} = \frac{5}{3} A^{3/5} \frac{(3bL/2)^{2/5}}{E_{i\,0}} \qquad L = d - L_s \qquad (7.3)$$

Here $E_{i\,0}$ is the strength for a steady-state corona ignition ($E_i \approx 30\,\text{kV cm}^{-1}$ at $r_a > 10$ cm in atmospheric air). The numerical example that follows makes use of the constants A and b from Section 6.10.3 and $L_s = U_{t\,min}/E_s$, where $U_{t\,min}$ is the leader head potential corresponding to $U_{0\,min}$ and $E_s \approx 4.65\,\text{kV cm}^{-1}$ for normal air. We will obtain the following values:

d (m)	15	20	30	50	75	100
r_{cr} (cm)	63	71	85	105	125	140

For $r_a > r_{cr}(d)$, the breakdown of a gap d will be determined by the corona voltage, as in a weakly nonuniform field. It is seen that the geometrical similarity is violated here: r_{cr} grows much slower than d.

The values of r_{cr} are overestimated, as compared with experimental data, because the calculation neglected the suspension knot effect considerably lowering the field (Section 4.9). For instance, in the limiting case when the suspension thread radius is equal to the sphere radius, we have $U_{i\,c} \approx 2r_a E_{i\,c}$ [formula (3.19)], decreasing the calculated value of r_{cr} by half. This is exactly what happens in reality, even though the suspension thread may be very thin. It displays a corona as the voltage rises, and the charge cover formed increases the equivalent thread radius, very much like it does the equivalent radius of a leader channel (Section 6.7.2). It is not surprising that the experimental critical radius for a sphere-plane gap of $10-17$ m long, $r_{cr} \approx 30-35$ cm, is nearly half the calculated value [7.12]. Nevertheless, the physical reason for the existence of a critical radius has been illustrated correctly.

The value of r_{cr} for a common wire-plane gap can also be found from the equation relating the cylindrical anode field to the voltage $U_{0\,min}$. Taking into account the wire image charge in the plane, we find

$$r_{cr} \ln \left(\frac{2d + r_{cr}}{r_{cr}} \right) = \frac{U_{0\,min}}{E_{i\,0}} \qquad (7.4)$$

For $d = 20$ m, $r_{cr} \approx 12$ cm is almost one fifth the value for a sphere-plane gap of the same length. This is clear because a 'cylindrical' field is more uniform than a 'spherical' field, requiring a higher voltage for the same $E_{i\,0}$. Note that the estimation accuracy of the critical radius of a wire from formula (7.4) is much higher than that of a sphere from formula (7.3), since the suspension knots of a long wire distort only the field at its ends.

Although the field in a long gap with $r_a \geq r_{cr}$ is strongly nonuniform, its insulation properties have much in common with those of a weakly nonuniform gap (Section 7.1). The withstand voltages are given, though less accurately, by the strength of a steady-state corona ignition, E_i: $U_{ws}(d) = E_i(r_a) f(r_a, d)$, where f is a geometrical factor equal to the ratio of the anode potential to its surface field. Local nonuniformities sporadically enhancing the anode field may lower the electrical strength to about the same extent as in gaps with a weak field nonuniformity. For example, experiments conducted in rain [7.13] show that the larger anode has no effect and that the gap strength appears to be the same as at $r_a < r_{cr}$.

The principal parameter, by which a strongly nonuniform gap with $r_a \geq r_{cr}$ can be distinguished from a weakly nonuniform one, is the electrical strength dependence on the pulse rise time. In the former, the initial corona charge is not taken by streamers out of the gap, so it slightly decreases the anode field. Voltage rise must compensate for this decrease for a leader to be excited; therefore, the corona will necessarily complete its development at the *pulse front*. In this situation, everything depends on the rate of voltage rise. If the rate is high, the voltage will rise considerably over the time of streamer flash propagation. Then a voltage higher than U_{ic} will determine the value of U_{iL} necessary for the leader start and for the gap breakdown.

A streamer propagates as long as the average field $U_0(t)/l_s$ along its length $l_s(t)$ exceeds $E_s \approx 5$ kV cm^{-1} in air, or while the rate of voltage rise is $A_u = dU_0/dt > E_s dl_s/dt = v_s E_s$. Since the minimum velocity of a streamer in air is $v_s \approx 10^7$ cm s^{-1}, it develops at $A_u > A_{u\,min} \approx 50$ kV μs^{-1}. When the pulse rise time t_f is about $U_0/A_{u\,min}$ and less, the breakdown voltage will grow with decreasing t_f. The rise time in megavolt experiments should be $t_f < 10-20$ μs. The experimental dependence of 50% breakdown voltage, $U_{50\%}$, on t_f is shown in Figure 7.1(a) [7.14]. In the experiments the pulse duration was maintained at the value of 2500 μs (0.5 level). As the rise time t_f was reduced from 90 to 1.7 μs, $U_{50\%}$ grew from 2.2 to 3.2 MV, while the average breakdown field $E_{50\%}$ rose to about 4 kV cm^{-1}. The average breakdown strength may be assumed to have an upper limit of 4.5-5 kV cm^{-1}, at which streamers are able to overlap the gap.

Figure 7.1(a) shows that the highest rate of strength variation with t_f is characteristic of the practically important range of $t_f = 1-10$ μs. This is a typical time for lightning current rise. The strength of air insulation at atmospheric overvoltages is usually tested by a standard pulse of $t_f = 1.2$ μs. The test results seem to be overestimated, especially in evaluations of stability to positive lightning discharges with the average rise time $t_f > 20$ μs. One should expect no effect of the rise time on the spark leader development, because at $r_a > r_{cr}$ no voltage rise is necessary, and the whole process occurs beyond the pulse front. Thus, in the case of a strongly

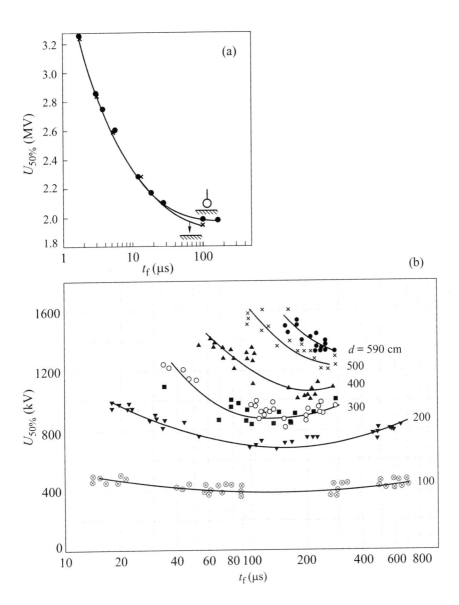

FIGURE 7.1
Breakdown voltage for long gaps with a strongly nonuniform field versus the
positive voltage pulse rise time: sphere-plane ($r_\mathrm{a} = 50$ cm) and rod-plane
($r_\mathrm{a} = 2.5$ cm) gaps of 8 m long with steep pulse fronts (a); rod-plane gaps with
smooth pulse fronts (b).

nonuniform gap with a supercritical anode radius, the breakdown voltage depends on voltages necessary for both the corona initiation and leader inception. The first condition relates U_b to the anode radius and the second to the rate of voltage pulse rise.

7.2.2   Minimum gap strength for a small anode

If the radius of a high voltage electrode is so small that we have $r_a \ll r_{cr}$ and $m_e \gg m_{e\,cr} \equiv m\,(r_{cr}/d)$, all external field is concentrated near the electrode. This is typical of a needle-plane or rod-plane gap. With such field nonuniformity, a corona flash and a leader following it may arise at a much lower voltage than $U_{0\,min}$ necessary for bridging a gap of length d (Section 6.10.3). Then the anode radius, to which the initial corona voltage is proportional, will not affect the breakdown.

To make a leader bridge a gap, the voltage must be raised during its propagation. This is a reason why breakdown voltage must depend on the pulse front duration t_f but for a longer time interval, because the initial leader stage lasts $100-1000$ μs or longer. The interest in smooth pulses is due to practical needs. Smooth pulses appear in prescribed and emergency commutations of power transmission lines; so, overvoltage arising in such circumstances is known as *switching overvoltage*. Let us first consider an optimal rate of voltage rise, at which a gap will have a minimum electrical strength. This situation is of special interest, because it may cause an emergency. Then we will find the gap strength at other pulse front durations.

The minimum voltage $U_{0\,min}\,(L)$ for a leader calculated in Section 6.10.3 corresponds to the most favorable voltage distribution between the leader channel and its head, as well as to optimal current i_{opt} and leader velocity $v_{L\,opt}$. These quantities have been derived for a given leader length L, irrespective of the leader evolution and the way the optimal condition was achieved. According to the derivation, the quantity $U_{0\,min}$ corresponds to the minimum 'static' current-voltage characteristic $U_0\,(i)$. With $i > i_{opt}$, the head potential is too large; with $i < i_{opt}$, too large are the field and voltage drop in the leader channel. In either case, we deal with $U_0 > U_{0\,min}$. In actual fact, the leader process is not steady-state but is described by a 'dynamic' relation of U_0 to i and L, which depends on the time evolution of both voltage and leader. But with the implementation of the voltage rise mode, in which the gap voltage will be equal to $U_{0\,min}\,(L)$ at any leader length L and at any moment of time, a gap of any length will possess a minimum strength, as compared to that attainable by any other voltage rise mode.

On the assumptions used for the derivation of the basic formula (6.24), the optimal leader velocity varies but little with its length, as $v_{L\,opt} \sim L^{1/5}$.

Indeed, experiments give the value $v_{\mathrm{L\,opt}} \approx \mathrm{const} \approx 1.8\ \mathrm{cm}\,\mu\mathrm{s}^{-1}$ in $5-15$ m gaps (Section 6.2). The optimal mode of leader development corresponds to the channel elongation law $dL/dt = v_{\mathrm{L\,opt}}(L)$, $L \sim t^{4/5}$ and to the voltage rise law $U_{0\,\mathrm{min}}(L) \sim L^{2/5} \sim t^{8/25} \approx t^{1/3}$. Obviously, the evaluation from formula (6.24) is approximate, and the experiment can hardly make voltage follow it exactly. However, one can choose an optimal pulse rise time for a gap of a given length d

$$t_{\mathrm{f\,opt}} \approx \frac{d - L_{\mathrm{s}}}{v_{\mathrm{L\,opt}}} \approx \frac{d}{v_{\mathrm{L\,opt}}} \qquad \text{at} \quad d \gg L_{\mathrm{s}}$$

in order to provide an approximately optimal leader velocity over the whole gap length. A voltage pulse with such a rise time will correspond to approximately optimal conditions for the gap overlap and to minimum breakdown voltage. Pulses must be sufficiently smooth, $t_{\mathrm{f\,opt}} \approx 150-800\ \mu\mathrm{s}$, at $v_{\mathrm{L\,opt}} \approx 1.8\ \mathrm{cm}\,\mu\mathrm{s}^{-1}$ in gaps with $d = 3-15$ m.

The relation of minimum breakdown voltage U_{b} to pulse rise time was studied experimentally [7.15–7.17] [Figure 7.1(b)]. The effect of t_{f} on the electrical strength of long gaps with variously shaped electrodes was later carefully tested [7.18–7.20] because of its great practical importance. No exceptions were found: all air gaps with a strongly nonuniform field possessed an optimal duration of the pulse front $t_{\mathrm{f\,opt}}$, at which the strength was minimal. The value of $t_{\mathrm{f\,opt}}$ increased with the gap length. The available data were generalized [7.20] by the empirical formula

$$t_{\mathrm{f\,opt}} \approx 50d\,[\mathrm{m}]\ \mu\mathrm{s} \tag{7.5}$$

This formula may be considered as reflecting the above theoretical conclusion that breakdown of a gap of length d occurs readily, if the voltage rise provides the leader with optimal velocity. Indeed, this formula gives $v_{\mathrm{L}} \approx d/t_{\mathrm{f\,opt}} = 2\ \mathrm{cm}\,\mu\mathrm{s}^{-1}$, a value close to the measured optimal velocity of $1.8\ \mathrm{cm}\,\mu\mathrm{s}^{-1}$.

7.2.3 Gap strength versus pulse rise time

The dependence of the gap strength on the rise time of the applied voltage pulse is quite appreciable in gaps with strongly nonuniform fields. The breakdown voltage rises by tens of percent over its minimum with increasing t_{f}, asymptotically approaching the value measured at 'constant' voltage, which was raised at a rate of $1-10$ kV s^{-1} during the tests. It is seen from Figure 7.2(a) that the breakdown voltage $U_{50\%}$ grows with d nearly linearly in constant voltage tests, corresponding to the average breakdown field $E_{50\%} = 4.5$ kV cm^{-1}. In tests with $t_{\mathrm{f\,opt}}$, the relationship $U_{50\%\,\mathrm{min}}(d)$ is strongly nonlinear, and the average breakdown field $E_{\mathrm{b\,min}}$ drops from

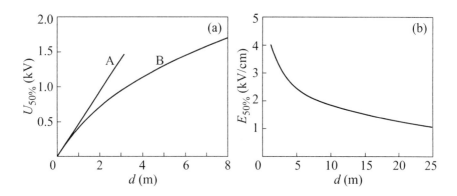

FIGURE 7.2
Electrical strength of a rod-plane gap at positive constant voltage (A) and at an optimal pulse front duration (B): 50% breakdown voltage (a); average breakdown field (b) [7.17, 7.20, 7.26].

4 kV cm^{-1} at $d = 1$ m to 2.3 kV cm^{-1} at $d = 6$ m. The tendency for $E_{b\ min}$ to decrease persists in longer gaps [Figure 7.2(b)]. Some empirical formulas have been suggested [7.21]

$$U_{50\%\ min} = \frac{3400}{1 + 8/d}\ \text{kV} \qquad d < 15\ \text{m} \tag{7.6}$$

$$U_{50\%\ min} = 1440 + 55d\ \text{kV} \qquad 15 \le d \le 30\ \text{m} \tag{7.7}$$

which provide a 5% accuracy. It seems that the lower limit of $E_{b\ min}$ may turn out to be longitudinal arc gradients corresponding to weak currents $i \lesssim 1$ A. In any case, the values $0.35-0.25$ kV cm^{-1} were registered for leader lengths of $100-200$ m under conditions far from being optimal.

Some suggestions can be made as for the physical mechanism of gap strength increase when t_f deviates from $t_{f\ opt}$. Let us go from smooth pulse fronts towards shorter rise times. The largest values of t_f correspond to the breakdown at constant voltage, if the voltage rises over the time of the leader propagation at minimum velocity $v_L \approx 1$ cm μs^{-1}. But when the voltage grows appreciably in time, a new component $C dU_0/dt$ appears in the leader current (Section 6.8.1). This current component charges additionally the available channel capacitance C. The leader current is exhausted on the way from the base to the head; so, the head current and potential do not change much. The average current in the channel is higher than that at constant voltage, while the longitudinal field and voltage drop in a channel with a descending current-voltage characteristic become smaller. The total voltage necessary for a leader of a given length also becomes lower. For instance, as t_f decreases from 'infinity' to $t_{f\ opt}$, the breakdown voltage drops from the value U_b at constant voltage to the minimum. This may account for the right-hand ascending branch of the curve $U_b(t_f)$.

The effect just described, which always tends to decrease U_b with lower t_f, seems to give way to the opposite effect on the left-hand branch with $t_f < t_{opt}$. With an abrupt voltage rise and strong current in the initial leader stage, too much positive charge is incorporated into the leader cover. Later, after the head has moved on, there appears an excessively strong reverse radial field in the old leader sites, giving rise to a more powerful negative corona, which neutralizes the excessive cover charge (Section 6.7.2). Its current flows in the direction opposite to the main leader current, that is, from the anode to the head. The total current decreases, while the voltage drop becomes greater, decreasing the head current and potential. In order to rectify this situation and make the leader move on, one has to raise the voltage above the optimal value. This probably accounts for the left-hand branch of the curve $U_b(t_f)$. The numerical calculations presented at the end of Section 6.10.3 do not contradict this reasoning.

A minimum in $U_{50\%}(t_f)$ is also observed for negative pulses [7.22, 7.23], but $t_{f\,opt}$ is 3–4 times less than for positive ones. The electrical strength is 1.5–1.7 times higher, but this difference vanishes in longer gaps.

The significance of strength decrease for applications can be demonstrated as follows. An air insulation gap of length $d = U/E_b \approx 3.3$ m withstands constant voltage $U' = 1.5$ MV. To withstand a voltage pulse with an optimal rise time, a gap must be at least 6.5 m long. A doubled gap length requires the use of larger supporting insulators, metallic supports, and other constructions, which entails additional expenditures. For this reason, strongly nonuniform fields are equalized by using metallic screens of a large radius.

Consider now the breakdown of a gap with a small anode, when voltage pulses have a very short rise time, $t_f = 1-10$ μs. The voltage rises so rapidly that the leader actually propagates at constant or even decreasing voltage, so that the effect of current variation along the channel is negligible. It might seem that the rise time cannot affect the gap electrical strength under these conditions, especially if we recall that at $r_a \ll r_{cr}$ a corona flash, whose charge is independent of t_f, arises at a voltage much lower than the breakdown voltage. In actual reality, however, the gap is considerably strengthened at smaller t_f. Tests with pulse rise times $t_f = 1.7-15$ μs showed that a rod-plane gap ($r_a = 2.5$ cm) and a sphere-plane gap ($r_a = 50$ cm $> r_{cr}$) of 8 m long had identical strengths [Figure 7.1(a)], whereas at a slower voltage rise the value of U_b in the former gap was lower than in the latter. The reason is that earlier corona streamers from a rod increase their length at higher gap voltage, as long as $A_u > A_{u\,min}$. They stop at the same voltage as streamers from a large sphere, when the rate of pulse

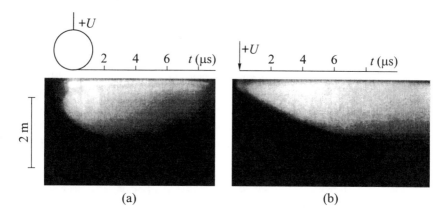

FIGURE 7.3
An initial corona flash in sphere-plane ($r_a = 50$ cm) (a) and rod-plane
($r_a = 2.5$ cm) (b) gaps of 8 m long at a positive voltage pulse of 10 μs.

rise becomes lower than $A_{u\ \min} \approx 50$ kV μs^{-1}. Streamers in both kinds
of gap have a nearly identical length (Figure 7.3), incorporate the same
charge, and have leaders starting at the same voltage in spite of differently
shaped electrodes. The strength of a gap with a small-radius anode, which
is lower at $A_u < A_{u\ \min}$, becomes equal to that of a gap of the same length
at $r_a > r_{cr}$. No special measures are required for equalizing the gap field—
this is accomplished by the corona flash. The effect of the anode radius is
detectable only at smooth voltage pulses with $A_u < A_{u\ \min}$.

The necessity to equalize the external field is especially acute, when the
pulse rise time is $t_f > 50$–100 μs. These are switching waves that are so
troublesome technically. Screening of high voltage equipment to enlarge the
radius of elements acting as an anode is a very costly measure. Besides, an
insulation construction with large-radius electrodes is subject to a harmful
effect of dust and rain droplets producing nonuniformities. Lately, the idea
of equalizing the gap field by pumping space charge by special electrode
coronas has become quite popular. The electrodes display no coronas at
the operating voltage, but as soon as an overvoltage pulse exceeds a certain
critical value, they start displaying intensive coronas.

Thus, electrical strength of a long gap with a strongly nonuniform field
and a small anode is largely determined by the voltage necessary for the
gap overlap. If the voltage pulses have a short rise time, the strength is
also affected by the leader inception conditions. It has a minimum at the
optimal rise time, increasing with the gap length.

7.3   The time-voltage characteristic

The time-voltage characteristic is the dependence of breakdown voltage U_b on the time t_b of the pulse action from the moment of the pulse application until the gap overlap. There are few ways of reducing the time of discharge evolution, if the leader is developing in its initial phase. The velocity of a long leader, $v_L \approx 10^6$ cm s^{-1}, increases slightly even at a doubled voltage (Section 6.2). In order to increase v_L and, thereby, reduce t_b considerably, the discharge must develop in the final jump mode from the very beginning. This means that the average field in a normal air gap must be higher than 4.5–5 kV cm^{-1}. The leader velocity will then immediately rise to $v_L \sim 10^7$ cm s^{-1} (Section 6.2), and the overlap time in a gap of length $d = 1$–5 m will drop from $t_b \sim 100$ μs, characteristic of an initial leader, to $t_b \sim 10$ μs.

Most time-voltage characteristics have been obtained for a standard pulse of 1.2/50 μs widely used for modeling overvoltages from lightning currents [Figure 7.4(a)]. The reduction in time is achieved by an abrupt voltage rise. For example, at $d = 3$ m, the pulse amplitude is to be increased from 1.7 to 3.3 MV in order to reduce t_b from 10 to 3 μs. Note for comparison that at smooth pulses, when the discharge formation takes more than 100 μs, the minimum electrical strength of a similar gap is equal only to 0.9 MV. If we plot the relationships $U_b(d)$ at $t_b = $ const, they all will look like

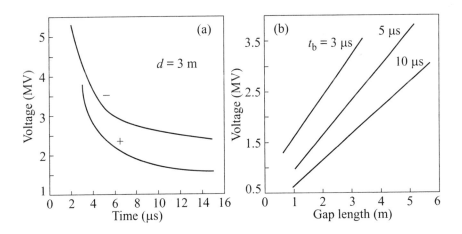

FIGURE 7.4
The relationship between the breakdown voltage and the discharge formation time at a voltage pulse of 1.2/50 μs: time-voltage characteristics at positive and negative polarities (a); the relationships $U_b(d)$ at a fixed time of discharge formation (b) [7.27].

straight lines [Figure 7.4(b)], indicating that the average field remains the same. This is also valid for negative voltages, with the only difference that the electrical strength is 1.5 times higher [Figures 7.4(a)].

At a fixed pulse amplitude, the breakdown of a gap of length d takes less time if the voltage beyond the pulse front decreases slower. For this reason, tests with any one type of pulse of fixed duration t_p are not particularly informative, and this is especially true of the standard pulse of 1.2/50 μs. Lightning current, whose overvoltages are to be simulated by a standard pulse, lasts from $t_p \approx 20$ μs to 1000 μs for lightning of both signs [7.24]. The actual electrical strength at $t_p > 50$ μs may be much lower than that obtained by the standard pulse. The following data illustrate this for a rod-plane gap of $d = 8$ m, for which $U_{50\%}$ was determined at positive pulses with $t_f = 10$ μs and various t_p:

t_p (μs)	80	100	160	560	2500
$U_{50\%}$ (MV)	3.73	3.57	3.43	3.3	2.35

Tests with a standard pulse of $t_p = 50$ μs under the same conditions yielded $U_{50\%} \approx 4.2$ MV. The difference is too large to be neglected, when designing insulation constructions.

7.4 Breakdown voltage deviation

What one must know when designing insulation elements is the voltage U_{ws} (rather than 50% breakdown voltage) that a gap can withstand for an infinitely long period of time or at an infinite number of pulses. Determination of U_{ws} requires a very large number of tests to be made. To reduce this number, one tries to find the relationship between the breakdown probability and the pulse amplitude, $\Psi(U_{max})$. In many cases, this probability can be satisfactorily approximated by the normal error law; its description requires the knowledge of only two experimental parameters— $U_{50\%}$ and standard deviation σ_u. The latter permits avoiding the many thousands of costly experiments that would be necessary for finding U_{ws}.

The reasons for the voltage deviation are poorly understood, and experimental data available are not quite reliable. Finding the deviation characteristics requires a very stable high voltage source to be used in an experiment conducted under strictly unvarying atmospheric conditions to avoid the effects of air density and humidity variation. It is very difficult to meet the latter requirement and nearly impossible to make a voltage source operate steadily. A 1% voltage variation is hard to maintain even in a zero-load regime. After the leader inception, the current loads the generator,

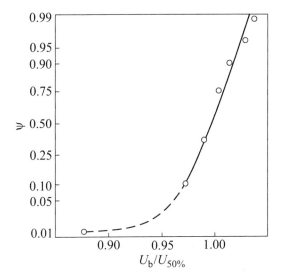

FIGURE 7.5

Integral curves for the breakdown voltage distribution in a 2 m sphere-plane gap ($r_a = 50$ cm) at voltage pulses $4500/6500$ μs.

which normally has a limited power and energy capacity. The voltage pulse becomes distorted, the distortion varying with the leader current and lifetime. It is not clear whether the deviation reflects the regular behavior of discharge or the voltage pulse distortion.

On the whole, experimental findings give the following picture. Minimum deviations correspond to weakly nonuniform gaps with $m_e < m_{e\,cr}$ (Section 7.1): $\sigma_u \approx 1\%$. It is quite likely that the actual value of σ_u is smaller, but this cannot be established with certainty, because a voltage source cannot operate with a smaller deviation. The deviation is unaffected by the pulse rise time, since the voltage does not have to be raised after the corona flash, and the leader process develops at a decreasing pulse. Pulse duration t_p would have an effect, if it were comparable with the statistical time lag of the corona, t_{lag}. In reality, however, we deal with $t_{lag} \ll t_p$, especially for electrodes with radius $r_a > 10$ cm, as is the case with long weakly nonuniform gaps: the probability for an active electron to appear on a larger electrode is higher. On the other hand, there is a greater probability for a local field on a large anode to be enhanced by a microfiber, web, and so on. For this reason, a discharge sometimes arises at 'anomalously' low voltage, producing a deviation from the normal law, $\Psi(U_b)$ (Figure 7.5). The deviation of breakdown voltage grows with increasing nonuniformity of the

TABLE 7.1
Breakdown standard deviation for different voltage pulses, σ (%)

	d (m)						
t_f/t_i	9.0	11.0	13.0	17.0	21.0	25.0	29.0
250/2500	-	5.0	6.5	6.5	6.6	6.2	6.6
1100/6000	6.0	-	4.3	3.4	6.6	6.0	-

external field. In the case of $m_e > m_{e\,cr}$ and electrodes with $r_a > r_{cr}$ (Section 7.3), this is associated with space charge variation of the corona flash. The charge delays the leader inception by decreasing the field near the electrode. The standard deviation σ_u goes up to 4–6% [7.25] for pulses with a short rise time, which excite an especially powerful corona. There seems to be a tendency for σ_u to grow with the rate of gap voltage rise. In strongly nonuniform rod-plane gaps, in which a leader arises at a much lower voltage than the breakdown value, the deviation is $\sigma_u \approx 4$–7%. Variation in U_b appears to be associated with the leader development rather than with the corona parameters. An attempt to relate the actual values of U_b with the leader path length failed, because the correlation between these parameters turned out to be too weak. No dependence of σ_u on the length of a rod-plane gap or on the pulse rise time in the range $100-1000\ \mu s$ was found [7.20]. This is demonstrated by Table 7.1.

To know the relationship between σ_u and the rise time is important for designing the insulation of power transmission lines, where switching overvoltages are quite frequent. An attempt was made to identify the effect of voltage rise time. For this, a rod-plane gap of length $d = 5$ m was tested by an almost linearly rising pulse, whose slope varied by less than 60% between the moment of its application and the breakdown. The standard deviation σ_u was observed to grow with decreasing pulse slope, especially remarkable at $A_u < 10\ \mathrm{kV}\,\mu s^{-1}$:

A_u ($\mathrm{kV}\,\mu s^{-1}$)	0.65	3.0	5.3	6.8	8.0	62
σ_u (%)	5.8	4.7	2.7	2.4	1.7	1.0

However, no well-defined dependence on the rise time has been found under usual test conditions, when the voltage pulse is formed by a capacitive circuit with $U(t) = U_0\left[\exp\left(-\alpha t\right) - \exp\left(-\beta t\right)\right]$ and the slope may change by several orders of magnitude over the discharge lifetime. We emphasize again that conventional testing techniques have failed to separate the parameter deviation of a voltage pulse from that of gas discharge processes that form a leader. This problem requires further investigation.

A

Supplement

What defines the radius and maximum field of the streamer head?
The answer to this question became clear after the writing of this book
had been completed. This is essentially one question, for the quantities r_m
and E_m are related roughly by the expression (3.21), $U \approx 2r_m E_m$, and the
streamer head potential U can be taken as a given parameter, since it is
described by the external conditions with respect to the head. The mechanism regulating the setting of r_m and E_m was discussed in Section 3.5.4. If
the value of r_m is smaller and that of E_m is larger than what is necessary
for the streamer, the head and the adjacent channel will expand due to
the ionization in an excessively strong radial field. But if r_m is larger and
E_m is smaller than the required values, a thinner channel with a stronger
field will start from the head front. These considerations were formulated
in [A.1] with reference to the work [A.2].

It follows from the discussion in [A.1] that the primary quantity 'chosen'
by the streamer is the field at the head front rather than the radius. We
will denote this field as E_m^0. Indeed, the fast streamer problem contains
the field scale, because it is implied by the dependence of the ionization
frequency ν_i on E, but it does not contain a direct scale for the radius.

There are essentially two physical factors which define the choice of E_m^0.
One is the 'quasi-threshold' nature of the function $\nu_i(E)$. The ionization
frequency is very small and rises sharply with E in relatively low fields.
But then, starting with some characteristic values of E, it becomes fairly
high while its growth remains relatively slow. The other factor, imposed
by electrostatics, is that the field normal to the streamer surface decreases
from the front point of the head towards its transition to the channel body
and then farther on down the channel.

To better understand the effects of these factors, assume the streamer to
be a perfectly conducting cylinder of length l with a hemispherical head of
radius r_m (Figure A.1) and let the function $\nu_i(E)$ have a truly threshold

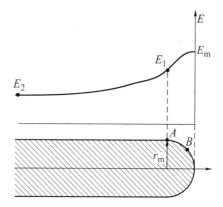

FIGURE A.1
Schematic streamer and the distribution of the field normal to its surface.

character: $\nu_i = 0$ at $E < E^*$ and, say, $\nu_i = a(E - E^*)$ at $E > E^*$. The values of $E^* = 125$ kV cm^{-1} and $a = 1.7 \times 10^{11}$ cm kV^{-1} s^{-1} provide an acceptable approximation of the curve for air (Figure A.2). Obviously, the threshold ionization field E^* will be eventually achieved at point A, where the hemisphere transforms to the cylinder; we denote the field at this point as E_1. Indeed, as long as we have $E_1 > E^*$, the head expands in the transverse direction. At $E_1 < E^* < E_m$, only the central head segment, which rests on the circumference where $E = E^*$ (point B in Figure A.1), moves forward. It is this segment that will form a new, thinner channel. The field at the front point of the new head is $E_m^0 = \beta E^*$, where β is defined by the surface charge distribution over the conductor of the final shape. At $l/r_m \sim 10^2$, we have $\beta \approx 1.5$; from expressions (3.18) and (3.21) the field at the channel center is $E_2 \approx 2E_m^0 / \ln(l/r_m) \approx 0.4E_m^0$, i.e. it is 1.7 times less than E_1, and the channel there does not expand.

With the real smooth $\nu_i(E)$ curve, we may expect the maximum value of E_m to be achieved when the velocity of the ionization wave at point A is much less than that of the head front point. The channel behind the head will expand all the same, and no strictly steady state will arise unless E_1 drops to the actual ionization threshold of 30 kV cm^{-1} for air, but this is hardly feasible. Let us make a rough estimation of the situation described. From expressions (3.7) and (3.8), the ionization wave velocity is $v_s \sim \nu_i(E)s \sim \nu_i(E)/E \equiv \gamma$, where $s \sim U/E$ is the size of the region at the surface where the field is still close to the surface value E. With allowance for $v_s(E_1) \ll v_s(E_m)$, we take in the first approximation $\gamma_m - \gamma_1 \approx \gamma_m \approx$

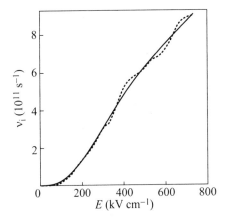

FIGURE A.2
Ionization frequency in air plotted from the data [A.3] on the ionization coefficient α and the electron drift velocity.

$(E_m - E_1)(d\gamma/dE)_{E=E_m}$ to obtain

$$\left(\frac{d\ln\nu_i}{d\ln E}\right)_{E=E_m} \approx \frac{2\beta - 1}{\beta - 1}$$

For the threshold function $\nu_i = a(E - E^*)$ at $E > E^*$, when $E_1 = E^*$, we have an exact equality

$$\left(\frac{d\ln\nu_i}{d\ln E}\right)_{E=E_m} = \frac{\beta}{1 - \beta} \qquad \beta = \frac{E_m}{E_1} \tag{A.1}$$

which can be used as an approximate expression for the real functions $\nu_i(E)$.

Reflecting the effect of the electrostatic factor, equation $(A.1)$ introduces an exactness in the qualitative criterion $(d\ln\nu_i/d\ln E)_{E=E_m} \sim 1$ suggested in [A.1] (if we take 3 or 1/3 as a value of the order of unit, we may have a deviation from the r_m estimate as large as one order of magnitude). In the work [A.1], the choice of the field by the streamer was attributed to the bending or saturation of the $\nu_i(E)$ curve. In our view, however, the reason for this choice should be the quasi-threshold character of this function rather than the bending. When the $\nu_i(E)$ function represents a straight line starting at the threshold point, it has neither a bending nor a saturation, whereas E_1 can be fixed with an utmost certainty, entailing the unambiguity of E_m^0 for a conductor of a steady state shape. Of course, finding the steady state and the electrostatic coefficient β is quite another matter, but the real range of the β values is not large: 1.3–2 [A.4].

The considerations above have been supported by calculations [A.4] for a simple streamer model. At constant potential, every point on the surface of a long perfect conductor moves along the normal s' with the velocity

$$v_s = \int_0^\infty \nu_i[E(s')]\, ds'$$

in keeping with the formula (3.7). At $t = 0$, we take a cylinder with a hemispherical end of $r = 0.1$ cm and $l = 5$ cm. The evolution of the conductor during its 4–8-fold elongation is examined at constant potential U taken to be equal to 20–40 kV. The radius of the channel front becomes smaller or larger with varying U, and the hemisphere becomes either flattened or elongated. In the calculation with the threshold $\nu_i(E)$ function, the field at the transition of the head to the channel becomes equal to $E_1 \approx E^* = 125$ kV cm^{-1}; $E_m^0 \approx 150$–210 kV cm^{-1}, $r_m \approx 0.6$–0.11 cm; $2E_m r_m/U \approx 0.95$–1.1, in agreement with the theoretical unity for a very long cylinder with a hemispherical end.

For a real $\nu_i(E)$ function, lower fields $E_m^0 \approx 90$–145 kV cm^{-1} are established, because the expansion also occurs at $E < 125$ kV cm^{-1}; $\beta = 1.5$–2; $r_m \approx 0.14$–0.20 cm; $2E_m r_m/U \approx 1.2$. There is a slight bending of the $\nu_i(E)$ function lying much higher than E_m^0, at 380 kV cm^{-1}, i.e. it is quite unimportant. There is also a continuous slow expansion of the channel, as in the more detailed calculations described in Section 5.6. Equation $(A.1)$ is fullfilled with a good accuracy.

Recently, a detailed calculation of the evolution of a short streamer with a length up to 0.8 cm has been performed in the work [A.5]. The ionization frequency was found from the same data as in [A.4]. A field $E_m^0 \approx 150$ kV cm^{-1} was achieved at $U \approx 20$ kV. The reader is also referred to the calculations of short streamers in nitrogen [A.6] and in air [A.7, A.8] performed in terms of a complete two-dimensional model.

To conclude, the idea that the maximum field E_m^0, more or less definite for a particular gas, is reached at the streamer front allows an experimental checkup. According to (3.11), the plasma density and conductivity immediately behind the head depend only on $\nu_{i\,m} \equiv \nu_i(E_m^0)$; hence, they should not depend on the voltage or streamer velocity. In principle, this can be registered experimentally. It should be noted that runaway electrons have been observed in some streamer experiments [A.9]. Their appearance requires much stronger fields than those derived from the above calculations [A.4–A.8, A.10]. Strong fields, probably, arise at the very early stages of the streamer development. This problem is, no doubt, worth further research effort.

References

Chapter 1

1.1 Schonland, B., *The Lightning Discharge* in Handbuch der Physik, Vol. 22, Springer-Verlag, Berlin, 1956.

1.2 Uman, M., *The Lightning Discharge*, Acad. Press, Orlando, 1987.

1.3 *Lightning*, Ed. R. H. Golde, Vol. 1 Physics of Lightning, Vol. 2 Lightning Protection, Acad. Press, Orlando, 1977.

1.4 Allibone, T. and Schonland, B., *Nature*, **134**, 3393, 1934.

1.5 Stekolnikov, I. S., Beljakov, A. P., and Mjakishev, I. N., *Elektrichestvo*, **8**, 51, 1937.

1.6 Komelkov, V. S., *Izv. Akad. Nauk SSSR, Tekhnika*, **8**, 955, 1947.

1.7 Loeb, L. B. and Meek, J. M., *The Mechanism of the Electric Spark*, Clarendon Press, Oxford, 1941.

1.8 Loeb, L. B., *Basic Processes of Gaseous Electronics*, Univ. of California Press, Berkeley, 1960.

1.9 Raether, H., *Electron Avalanches and Breakdown in Gases*, Butterworths, London, 1964.

1.10 Meek, J. M. and Craggs, I. D., *Electrical Breakdown of Gases*, Clarendon Press, Oxford, 1953.

Chapter 2

2.1 Raizer, Yu. P., *Gas Discharge Physics*, Springer-Verlag, Berlin, New York, 1991.

2.2 Nielsen, R. A. and Bradbury, N. E., *Phys. Rev.*, **51**, 69, 1937.

2.3 Crompton, R. W., Huxley, L. C., and Sutton, D. J., *Proc. Roy. Soc. A*, **218**, 507, 1953; Crompton, R. W. and Sutton, D. J., *Proc. Roy. Soc. A*, **215**, 467, 1952.

2.4 Dutton, J., Harris, E. M., and Elewellyn Lones, F., *Proc. Phys. Soc.*, **81**, 52, 1963.

2.5 Von Engel, A., Handbuch der Physik, Vol. 21, Springer-Verlag, Berlin, 1956.

2.6 Lin, S. C. and Teare, J. D., *Phys. Fluids*, **6**, 355, 1963.

2.7 Park, C., *J. Thermophysics*, **3**, 233, 1989.

2.8 Kossyi, I. A., Kostinsky, A. Yu., Matveyev, A. A., and Silakov, V. P., *Plasma Sources Sci. Technol.*, **1**, 207, 1992.

2.9 Aleksandrov, N. L., Bazelyan, A. E., and Bazelyan, E. M., *Plasma Physics Reports*, **21**, 57, 1995.

2.10 Rodriguez, A. E., Morgan, W. L., Touryan, K. J., Moeny, W. M., and Martin, T. H., *J. Appl. Phys.*, **70**, 2015, 1991.

2.11 Aleksandrov, N. L., *Zh. Tekh. Fiz.*, **56**, 1411, 1986; *Chem. Phys. Lett.*, **212**, 409, 1993.

2.12 Aleksandrov, N. L., Vysikailo, F. I., Islamov, R. Sh., Kochetov, I. V., Napartovich, A. P., and Pevgov, V. G., *High Temperature*, **19**, 21, 1981.

2.13 Cristiphorou, L. G., *Contrib. Plasma Phys.*, **27**, 237, 1987.

2.14 Aleksandrov, N. L., *Sov. J. Tech. Phys.*, **56**, 835, 1986.

2.15 Guthrie, J. A., Chaney, R. C., and Cunningham, A. J., *J. Chem. Phys.*, **95**, 930, 1991.

2.16 Bohringen, H., Arnold, F., Smith, D., and Adams, N., *Int. J. Mass Spectrom. Ion Phys.*, **52**, 25, 1983.

2.17 Cao, Y. S. and Johnsen, R., *J. Chem. Phys.*, **95**, 7356, 1991.

2.18 Bortnik, I. M., Kushko, A. N., and Lobanov, A. N., *Proc. II All-Union Conf. 'Physics of Electrical Breakdown of Gases'*, Tartu, p. 270, 1984.

2.19 Khvorostovskaya, L. E. and Yankovsky, V. A., *Contrib. Plasma Phys.*, **31**, 71, 1991.

2.20 Finkelburg and Maecker, H., *Electrische Bogen und Thermisches Plasma* in Handbuch der Physik, Vol. 22, Springer-Verlag, Berlin, 1956.

Chapter 3

3.1 Loeb, L. B., *Science*, **148**, 1417, 1965.

3.2 Lozansky, E. D. and Firsov, O. B., *Theory of Spark*, Atomizdat, Moscow, 1975 (in Russian).

3.3 Diakonov, M. I. and Kachorovsky, V. Yu., *Zh. Eksp. Teor. Fiz.*, **94**, 32, 1988; **95**, 1850, 1989.

3.4 Shveigert, V. A., *Teplofiz. Vys. Temp.*, **28**, 1056, 1990.

3.5 Bazelyan, E. M. and Raizer, Yu. P., *Teplofiz. Vys. Temp.*, **35**, 181, 1997.

3.6 Raizer, Yu. P. and Simakov, A. N., *Fiz. Plazmy*, **22**, 668, 1996.

3.7 Mnacakanjan, A. H. and Naidis, G. V., in *Plasma Chemistry*, Ed. B. M. Smirnov, Energoatomizdat, Moscow, **14**, 227, 1987 (in Russian).

3.8 Dhali, S. K. and Williams, P. F., *Phys. Rev. A*, **31**, 1219, 1985.

Chapter 4

4.1 Schwab, A., *Hochspanungmesstechnik (Messgerate und Messverfahren)*, Springer-Verlag, Berlin, 1969.

4.2 Bazelyan, E. M., *Zh. Tekh. Fiz.*, **34**, 474, 1964.

4.3 Bazelyan, E. M., *Zh. Tekh. Fiz.*, **36**, 365, 1966.

4.4 *Positive Discharges in Air Gaps at Les Renardieres—1975*, Electra, **53**, 31, 1977.

4.5 Hidaka, K. and Nurooka, Y., *Proc. IEEE*, **132**, Pt. A, 139, 1985.

4.6 Chernov, E. N., Lureiko, A. V., and Petrov, N. I., *Proc. VII Int. Symp. on High Voltage Engin.*, Dresden, p. 141, 1991.

4.7 Norinder, H. and Salka, O., *Arkiv for Fysik.*, Band 3, Nr. 19, 347, 1951.

4.8 Gorin, B. N. and Inkov, A. Ya., *Zh. Tekh. Fiz.*, **32**, 329, 1962.

4.9 Brago, E. N. and Stekolnikov, I. S., *Izv. Akad. Nauk SSSR, Tekhnika*, **11**, 50, 1958.

Chapter 5

5.1 Norinder, H. and Salka, O., *Arkiv for Fysik.*, Band 3, Nr. 19, 347, 1951.

5.2 Bazelyan, E. M., Goncharov, V. A., and Gorjunov, A. Yu., *Izv. Akad. Nauk SSSR, Energetika transp.*, **2**, 154, 1985.

5.3 Rudenko, N. S. and Smetanin, V. I., *Zh. Eksper. Teor. Fiz.*, **61**, 146, 1971.

5.4 *Research on Long Air Gap Discharges at Les Renardieres—1973 results. By the 'Les Renardieres Group'*, Electra, **35**, 49, 1974.

5.5 *Positive Discharges in Air Gaps at Les Renardieres—1975*, Electra, **53**, 31, 1977.

5.6 Gallimberti, I., *J. de Physique Coll. C7 Suppl.*, **40**, 193, 1979.

5.7 Marode, E. J., *J. Appl. Phys.*, **46**, 2065, 1975.

5.8 Meek, J. M. and Craggs, J. D., *Electrical Breakdown of Gases*, John Wiley and Sons, New York, 1978.

5.9 Bazelyan, E. M., Burmistrov, M. V., et al., *Izv. Akad. Nauk SSSR, Energetika transp.*, **2**, 99, 1978.

5.10 Brago, E. N., *Vestn. Akad. Nauk SSSR*, **3**, 31, 1959.

5.11 Stekolnikov, I. S., *Long Spark Nature*, Russian Acad. Sci. Press, 1960 (in Russian).

5.12 Bazelyan, E. M. and Gorjunov, A. Yu., *Elektrichestvo*, **11**, 27, 1986.

5.13 Bazelyan, E. M. and Gorjunov, A. Yu., *Izv. Akad. Nauk SSSR, Energetika transp.*, **4**, 75, 1982.

5.14 Vasilyak, L. M., Kostjuchenko, S. V., et al., *Usp. Fiz. Nauk*, **164**, 263, 1994.

5.15 Bazelyan, E. M. and Razhansky, I. M., *Air Spark Discharge*, Nauka, Siberia, 1988 (in Russian).

5.16 *Negative Discharges in Long Air Gaps at Les Renardieres*, Electra, **74**, 67, 1981.

5.17 Pulavskja, I. G., *Electrical Breakdown of Air Isolation*, Trans. of Krzhizhanovsky Power Engineering Institute, Moscow, p. 39, 1982.

5.18 Gorin, B. N. and Shkilev, A. V., *Elektrichestvo*, **6**, 31, 1976.

5.19 Gorin, B. N. and Shkilev, A. V., *Elektrichestvo*, **2**, 29, 1974.

5.20 Aleksandrov, D. S., Bazelyan, E. M., and Bekzhanov, B. I., *Izv. Akad. Nauk SSSR, Energetika transp.*, **2**, 120, 1984.

5.21 Allen, N. L. and Ghaffar, A., *J. Phys. D.*, **28**, 331, 1995.

5.22 Stekolnikov, I. S. and Shkilev, A. V., *Dokl. Akad. Nauk SSSR*, **145**, 782, 1962.

5.23 Bazelyan, E. M., *Zh. Tekh. Fiz.*, **34**, 474, 1964.

5.24 Park, Y. and Cones, H., *J. of Research of Nat. Bureau of Standard*, **56**, 201, 1959.

5.25 Brago, E. N. and Stekolnikov, I. S., *Izv. Akad. Nauk SSSR, Tekhnika*, **11**, 50, 1958.

5.26 Dawson, G. and Winn, W., *J. Phys.*, **183**, 159, 1965.

5.27 Bazelyan, E. M., *Zh. Tekh. Fiz.*, **36**, 365, 1966.

5.28 Bazelyan, E. M., *Izv. Akad. Nauk SSSR, Energetika transp.*, **3**, 82, 1982.

5.29 Wu, C. and Kunhardt, E. E., *Phys. Rev. A*, **37**, 4396, 1988.

5.30 Wang, M. and Kunhardt, E. E., *Phys. Rev. A*, **42**, 2366, 1990.

5.31 Jing-Ming Guo and Chwan-Hwa John Wu, *IEEE Trans. Plasma Sci.*, **21**, 684, 1993.

5.32 Dhali, S. K. and Williams, P. F., *Phys. Rev. A*, **31**, 1219, 1985.

5.33 Jing-Ming Guo and Chwan-Hwa John Wu, *J. Phys. D.*, **26**, 487, 1993.

5.34 Evlachov, N. V., Kachorovsky, V. Yu., and Chistjakov, V. M., *Zh. Eksp. Teor. Fiz.*, **102**, 59, 1992.

5.35 Naidis, G. V., *J. Phys. D.*, **29**, 779, 1996.

5.36 Kulikovsky, A. A., *J. Phys. D.*, **27**, 2556, 1994.

5.37 Vitello, P. A., Penetrante, B. M., and Bardsley, J. N., *Phys. Rev. E*, **49**, 727, 1994.

5.38 Kulikovsky, A. A., *XXII Int. Conf. on Phenomena in Ionized Gases*, Hoboken, USA, **1**, 43, 1995.

5.39 Djermoude, D., Marode, E., and Segur, P., *XXII Int. Conf. on Phenomena in Ionized Gases*, Hoboken, USA, **1**, 33, 1995.

5.40 Davies, A. J. and Al-Hussany, A., *XXII Int. Conf. on Phenomena in Ionized Gases*, Hoboken, USA, **4**, 185, 1995.

5.41 Vitello, P. A., Penetrante, B. M., and Bardsley, J. N., *Non-Thermal Plasma Techniques for Pollution Control*, NATO ASI Series, **34A**, 249, Springer-Verlag, Berlin, 1993.

5.42 Guo, J. M. and Wu, C. H., *Non-Thermal Plasma Techniques for Pollution Control*, NATO ASI Series, **34A**, 287, Springer-Verlag, Berlin, 1993.

5.43 Kulikovsky, A. A., Mnatsakanian, A. Kh., Naidis, G. V., and Solozobov, Yu. M., *XXI Int. Conf. on Phenomena in Ionized Gases*, Bochum, 297, 1993.

5.44 Babaeva, N. Yu. and Naidis, G. V., *J. Phys. D*, **29**, 2423, 1996.

5.45 Babaeva, N. Yu. and Naidis, G. V., *Phys. Lett. A*, **215**, 187, 1996.

5.46 Braun, D., Gibalov, V., and Pietsch, G., *Plasma Sources Sci. Technol.*, **1**, 166, 1992.

5.47 Ganesh, S., Rajabooshanam, A., and Dhali, S. K., *J. Appl. Phys.*, **72(9)**, 3957, 1992.

5.48 Tochibuto, F., Miyamoto, A., and Watanabe, T., *XI Int. Conf. on Gas Discharge and their Appl.*, Tokio, **I-168**, 1995.

5.49 Kato, S., *Proc. VII Int. Conf. 'Gas Discharges and their Applic.'*, London, p. 328, 1982.

5.50 Morrow, R., *Phys. Rev.*, **32**, 1799, 1985.

5.51 Gallimberti, I., *Pure Appl. Chem.*, **60**, 663, 1988.

5.52 Bazelyan, A. E. and Bazelyan, E. M., *Teplofiz. Vys. Temp.*, **31**, 867, 1994.

5.53 Bazelyan, A. E. and Bazelyan, E. M., *Teplofiz. Vys. Temp.*, **32**, 35, 1994.

5.54 Aleksandrov, N. L., Bazelyan, A. E., and Bazelyan, E. M., *Fiz. Plazmy*, **21**, 60, 1995.

5.55 Aleksandrov, N. L. and Bazelyan, E. M., *J. Phys. D.*, **29**, 740, 1996.

5.56 Meleshko, V. P. and Shveigert, V. A., *Preprint 14–89*, Institute for Theoretical and Applied Mechanics, Russian Academy of Science, Novosibirsk, 1989 (in Russian).

5.57 Kunhardt, E. E. and Wu, C., *Comp. Phys.*, **68**, 127, 1987.

5.58 Kossyi, I. A., Kostinsky, A. Yu., Matveev, et al., *Plasma Sources*, **1**, 207, 1992.

5.59 Aleksandrov, N. L. and Bazelyan, E. M., *J. Phys. D.*, **29**, 740, 1996.

5.60 Aleksandrov, N. L. and Bazelyan, E. M., *Fiz. Plazmy*, **22**, 60, 1996.

5.61 Mnacakanjan, A. H. and Naidis, G. V., in *Plasma Chemistry*, Ed. B. M. Smirnov, Energoatomizdat, Moscow, **14**, 227, 1987 (in Russian).

5.62 Lozansky, E. D. and Firsov, O. B., *Theory of Spark*, Atomizdat, Moscow, 1975 (in Russian).

Chapter 6

6.1 Bazelyan, E. M., Gorin, B. N., and Levitov, V. I., *Izv. Akad. Nauk SSSR, Energetika transp.*, **5**, 30, 1975.

6.2 Bazelyan, E. M., *Zh. Tekh. Fiz.*, **36**, 365, 1966.

6.3 Bazelyan, E. M., Burmistrov, M. V., et al., *Izv. Akad. Nauk SSSR, Energetika transp.*, **2**, 99, 1978.

6.4 *Positive Discharges in Air Gaps at Les Renardieres—1975*, Electra, **53**, 31, 1977.

6.5 Anisimov, E. I., Bogdanov, O. V., et al., *Elektrichestvo*, **11**, 55, 1988.

6.6 Lupeiko, A. V., Miroshnizenko, V. P., et al., *Proc. II All-Union Conf. 'Physics of Electrical Breakdown of Gases'*, Tartu, p. 259, 1984.

6.7 Baikov, A. P., Bogdanov, O. V., et al., *Elektrichestvo*, **9**, 60, 1988.

6.8 Waters, R. T., *Proc. IEEE*, **128**, Pt. A, 319, 1981.

6.9 Kekez, M. and Savich, P., *Proc. IV Int. Symp. on High Voltage Engin.*, Athens, Rep. No. 45.04, 1983.

6.10 Allibone, T. and Schonland, B., *Nature*, **134**, 3393, 1934.

6.11 Stekolnikov, I. S., Beljakov, A. P., and Mjakishev, I. N., *Elektrichestvo*, **8**, 51, 1937.

6.12 Bazelyan, E. M. and Razhansky, I. M., *Air Spark Discharge*, Nauka, Siberia, 1988 (in Russian).

6.13 Bazelyan, E. M., Valamat-Zade, T. G., and Skilev, A. V., *Izv. Akad. Nauk SSSR, Energetika transp.*, **6**, 149, 1975.

6.14 Petrov, N. I., Avansky, V. R., and Bombenkova, N. V., *Zh. Tekh. Fiz.*, **64**, 50, 1994.

6.15 Komelkov, V. S., *Izv. Akad. Nauk SSSR, Tehknika*, **8**, 955, 1947.

6.16 Stekolnikov, I. S., *Izv. Akad. Nauk SSSR, Tehknika*, **5**, 133, 1957.

6.17 Larionov, V. P., *Elektrichestvo*, **4**, 80, 1957.

6.18 Gorin, B. N. and Shkilev, A. V., *Elektrichestvo*, **2**, 29, 1974.

6.19 Bazelyan, E. M., Levitov, V. I., and Ponizovsky, A. Z., *Proc. III Int. Symp. on High Voltage Engin.*, Milan, Rep. No. 51.09, 1979.

6.20 Stekolnikov, I. S. and Shkilev, A. V., *Gas Discharges and the Electricity Supply Industry*, Leatherhead, England, p. 242, 1962.

6.21 Stekolnikov, I. S. and Shkilev, A. V., *Dokl. Akad. Nauk SSSR*, **141**, 1079, 1962.

6.22 Stekolnikov, I. S. and Shkilev, A. V., *Dokl. Akad. Nauk SSSR*, **151**,

837, 1963.

6.23 Komelkov, V. S., *Izv. Akad. Nauk SSSR, Tekhnika*, **6**, 856, 1950.

6.24 Gorin, B. N. and Inkov, A. Ya., *Zh. Tekh. Fiz.*, **32**, 329, 1962.

6.25 *Negative Discharges in Long Air Gaps at Les Renardieres*, Electra, **74**, 67, 1981.

6.26 Park, Y. and Cones, H., *J. of Research of Nat. Bureau of Standard*, **56**, 201, 1959.

6.27 Bogdanova, N. B. and Popkov, V. I., *Izv. Akad. Nauk SSSR, Energetika transp.*, **1**, 79, 1968.

6.28 Bogdanova, N. B., Pevchev, B. G., and Popkov, V. I., *Izv. Akad. Nauk SSSR, Energetika transp.*, **1**, 96, 1978; *Proc. IV Int. Conf. on Gas Discharges*, London, 1976.

6.29 Bazelyan, E. M., *Izv. Akad. Nauk SSSR, Energetika transp.*, **3**, 82, 1982.

6.30 Bazelyan, E. M. and Gorjunov, A. Yu., *Izv. Akad. Nauk SSSR, Energetika transp.*, **4**, 75, 1982.

6.31 Aleksandrov, D. S., Bazelyan, E. M., and Bekzhanov, B. I., *Izv. Akad. Nauk SSSR, Energetika transp.*, **2**, 120, 1984.

6.32 Bazelyan, E. M., Goncharov, V. A., and Gorjunov, A. Yu., *Izv. Akad. Nauk SSSR, Energetika transp.*, **2**, 154, 1985.

6.33 Gallimberti, I., *The Characteristics of the Leader Channel in Long Gaps*, II World Electrotech. Symp., Moscow, 1977; *J. Phys. (France)*, **40**, C7-7, p. 193, 1979.

6.34 Gallimberti, I. and Bondiou, A., *J. Phys. D.*, **27**, 1252, 1994.

6.35 Raizer, Yu. P., *Gas Discharge Physics*, Springer-Verlag, Berlin, New York, 1991.

6.36 Norinder, H. and Salka, O., *Arkiv for Fysik*, Band 3, Nr. 19, 347, 1951.

6.37 Zeldovich, Ya. B. and Raizer, Yu. P., *Physics of Shock Waves and High Temperature Hydrodynamic Phenomena*, Academic Press, New York, 1968.

6.38 Aleksandrov, N. L. and Bazelyan, E. M., *J. Phys. D.*, **29**, 740, 1996.

6.39 Gorin, B. N. and Shkilev, A. V., *Elektrichestvo*, **6**, 31, 1976.

6.40 Stekolnikov, I. S. and Shkilev, A. V., *Dokl. Akad. Nauk SSSR*, **145**, 1962.

6.41 Stekolnikov, I. S. and Shkilev, A. V., *Dokl. Akad. Nauk SSSR*, **151**, 1085, 1963; *Int. Conf.*, Montreux, p. 466, 1963.

6.42 Schonland, B., *The Lightning Discharge* in Handbuch der Physik,

Vol. 22, Springer-Verlag, Berlin, 1956.

6.43 Bazelyan, E. M., *Elektrichestvo*, **11**, 27, 1991.

6.44 Bazelyan, E. M., Chlapov, A. V., and Shkilev, A. V., *Elektrichestvo*, **9**, 19, 1992.

Chapter 7

7.1 Aleksandrov, G. N., Ivanov, V. L., and Kizewetter, V. E., *Dielectric Strength High Voltage Outside Insulation*, Energia, Leningrad, 1969 (in Russian).

7.2 Aleksandrov, G. N., *Elektrichestvo*, **8**, 15, 1975.

7.3 Lemke, E. E., *Wissenschaft Zeit. der Techn. Univer.*, Dresden, **26**, 1, 1977.

7.4 Bazelyan, E. M., *Elektrichestvo*, **7**, 22, 1977.

7.5 Gallimberti, I., Goldin, M., and Poli, E., *Proc. IV Int. Symp. on High Voltage Engin.*, Athens, Rep. 42.08, 1983.

7.6 Gallimberti, I. and Bondiou, A., *J. Phys. D.*, **27**, 1252, 1994.

7.7 Bazelyan, E. M., *Elektrichestvo*, **5**, 20, 1987.

7.8 Bazelyan, E. M. and Gorjunov, A. Yu., *Izv. Akad. Nauk SSSR, Energetika transp.*, **4**, 75, 1982.

7.9 Aleksandrov, D. S., Bazelyan, E. M., and Bekzhanov, B. I., *Izv. Akad. Nauk SSSR, Energetika transp.*, **2**, 120, 1984.

7.10 Carrara, G. and Thione, L., *IEEE Summer Meeting*, Paper CH 0910-0-PWR, **41**, 1974.

7.11 Carrara, G. and Thione, L., *IEEE Trans.* **PAS-95**, 512, 1976.

7.12 Pigini, A., Rizzi, G., Brambila, R., et al., *Proc. III Int. Symp. on High Voltage Engin.*, Milan, Rep. 52.15, 1979.

7.13 Schneider, H. and Turner, F., *IEEE Trans. on Power Appar. and Systems*, **94**, 551, 1975.

7.14 Bazelyan, E. M., Burmistrov, M. V., et al., *Izv. Akad. Nauk SSSR, Energetika transp.*, **3**, 122, 1979.

7.15 Stekolnikov, I. S., Brago, E. N., and Bazelyan, E. M., *Dokl. Akad. Nauk SSSR*, **133**, 550, 1960.

7.16 Stekolnikov, I. S., Brago, E. N., and Bazelyan, E. M., *Gas Discharges and the Electricity Supply Industry*, Leatherhead, England, p. 139, 1962.

7.17 Bazelyan, E. M., Brago, E. N., and Stekolnikov, I. S., *Zh. Tekh. Fiz.*, **32**, 993, 1962.

7.18 Menemenlis, G. and Harbec, G., *IEEE Trans. Winter Power Meeting*, New York, p. 225, 1973.

7.19 Paris, L., *IEEE Trans.* **PAS-86**, 936, 1967.

7.20 Barnes, H. and Winters, D., *IEEE Trans.*, **PAS-90**, 1579, 1981.

7.21 Gallet, G. and Leroy, J., *IEEE Conf.*, Paper C73-408-2, 1973.

7.22 Hughes, R. and Roberts, W., *Proc. IEEE*, **112**, 198, 1965.

7.23 Gorin, B. N. and Shkilev, A. V., *Elektrichestvo*, **6**, 31, 1976.

7.24 Uman, M., *The Lightning Discharge*, Academic Press, New York, 1987.

7.25 Bazelyan, E. M. and Razhansky, I. M., *Air Spark Discharge*, Nauka, Siberia, 1988 (in Russian).

7.26 Tichodeev, N. N. and Tushnov, A. N., *Elektrichestvo*, **3**, 37, 1958.

7.27 Udo, T., *IEEE Trans. on Power Appar. and Systems*, **PAS-64**, 471, 1964.

Supplement

A.1 Dyakonov, M. I. and Kachorovsky, I. Yu., *Zh. Eksp. Teor. Fiz.*, **94**, 32, 1988; **95**, 1850, 1989; **98**, 897, 1990.

A.2 Cravath, A. M. and Loeb, L. B., *Physics* (now *J. Appl. Phys.*), **6**, 125, 1935.

A.3 Dutton, J. A., *J. Phys. Chem. Ref. Data*, **4**, No. 3, 577–856, 1975.

A.4 Raizer, Yu. P. and Simakov, A. N., *Fiz. Plasmy* (in print).

A.5 Babaeva, N. Yu. and Naidis, G. V., *J. Phys. D*, **29**, 2423, 1996.

A.6 Vitello, P. A., Penetrante, B. M., and Bardsley, J. N., *Phys. Rev. E*, **49**, 5574, 1994.

A.7 Morrow, R. and Lowke, J. J., *J. Phys. D*, **30**, 614, 1997.

A.8 Kulikovsky, A. A., *J. Phys. D*, **30**, 441, 1997.

A.9 Babich, L. P., Loiko, T. V., and Tsukerman, V. A., *Usp. Fiz. Nauk*, **160**, No. 7, 49, 1990.

A.10 Dhali, S. K. and Williams, P. F., *Phys. Rev. A*, **31**, 1219, 1985.

Index